Nanoelectromechanics in Engineering and Biology

Nano- and Microscience, Engineering, Technology, and Medicine Series

Series Editor
Sergey Edward Lyshevski

Titles in the Series

MEMS and NEMS:
Systems, Devices, and Structures
Sergey Edward Lyshevski

Microelectrofluidic Systems: Modeling and Simulation
Tianhao Zhang, Krishnendu Chakrabarty,
and Richard B. Fair

Nano- and Micro-Electromechanical Systems: Fundamentals
of Nano- and Microengineering
Sergey Edward Lyshevski

Nanoelectromechanics in Engineering and Biology
Michael Pycraft Hughes

Microdrop Generation
Eric R. Lee

Micro Mechatronics: Modeling, Analysis, and Design
with MATLAB®
Victor Giurgiutiu and Sergey Edward Lyshevski

Nanoelectromechanics in Engineering and Biology

Michael Pycraft Hughes

CRC PRESS

Boca Raton   London   New York   Washington, D.C.

Library of Congress Cataloging-in-Publication Data

Hughes, Michael Pycraft, 1970–
    Nanoelectromechanics in engineering and biology / Michael Pycraft Hughes.
          p. cm. -- (Nano- and microscience, engineering, technology, and medicine series)
    Includes bibliographical references and index.
    ISBN 0-8493-1183-7 (alk. paper)
    1. Nanotechnology. 2. Biotechnology. 3. Biomedical engineering. I. Title. II. Series.

T174.7 .H84 2002
620'.5--dc21   2002073652

Visit the CRC Press Web site at www.crcpress.com

© 2003 by CRC Press LLC

No claim to original U.S. Government works
International Standard Book Number 0-8493-1183-7
Library of Congress Card Number 2002073652
Printed in the United States of America 1 2 3 4 5 6 7 8 9 0
Printed on acid-free paper

Dedication

_____

For

Alis, my parents, and Svipp

"Begin with the possible, and gradually move towards the impossible"

—*Robert Fripp*

Preface

The subject matter of this book describes the coming together of two fields of science, with histories stretching back to the middle of the twentieth century, that have only united in its closing years. Particle electrokinetics — the manipulation of particles, particularly with nonuniform electric fields — was first studied in significant depth by Herbert Pohl in the 1950s and beyond. Since then, it has become a tool for manipulating, separating, and studying all manner of objects on the scale of one micron or above, most particularly in the study of biological cells by, among others, Nobel laureate Albert Szent-Györgyi. Nanotechnology's origins are difficult to trace, but they can be attributed in concept to a talk by another Nobel laureate, Richard Feynman, who described the concept of building machines capable of manipulating objects on the nanometer scale. Over the last few years, the field has grown into a multi-billion-dollar industry driven by the constant scaling down of electronic devices and the promise of fantastic science-fiction outcomes such as nanosized robots swimming around the bloodstream and repairing cells from within. In the early 1990s, these two fields finally met at a laboratory in Japan, with the manipulation of DNA and protein molecules by Professor Masao Washizu and colleagues. The intervening period has seen the development of the field into a mature technology where researchers can trap single viruses, stretch DNA molecules, and build nanocircuits using molecular components.

The aim of this book is to give a comprehensive description of how electrokinetic techniques can be used to manipulate particles on the nanometer scale. The subject matter can truly be described as interdisciplinary; it offers benefits to groups as diverse as colloid chemists, virologists, electronic engineers, and biophysicists. A nanoparticle suspended in liquid behaves in much the same way whether it is of interest to a biologist — as a fragment of DNA would be, for example — or an engineer making devices with nanotubes. That behavior is dictated by a mix of electronics, physics, and chemistry. Since practitioners from any of these disciplines may find something here of use, I have tried to make this book's scope as broad as possible; it deals with everything from manipulating molecules to determining the biophysical state of virus membranes and from separating particles to examining hypothetical nanometer-scale motors that look at how electro-mechanical forces might be applied in the future.

The structure of this book is loosely divided into three sections. The first three chapters deal with the background, including a historical review of the subject, a review of electrokinetic theory, and a brief introduction into the science of colloids and surfaces. The following four chapters deal directly with the manipulation of nanoparticles; these increase in complexity from simple solid spheres to hypothetical nanomachines, and the theories in the first three chapters are developed and applied to these specific cases. Then, in the final four chapters, the practicalities of the subject are discussed. Here the reader can find how-to descriptions of building particle separators, electrodes, and laboratories on a chip; simulating electric fields; and analyzing data for mathematical methods.

The book is aimed at final-year undergraduates and graduate students, as well as researchers and the curious; however, since the field (as is evident from the title) is assumed to be quite interdisciplinary, I have attempted to introduce concepts such as electric fields from the ground up, building to derivation of the dielectrophoretic force equation and beyond. Similarly, I have introduced viruses and proteins assuming no prior knowledge of their structure. The mathematics involved is not too complex, rarely involving anything more complicated than basic calculus or cross-products, although complex numbers are used throughout. Those wishing to use this book to teach a course should feel free to adapt it to fit the interests and background of their target audience; for example, they may wish to make Chapter 11 a practical session rather than a lecture.

Acknowledgments

I am indebted to my teachers, including Gareth Jones, Gorwel Owen, Goronwy Williams, Peter Gascoyne, Hywel Morgan, John Bloom Roberts, and, in particular, Ronald Pethig, who introduced me to the field and whose approach to and enthusiasm for science has been a great inspiration to me. I also acknowledge his help in the preparation of my doctoral thesis, parts of which have made their way into Chapter 10. Similarly, many of the experiments and photographs in Chapters 4 to 6 were produced during my time working at Hywel Morgan's lab, for which I acknowledge his help and advice.

I wish to thank my colleagues for their encouragement in writing this book, in particular Professors Peter Macdonald and Paul Smith, and all my colleagues and students in the Centre for Biomedical Engineering. I am also grateful to Professor Thomas B. Jones and Doctors Jamie Cleaver and Kai Hoettges for their feedback on chapters of this book. I would like to thank series editor Sergey Lyshevski for his kind invitation to write this book, and Nora Konopka and Helena Redshaw at CRC Press for their support, help, and tremendous patience!

Finally, I wish to thank my parents Eric and Ceri for their support, my dog Svipp for an hour of uninterrupted thinking (sorry, walking) time per day, and most especially my wife Alis for her love, patience, and encouragement during the long days spent writing this book.

About the author

Michael Pycraft Hughes was born on Holy Island off the coast of Wales in 1970. He attended the University of Wales at Bangor for both his master of engineering (1992) and Ph.D. (1995) studies and has since worked in laboratories in the University of Glasgow (Scotland) and the M.D. Anderson Cancer Center, Houston (USA). In 1999, he was appointed lecturer in microengineering at the Centre for Biomedical Engineering, in the School of Engineering at the University of Surrey (England). He has authored or co-authored over 60 journal papers, conference papers, and book chapters. Dr. Hughes's research activities include dielectrophoresis applied to cancer biology and biosensors, laboratory-on-a-chip technology, applications of electrokinetics to nanotechnology, colloid and surface science, microengineered neural implants, neural signal processing, and neural computing. His teaching activities include instrumentation, sensors, artificial intelligence, biological safety, physiological measurement, and microengineering in medicine and biology.

Contents

chapter one

Movement from electricity

1.1 Introduction

It has been known since the discovery in antiquity that electrostatic inter-actions between objects (such as a rubbed material picking up small items) can induce a force, either attractive or repulsive. In the intervening millennia (but mostly in the last few years) we have learned to use this force to actuate printers and hence produce the written word, to separate and sequence strands of DNA and hence diagnose diseases, to flip mirrors the size of blood cells and hence make data projectors work. The list is endless. Furthermore, as the size of the object being manipulated is decreased, so electrostatic interactions become one of the dominant forces acting on the object. This is important, since the manipulation of ever-smaller objects has increasingly become the cornerstone of technological development. Technology has only recently begun to allow mankind the ability to exert its will over particles so small they may consist of a single molecule, thus allowing us the ability to manipulate, structure, and construct, or to study, discriminate, and separate, the fabric of materials on the level of the molecules from which those materials are made, or the fundamental biological structures that make life work.

Nanoelectromechanics — from the Greek *nanos* (dwarf), *electro* (from the goddess Electra, believed in ancient times to be the source of electric charge), and *mechanics* (the study of forces and their effects on bodies) — is the study of forces exerted on small objects, nanometer-scale particles such as viruses, proteins, nanotubes, and DNA, by the application of electric fields. These studies occupy the space between the quantum world of atoms and the microscopic world of cells, the space that contains nanometer-scale particles, which possess complex properties in both how they work and how they interact with their environment. Moving particles with precision on such scales requires new challenges to be overcome and new insights into the physics of the interaction between electric fields, nanoparticles, and the molecules that surround them. This book will examine, in language accessible to engineers, physicists, and biologists, how these factors can be addressed to use nanodynamics both as an investigative tool, for example in studying the interiors of single viruses without harming them, or as a manipulation tool for nanoparticle separation or molecular manufacturing. This book is concerned with the application of nanomanipulation to present and future problems in nanoscale engineering, physics, chemistry, and biology. The manipulation of particles on the nanometer scale is a key technique in the exploitation of nanotechnology, and this book will study the nanodynamics of nanotechnological devices such as molecular motors and computers.

These disparate fields all need to perform the same tasks — to selectively identify, manipulate, and separate molecules and other nanoparticles from solution. This book will review the current techniques available for this purpose, presenting the range of techniques being developed but concentrating on electrostatic techniques, which dominate the field. This book describes the first major application of what is commonly referred to as nanotechnology (the precise manipulation of nanometer-scale structures) and its use in microbiology, biochemistry, and nanoelectronics. For example, many major technology companies have described the biochip market as the key technology industry of the twenty-first century. Such a market will require miniaturized, analytical methods of identifying and separating proteins, DNA, viruses, and other nanomaterial. Similarly, drug companies and forensic scientists need devices to provide rapid biochemical analysis of tiny samples. At the same time, molecular technologists require methods to position components such as nanowires and fullerenes to form molecular diodes and transistors.

A number of approaches have been taken to the study of the dynamic interactions between moving objects on the molecular scale, which form the basis of the science of *molecular* dynamics. The work presented in this book concentrates firmly on the scale of the macromolecular and the supramolecular — larger molecules, of the orders of nanometers across and larger — and nanometer-scale objects consisting of many molecules, such as colloids, viruses, and nanowires (as shown in Figure 1.1). Similarly, there are a number of different approaches that may be taken to impart force to nanometer-scale objects with high precision. These have included the manipulation of atoms on a dry surface using atomic force microscope tips or the manipulation of molecules in suspension using a focused laser. However, one method of precision manipulation has demonstrated great potential for trapping, positioning, or studying nanometer-scale particles; this is the manipulation by controlling the electrostatic interactions between an object and its environment — a science known variously as *electrokinetics, electromechanics,* and the study of *ponderomotive* forces. From its origins in antiquity, the subject was first explored mathematically in the eighteenth century and was later described by luminaries such as James Clerk Maxwell but was the subject of significant study only in the latter part of the twentieth century for the study of micrometer particles such as biological cells — and subsequently submicrometer particles such as those described here. In particular, this book will focus on the manipulation of particles using magnitude-variant or phase-variant electric fields, generally known (since the early 1990s) as AC electrokinetics.

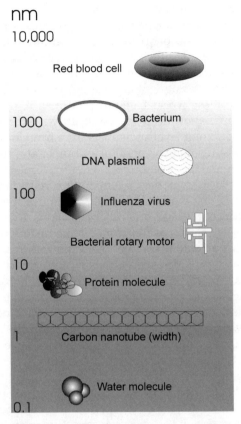

Figure 1.1 A diagram showing the relative sizes of a range of particles on a logarithmic scale. Particles in the nanometer range, between 1 nm and 1,000 nm, demonstrate particular properties separate from those of bulk matter (on the micrometer scale) and individual small molecules (on the atomic scale). This book is concerned with the manipulation of particles on this scale.

1.2   The promise of nanotechnology

Taniguchi[1] invented the term nanotechnology in 1974 to describe the precision machining of surfaces. Since then it has grown to encompass a vast array of different technologies and sciences. Nanotechnology first caught the attention of the general public in 1986, when K. Eric Drexler published the popular-science book *Engines of Creation*,[2] in which he described how machines, micrometers across, operating with atomic precision might one day revolutionize the world; these ideas were explored with considerable rigor in Drexler's second work on the subject, *Nanosystems: Molecular Machinery, Manufacturing and Computation*.[3] In fact, such ideas can be traced back to Nobel laureate Richard Feynman, who first produced these ideas in his lecture "There's Plenty of Room at the Bottom" in 1960.[4] In this, he considered the idea that by developing a scalable manufacturing system, a device could be made that could make a miniature replica of itself, which could in turn replicate itself in miniature, and so on down to molecular scale. He later revisited the subject in a subsequent lecture, "Infinitesimal Machinery" in 1983,[5] in which he postulated the now-famous idea of swimming machines in the human blood stream repairing damaged tissues, an idea he attributed to Al Hibbs.

However, since these works came to prominence, the subjects encompassed by the term nanotechnology have grown immeasurably; this is because, on many different levels, scientists have been manipulating objects on a molecular level for many years. For example, at a fundamental level, chemistry is the original nanotechnology, where custom molecules are delivered to order. Similarly, materials science often relies on molecular-scale arrangements of different materials in order to control specific properties of the ensemble. In recent years, microscopists have discovered that scanning-probe microscopes such as the atomic force microscope (AFM) can be used to push atoms around a surface.[6] And beyond this, nature itself has over billions of years provided us with examples of what can be achieved by developing rotary motors (functionally the same as electric stepper motors) nanometers in diameter to provide locomotion to swimming bacteria, linear motors to provide the basis for our muscles, and a method of data storage and retrieval powerful enough to describe a complete living entity, but compact enough that a complete copy resides on a molecular punch tape 2-nm wide inside almost every cell in the body: DNA.

With so many applications for the term, a new definition for nanotechnology needs to be formed that encompasses them all without being so general as to be meaningless; one current definition is *the study of structures with at least one dimension on the nanometer scale*. Even the definition of nanometer scale (or nanoscale for short) is vague, with the threshold between micrometer scale and nanometer scale falling either at 30 nm (the halfway point between the two on a logarithmic scale) or 100 nm (where 0.1 μm is considered sufficiently submicrometer to warrant the nanoscale label); often objects with minimum dimensions of 2–300 nm are considered, especially where they form part of a family of objects that extends downward in size; for example, herpes simplex viruses are over 200 nm in diameter but are still applicable here since they represent the largest of viruses, a class of organism that extends downward in size to some examples that are only 5 nm in diameter.

Where, in a subject so broad as to contain a hundred volumes, does this book fit in? As stated previously, it is concerned with the manipulation of particles on the nanoscale using the force that is most dominant at this scale, that of electrostatics. Drexler has divided the methods for the manipulation of nanoscale particles (nanoparticles) into two categories: top down, where larger devices are used to move smaller ones, and bottom up, where small structures self-assemble into larger ones. Here we will examine the application of electrostatic interactions with particles in solution, for the manipulation and assembly (and in some cases, self-assembly) of nanoparticles. The techniques can be used either as tools for assembling particles (that is, for engineering) or for the determination of the electrical properties of the particles being investigated. Since there are many biological objects on this scale whose biophysical state cannot be determined by any other means, a section of this work is devoted to the study of viruses and macromolecules, and methods of devising integrated systems for detecting and separating them in so-called laboratories on a chip. Hence, we have the study of nanodynamics in engineering and biology.

1.3   Electrokinetics

As stated previously, electrostatic forces can manifest themselves in a large number of forms and have a large number of applications. However, in this book we will be dealing specifically with those forces experienced by sub-micrometer-scale particles that we have the ability to control — these are the forces that relate to the interaction between particles and an electric field that we may choose to apply. The best known of these is *electrophoresis*, a force imparted on a charged object due to the attraction between the electrode and the charges on the particle, causing the particle to move toward the electrode of opposite polarity. Electrophoresis was developed in the late 1930s by Arne Tiselius[7] of Uppsala University in Sweden for the physical separation of colloidal mixtures and later proteins; he was awarded the Nobel Prize for chemistry for this in 1948. The principle of electrophoresis is that charged particles move through a nonmoving liquid in an electric field at a speed proportional to their size and electrical charge, although typically one selects separation to be principally dictated by one or the other by careful choice of experimental conditions (such as pH and medium viscosity). Two-dimensional electrophoresis techniques actually allow for separation according to both parameters (or indeed, combinations of other parameters) independently with, for example, electrical charge being first applied as a classifier along one axis and then particle size being the classifier along the other. Controlled electrophoresis experiments require the application of an electric field that acts in one direction only. Furthermore, it should have constant electric field magnitude, known as a direct-current (DC) electric field, such that particles maintain a constant velocity independent of time or position on the matrix. Since the magnitude and direction of the force are proportional to the applied electric field, an alternating (AC) electric field would cause the particle to wobble around an axis but not move from that average point over time.

Although electrophoresis is an important force for the manipulation of nanoscale particles such as proteins and DNA, it has somewhat less promise for the *precise* manipulation of particles on this scale; it operates best on larger scales such as the now-famous stripes of DNA electrophoresis gels commonly used in medicine and forensic science for the determination of identity. For our study here, we require forces that are capable of manipulating submicrometer particles on at least a scale of the order of a few micrometers. In order to achieve this, another class of electrokinetic methods can be introduced.

The second of these electrokinetic forces, known as *dielectrophoresis,* is the translational motion of particles induced by polarization effects in non-uniform electric fields, described in some detail in texts by Pohl,[8] Jones,[9] and Zimmermann and Neil.[10] Certain types of particle, when subjected to an electric field, will polarize; the inherent charges separate and form a dipole (and, as described in Chapter 2, higher orders such as quadrupoles and octopoles, though they only contribute significantly under certain circumstances). The poles interact with the electric field and generate electrostatic forces. If the field is nonuniform, the greater electric field strength across one side of the particle means that the force generated on that side is greater than the force induced on the opposing side of the particle and a net force is exerted toward the region of highest electric field. Moreover, this force will act toward the region of greatest electric field *regardless of the orientation of the electric field* and will thus also be present when an AC electric field is applied between the electrodes. This motion of the particle is termed *positive dielectrophoresis.* However, if the particle is suspended in a medium more polarizable than it, the electric field will be distorted around the particle, the induced dipole will orient in the opposite direction, and the force on the particle will be directed away from the high-field regions toward the low-field regions. This motion is referred to as *negative dielectrophoresis.* The polarizability of the particle and medium is dependent on the frequency of the electric field, and it is possible for a particle to experience either positive or negative dielectrophoresis according to the frequency of the applied electric field. This is because the orientation of the dipole depends largely on the accumulation of charge on either side of the particle/medium interface (called a *Maxwell–Wagner interfacial polarization*). The relative amount of charge accumulated depends on the impedance of these materials and hence the frequency of the applied field. As frequency changes, the relative dielectric behavior of the particle and medium change; in a given frequency window (called the dielectric dispersion), the net behavior of the system changes from being dominated by the particle to being dominated by the medium, and the particle goes from experiencing positive dielectrophoresis to experiencing negative dielectrophoresis.

Dielectrophoresis is commonly referred to as a component of AC electrokinetics, but this title is misleading; it can be observed equally in AC and DC fields. Strictly speaking, AC electrokinetics describes the interactions between an electric field and an induced dipole in the particle rather than its inherent charge, not necessarily implying use of AC fields. These interactions are due to different principles than those governing electrophoresis, and a particle may experience both DC electrokinetic (electrophoretic) and dipolar forces simultaneously. However, since the use of AC fields causes the electrophoretic force to average to zero, AC electrokinetics also describes the only forces acting when AC electric fields are used.

There are a number of early observations of the dielectrophoretic force; among the first experimental observations of the motion of particles in non-uniform electric fields were undertaken by Hatschek and Thorne[11] in the study of nickel suspended in toluene and benzene. The phenomenon was named dielectrophoresis by Herbert Pohl in 1951,[12] who later published an in-depth treatise on the subject in his 1978 book *Dielectrophoresis*.[8] Pohl's work advanced the use of dielectrophoresis for investigating the properties of suspensoids and for providing a means of separating particles from suspension. Similar investigations have been conducted using a frequency-based examination of dielectrophoretic response of populations of cells (e.g., Gascoyne et al.[13] and Kaler and Jones[14]), yeast (e.g., Pohl and Hawk[15] and Huang et al.[16]), and bacteria (e.g., Hughes and Morgan[17]), including work by Nobel laureate Albert Szent-Györgyi. Practical applications of dielectrophoresis have included the collection of cells for cellular fusion in biological experiments.[18–20]

Positive and negative dielectrophoresis has been used to separate mixtures of viable and nonviable yeast cells[15,21] and mixtures of healthy and leukemic blood cells.[22] Work by Rousselet et al.[23] and others applied dielectrophoresis to the induction of continuous linear motion of particles, expanding on the basic concept of dielectrophoresis as a means of trapping particles in a specific region in space. An important class of electrokinetic particle manipulator is the levitator — a device used to propel a particle against gravity, resulting in it hovering in midsolution (or midair) at a height governed by its dielectric properties, allowing those properties to be measured, and allowing those particles to be selected and trapped.[24,25] Early experiments used electric fields generated by (relatively) large electrodes and high voltages to trap particles (as described by Pohl[8]); more recently, electrode structures have been fabricated using techniques borrowed from the computer industry (e.g., Huang et al.,[16] Markx and Pethig,[21] and Rousselet et al.[23]) to manipulate much smaller particles at much lower voltages; we will look at the applications of this to nanoparticles in Section 1.4.

Another form of electrokinetics is that of *electrorotation*, the continuous rotation of particles suspended within rotating electric fields; although this phenomenon produces quite different particle behavior than dielectrophoresis, the two are closely related in origin.[26,27] Cell rotation was observed and reported by experiments on AC dielectrophoresis (e.g., Teixeira-Pinto et al.[28]) and was later suggested to be the result of the dipole–dipole interaction of neighboring cells.[29] This led Arnold and Zimmerman[30] to the principle of suspending single particles in a rotating field and thus to a more amenable means of studying the phenomenon. Electrorotation occurs when a dipole is induced by a rotating electric field. The dipole takes a finite time (the relaxation time) to form, by which time the electric field has rotated slightly. There is a lag between the orientation of the electric field and that of the dipole moment, and thus a torque is induced as the dipole moves to reorient itself with the electric field. Owing to the continuous rotation of the electric field, the torque is induced continually and the cell rotates. The direction of rotation is determined by the angle between the dipole moment and the electric field; if the phase lag is less than 180°, the particle rotation will follow that of the applied field, referred to as cofield rotation. If the phase angle is greater than 180°, the shortest way in which the dipole can align with the electric field is by rotating in a contrary direction to that of the electric field, and hence particle rotation will act in this direction (antifield rotation). As with dielectrophoresis, the rate and direction of cell rotation is related to the dielectric properties of both the particle and the suspending medium. The technique can thus be used as an investigative technique for studying these properties. Electrorotation has been used to study the dielectric properties of matter, such as the interior properties of biological cells and biofilms (e.g., Arnold and Zimmermann[31] and Zhou et al.[32]). As with dielectrophoresis, electrorotation is commonly listed in AC electrokinetics. A DC version (called Quinke rotation[9]) does exist; however, this is far more likely to be observed as a result of other work than specifically used for analysis.

The final example of electrokinetics discussed here is that of *traveling-wave dielectrophoresis*. The phenomenon was first reported Batchelder[33] and subsequently by Masuda et al.,[34,35] where the electric fields travel along a series of bar-shaped electrodes where low frequency (0.1 Hz to 100 Hz) sinusoidal potentials, advanced 120° for each successive electrode, were applied. This was found to induce controlled translational motion in lycopodium particles[35] and red blood cells.[27] At low frequencies, the translational force was largely electrophoretic, and it was proposed that such traveling fields could eventually find application in the separation of particles according to their size or electrical charge. However, later work by Fuhr and co-workers,[36] using applied traveling fields at much higher frequency ranges (10 kHz to 30 MHz), demonstrated the induction of linear motion in pollen and cellulose particles and also demonstrated that the mechanism inducing traveling motion at these higher frequencies is dielectrophoretic, rather than electrophoretic, in origin. Since then, Huang et al.[37] and others have, for example, used traveling fields to linearly move yeast cells and separate them from a heterogeneous population of yeast and bacteria.

Traveling-wave dielectrophoresis is effectively an extension of the principle of electrorotation to include a linear case. An AC electric field is generated that travels linearly along a series of electrodes. Particles suspended within the field establish dipoles that, due to the relaxation time, are displaced from the regions of the high electric field. This induces a force in the particle as the dipole moves to align with the field. If the dipole lags within half a cycle of the applied field, net motion acts in the direction of the applied field, while a lag greater than this results in motion counter to the applied field.

The means by which the nonuniform, time-variant electric fields described in this book may be generated should be noted. The bulk of the work described here employs microelectrode structures of one form or another; however, the same effects are generated when a focused beam of electromagnetic radiation is used on an object. For example, an object (on the micrometer scale or smaller) exposed to a laser beam will experience two forces. The first is optical pressure, where light is diverted through a transparent object (causing the particle to move toward the center of the beam) or away from an opaque object (causing it to be deflected away from the center of the beam). The second force is the gradient force, caused by interactions with the gradient of light within the laser beam. This second effect is not just an analog of dielectrophoresis, it is the *same effect*, actuated through a different means of electric field delivery and using electric field frequencies many orders of magnitude higher than those described for dielectrophoretic forces. Furthermore, it has been shown that through the use of rotating light modes within the laser, it is actually possible to spin the particle, in a direct analogue of electrorotation. These phenomena are, like dielectrophoresis, gaining acceptance for use both in nanomanipulation and in biological investigation; the common terms for the optical equivalents of dielectrophoresis and electrorotation are "laser tweezers" and "laser spanners," respectively.

Of the methods of inducing forces with electric fields described here — electrophoresis, laser tweezers, and the forces coming under the umbrella term of AC electrokinetics — the former two are already covered by existing literature; in particular, electrophoresis is an extensively used tool throughout modern biochemistry, and many volumes already exist to cover it and its subsidiary techniques, such as capillary electrophoresis. Similarly, those wishing to learn more about the applications of laser tweezers may read books such as Sheetz's.[38] However, in this book we will still encounter both these techniques where they impinge on the study of AC electrokinetic techniques, since at the nanometer scale there is considerable overlap between them.

1.4 Electrokinetics and nanoparticles

We have seen that AC electrokinetic techniques such as dielectrophoresis and electrorotation[4,5] have been used for many years for the manipulation, separation, and analysis of objects with lengths of the order 1 µm–1 mm in solution. Since the force experienced by a particle undergoing dielectrophoresis scales as a function of the particle volume, it was believed for many years that a lower threshold of particle size existed, below which the dielectrophoretic force would be overcome by Brownian motion. It was held that to increase the force would require electric fields of such magnitude that local medium heating would increase local fluid flow, again acting to prevent dielectrophoretic manipulation. Since electrode fabrication techniques were relatively crude, generating electric fields of sufficient nonuniformity required very large potentials to be applied across relatively large interelectrode volumes, and consequently particles with diameters less than about 1 µm could not be trapped. Indeed, Pohl[8] speculated that particles smaller than 500 nm would require excessively large electric fields to trap against the action of Brownian motion.

However, with the use of very small electrodes (usually formed by top-down methods), it is possible to generate electric fields with such complex local geometry that the manipulation of single molecules in solution becomes achievable — and these fields can be manipulated so that the particles can be steered across the electrodes and used to assemble miniature electric circuits.[39] Recent advances in semiconductor manufacturing technology have enabled researchers to develop electrodes for manipulating proteins,[40,41] to concentrate 14-nm beads from solution,[42] or to trap single viruses and 93-nm-diameter latex spheres in contactless potential energy cages.[43]

The first group to break this threshold was that of Washizu and co-workers,[40] who used positive dielectrophoresis to precipitate DNA and proteins as small as 25 kDa. This step downward in size was accelerated by improvements in technologies for electrode fabrication, principally the use of electron beam fabrication. This renewed interest in manipulation of submicron particles, and subsequent work by Schnelle et al.,[44] Müller et al.,[45] Morgan and Green,[46] and Green et al.,[47] demonstrated that viruses of 100-nm diameter could be manipulated using *negative* dielectrophoresis. It was also demonstrated by Müller et al. that latex spheres of 14-nm diameter could be trapped by either positive or negative dielectrophoresis.[42] Subsequent work by Hughes and co-workers demonstrated that by varying the frequency of the applied electric field, herpes viruses can be trapped using either positive or negative electric fields.[43] Another study demonstrated that molecules of the 68 kDa-protein avidin can be concentrated from solution by both positive and negative dielectrophoresis.[49]

The most obvious nanotechnological application of this technique is in the concentrating of parts of molecular machinery to one site. This was suggested by Hughes[50] shortly before the first demonstration of the technique by Huang et al.[51,53] and Cui and Lieber,[52] who used dielectrophoretic assembly to construct nanoscale circuits using semiconductor nanowires. Similar techniques were used by Velev and Kaler[54] to construct microscopic biosensors from solution by stacking particles of different types and to self-assemble nanowires from colloids.[55] Self-assembling computers using nanocomponents manipulated by electrokinetic forces is perhaps a long way off, but its prospects appear good. Electrokinetic techniques are simple and cheap and require no moving parts; they rely entirely on the electrostatic interactions between the particle and dynamic electric field.

1.5   A note on terminology

The intention of this book is to draw together a series of effects caused by the interactions of electric fields and the dipoles induced in particles exposed to these fields, which result in the induced motion of those particles. The phenomena described include dielectrophoresis (by far the best known of the group), electrorotation, traveling-wave dielectrophoresis, electro-orientation, and so on. These techniques are collectively referred to as AC (alternating

Table 1.1 A Summary of the Electrokinetic Forces That Can Be Exerted on Particles, Whether the Force Is Most Significant in AC or DC Fields, and the Origin of the Effect

Force	AC or DC	Origin
Electrophoresis	DC	Caused by charge in electric field
Dielectrophoresis	AC/DC	Caused by induced dipole in nonuniform field
Electro-osmosis	AC/DC	Caused by interaction between free charge in electrical double layer and tangential electric field
Electrorotation	AC	Caused by dipole lag in rotating electric fields
Traveling-wave dielectrophoresis	AC	Caused by dipole lag in traveling electric fields
Electro-orientation	AC/DC	Caused by interaction between dipole and electric field

current) electrokinetics, to distance them from other effects such as electrophoresis and electro-osmosis; these are summarized in Table 1.1. However, AC electrokinetics is something of a misnomer, since although we generally use alternating (oscillating, rotating, or traveling) electric fields to manipulate particles, it need not actually be so; dielectrophoresis can take place with either an alternating or stable (DC) electric field. Some have attempted to use other titles, such as electromechanics or electromanipulation, but these are sufficiently broad to take in all the techniques described above, including those dependent on the interaction of charges attracted to electrodes, such as electrophoresis, which are beyond the scope of this text; we are specifically concerned with the interaction of induced dipoles with an electric field. In many ways, dielectrophoresis is perhaps the most fitting title for these phenomena. Coined by Pohl in 1951, the term is derived from the Greek word meaning to force, with the prefix *dielectro-* to remind us that the origin of this force is dielectric and dipolar.

AC electrokinetic effects have been applied to the study of cells for some considerable time; from the 1950s onward there has been a steady production of scientific work where the techniques were employed in the investigation of the dielectric — and by implication, biophysical — properties of cells and other particles on the micrometer scale. However, it is only in more recent times that the technology has been available to particles smaller than these so-called "nanoparticles." As with AC electrokinetics, there is some dispute concerning the point at which particles can be said to be on the nanometer, rather than micrometer, scale. For example, one might argue that the nanometer scale concerns measurements between 1 nm and 1 μm. Alternatively, it could be argued that on a logarithmic scale, with 1 nm as its center, nanometer scale might refer to particles smaller than 30 nm (the halfway point between 1 nm and 1 μm). In order to circumvent this, we can define our sphere of interest as particles that can be described as *colloids* — a term originally used to describe particles for which Brownian motion is more significant than sedimentation forces and that occupy a state between molecules and bulk matter.

Finally, a number of workers in these fields have adopted their own abbreviation systems to refer to the techniques described here. For example, some abbreviate the word dielectrophoresis to DEP, others to DP, some refer to it as conventional dielectrophoresis and abbreviate it cDEP, while others discriminate between the positive and negative forms by using pDEP or +DEP, and nDEP or –DEP, respectively. In order to overcome confusion arising from this, such terms have not been abbreviated in this book, except where the abbreviation is so widely accepted as to be beyond confusion — such as DNA, for example.

References

1. Taniguchi, N., On the basic concept of nanotechnology, *Proc. Int. Conf. Prod. Eng.*, p. 18, 1974.
2. Drexler, K.E., *Engines of Creation*, Anchor Books/Doubleday, New York, 1986.
3. Drexler, K.E., *Nanosystems: Molecular Machinery, Manufacturing and Computation*, Wiley, New York, 1992.
4. Feynman, R., There's plenty of room at the bottom, 1960; reprinted in *J. Microelectromech. Syst.*, 1, 60, 1992.
5. Feynman, R., Infinitesimal machinery, 1983; reprinted in *J. Microelectromech. Syst.*, 2, 4, 1993.
6. Eigler, D.M. and Schweizer, E.K., Positioning single atoms with a scanning tunnelling microscope, *Nature*, 344, 524, 1990.
7. Tiselius, A., A new apparatus for electrophoretic analysis of colloidal mixtures, *Trans. Faraday Soc.*, 33, 524, 1937.
8. Pohl, H.A., *Dielectrophoresis*, Cambridge University Press, Cambridge, 1978.
9. Jones, T.B., *Electromechanics of Particles*, Cambridge University Press, Cambridge, 1995.
10. Zimmermann, U. and Neil, G.A., *Electromanipulation of Cells*, CRC Press, Boca Raton, FL, 1995.
11. Hatschek, E. and Thorne, P.C.L., Metal sols in non-dissociating liquids. I. nickel in toluene and benzene. *Proc. R. Soc. London*, 103, 276, 1923.
12. Pohl, H.A., The motion and precipitation of suspensoids in divergent electric fields, *J. Appl. Phys.*, 22, 869, 1951.
13. Gascoyne, P.R.C., Pethig, R., Burt, J.P.H., and Becker, F.F., Membrane changes accompanying the induced differentiation of Friend murine erythroleukemia cells studied by dielectrophoresis, *Biochim. Biophys. Acta*, 1149, 119, 1993.
14. Kaler, K.V.I.S. and Jones, T.B., Dielectrophretic spectra of single cells determined by feedback-controlled levitation, *Biophys. J.*, 57, 173, 1990.
15. Pohl, H.A. and Hawk, I., Separation of living and dead cells by dielectrophoresis, *Science*, 152, 647, 1966.
16. Huang, Y., Hölzel, R., Pethig, R., and Wang, X.B., Differences in the AC electrodynamics of viable and non-viable yeast cells determined through combined dielectrophoresis and electrorotation studies, *Phys. Med. Biol.*, 37, 1499, 1992.
17. Hughes, M.P. and Morgan, H., Determination of bacterial motor force by dielectrophoresis, *Biotechnol. Prog.*, 245, 15, 1999.
18. Zimmermann, U., Vienken, J., and Scheurich, P., Electric field induced fusion of biological cells, *Biophys. Struct. Mech.*, 6, 86, 1980.

19. Zimmermann, U., Electric field-mediated fusion and related electrical phenomena, *Biochim. Biophys. Acta,* 694, 227, 1982.
20. Abidor, I.G. and Sowers, A.E., Kinetics and mechanism of cell-membrane electrofusion, *Biophys. J.,* 61, 1557, 1992.
21. Markx, G.H. and Pethig, R., Dielectrophoretic separation of cells: continuous separation, *Biotechnol. Bioeng.,* 45, 337, 1994.
22. Gascoyne, P.R.C., Huang, Y., Pethig, R., Vykoukal, J., and Becker, F.F., Dielectrophoretic separation of mammalian cells studied by computerized image analysis, *Meas. Sci. Technol.,* 3, 439, 1992.
23. Rousselet, J., Salome, L., Ajdari, A., and Prost, J., Directional motion of Brownian particles induced by a periodic asymmetric potential, *Nature,* 370, 446, 1994.
24. Jones, T.B. and Bliss, G.W., Bubble dielectrophoresis, *J. Appl. Phys.,* 48, 1412, 1977.
25. Lin, I.J. and Jones, T.B., General conditions for dielectrophoretic and magnetohydrostatic levitation, *J. Electrostatics,* 15, 53, 1984.
26. Wang, X.-B., Huang, Y., Becker, F.F., and Gascoyne, P.R., A unified theory of dielectrophoresis and traveling-wave dielectrophoresis, *J. Phys. D: Appl. Phys.,* 27, 1571, 1994.
27. Jones, T.B. and Washizu, M., Multipolar dielectrophoretic and electrorotation theory, *J. Electrostatics,* 37, 121, 1996.
28. Teixeira-Pinto, A.A., Nejelski, L.L., Cutler, J.L., and Heller, J.H., The behaviour of unicellular organisms in an electromagnetic field, *J. Exp. Cell Res.,* 20, 548, 1960.
29. Holzapfel, C., Vienken, J., and Zimmermann, U., Rotation of cells in an alternating electric-field: theory and experimental proof, *J. Membrane Biol.,* 67, 13, 1982.
30. Arnold, W.M. and Zimmermann, U., Rotating-field-induced rotation and measurement of the membrane capacitance of single mesophyll cells of *Avena sativa, Z. Naturforsch.,* 37c, 908, 1982.
31. Arnold, W.M. and Zimmermann, U., Electro-rotation: development of a technique for dielectric measurements on individual cells and particles, *J. Electrostatics,* 21, 151, 1988.
32. Zhou, X.-F., Markx, G.H., Pethig, R., and Eastwood, I.M., Differentiation of viable and non-viable bacterial biofilms using electrorotation, *Biochim. Biophys. Acta,* 1245, 85, 1995.
33. Batchelder, J.S., Dielectrophoretic manipulator, *Rev. Sci. Instrum.,* 54, 300, 1983.
34. Masuda, S., Washizu, M., and Iwadare, M., Separation of small particles suspended in liquid by nonuniform traveling field, *IEEE Trans. Ind. Appl.,* 23, 474, 1987.
35. Masuda, S., Washizu, M., and Kawabata, I., Movement of blood-cells in liquid by nonuniform traveling field, *IEEE Trans. Ind. Appl.,* 24, 217, 1988.
36. Fuhr, G., Hagedorn, R., Müller, T., Benecke, W., Wagner, B., and Gimsa, J., Asynchronous traveling-wave induced linear motion of living cells, *Stud. Biophys.,* 140, 79, 1991.
37. Huang, Y., Wang, X.B., Tame, J., and Pethig, R., Electrokinetic behaviour of colloidal particles in travelling electric fields: studies using yeast cells, *J. Phys. D: Appl. Phys.,* 26, 312, 1993.
38. Sheetz, M.P., *Laser Tweezers in Cell Biology,* Academic Press, New York, 1997.
39. Bezryadin, A., Dekker, C., and Schmid, G., Electrostatic trapping of single conducting nanoparticles between nanoelectrodes, *Appl. Phys. Lett.,* 71, 1273, 1997.

40. Washizu, M., Suzuki, S., Kurosawa, O., Nishizaka, T., and Shinohara, T., Molecular dielectrophoresis of biopolymers, *IEEE Trans. Ind. Appl.*, 30, 835, 1994.

41. Bakewell, D.J.G., Hughes, M.P., Milner, J.J., and Morgan, H., Dielectrophoretic Manipulation of Avidin and DNA, paper presented at Proceedings of the 20th Annual International Conference of the IEEE Engineering in Medicine and Biology Society, 1998.

42. Müller, T., Gerardino, A., Schnelle, T., Shirley, S.G., Bordoni, F., DeGasperis, G., Leoni, R., and Fuhr, G., Trapping of micrometre and sub-micrometre particles by high-frequency electric fields and hydrodynamic forces, *J. Phys. D: Appl. Phys.*, 29, 340, 1996.

43. Hughes, M.P. and Morgan, H., Dielectrophoretic manipulation of single sub-micron scale bioparticles, *J. Phys. D: Appl. Phys.*, 31, 2205, 1998.

44. Schnelle, T., Müller, T., Fiedler, S., Shirley, S.G., Ludwig, K., Herrmann, A., Fuhr, G., Wagner, B., and Zimmermann, U., Trapping of viruses in high-frequency electric field cages, *Naturwissenschaften*, 83, 172, 1996.

45. Müller, T., Fiedler, S., Schnelle, T., Ludwig, K., Jung, H., and Fuhr, G., High frequency electric fields for trapping of viruses, *Biotechnol. Tech.*, 4, 221, 1996.

46. Morgan, H. and Green, N.G., Dielectrophoretic manipulation of rod-shaped viral particles, *J. Electrostatics*, 42, 279, 1997.

47. Green, N.G., Morgan, H., and Milner, J.J., Manipulation and trapping of sub-micron bioparticles using dielectrophoresis, *J. Biochem. Biophys. Meth.*, 35, 89, 1997.

48. Hughes, M.P., Morgan, H., Rixon, F.J., Burt, J.P.H., and Pethig, R., Manipulation of herpes simplex virus type 1 by dielectrophoresis, *Biochim. Biophys. Acta*, 1425, 119, 1998.

49. Hughes, M.P. and Morgan, H., Positive and negative dielectrophoretic manipulation of avidin, paper presented at the Proceedings of the First European Workshop on Elelectrokinetics and Electrohydrodynamics, Glasgow, 2001.

50. Hughes, M.P., AC electrokinetics: applications for nanotechnology, *Nanotechnology*, 11, 124, 2000.

51. Huang, Y., Duan, X.F., Wei, Q.Q., and Lieber, C.M., Directed assembly of one-dimensional nanostructures into functional networks, *Science*, 291, 630, 2001.

52. Cui, Y. and Lieber, C.M., Functional nanoscale electronic devices assembled using silicon nanowires as building blocks, *Science*, 291, 851, 2001.

53. Huang, Y., Duan, X.F., Cui, Y., Lauhon, L.J., Kim, K.H., and Lieber, C.M., Logic gates and computation from assembled nanowire building blocks, *Science*, 294, 1313, 2001.

54. Velev, O.D. and Kaler, E.W., *In situ* assembly of colloidal particles into miniaturised biosensors, *Langmuir*, 15, 3693, 1999.

55. Hermanson, K.D., Lumsdon, S.O., Williams, J.P., Kaler, E.W., and Velev, O.D., Dielectrophoretic assembly of electrically functional microwires from nanoparticle suspensions, *Science*, 294, 1082, 2001.

chapter two

Electrokinetics

2.1 The laws of electrostatics

In order to understand the interactions between particles on the nanometer scale, it is worthwhile to examine the underlying processes from which these forces arise. The forces that we will later use for particle manipulation and study form part of *electromagnetics*, that is, the study of the interaction between electric charges; however, for charges of the size considered here, we can disregard magnetic effect and concentrate on electric fields.

Electric charge is a fundamental quantity of subatomic particles, along with mass and other forces that govern interactions between these particles. Of most significance are the proton and the electron, which carry positive and negative charges with values approximately 1.6×10^{-19} Coulombs (C). Objects carrying equal numbers of protons and electrons are *charge neutral*; those carrying an excess of one or other will be *charged*. Charged atoms and molecules in a solution are called *ions*.

It is common knowledge that particles of opposite sign attract each other, and opposite signs repel; electric charges can be either *dipolar* or *monopolar* (or, indeed, have higher orders of poles), each *pole* can have either positive or negative *polarity*. The way that the charges interact with each other can influence this, even to the extent of inducing dipoles where none existed previously. In order to examine this we must look at the fundamental aspects governing electrostatics, beginning with Coulomb's law.

2.2 Coulomb's law, electric field, and electrostatic potential

As stated above, there is an attraction between opposing charge signs and a repulsion between like charges. In 1785, the French scientist Charles-Augustin de Coulomb (after whom the unit of charge is named) determined that the magnitude and direction of this force was proportional to the product of the magnitudes of the two charges multiplied together, inversely proportional to the square of the distance between them, and acts along the vector running through the centers of the charges. Subsequent work has demonstrated that the force can be expressed mathematically as

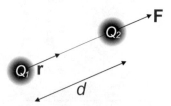

Figure 2.1 Two charges, a distance d apart along vector $\mathbf{r}$, will experience a force in accordance with Coulomb's law. This force acts along vector $\mathbf{r}$ and is repulsive when the charges are alike (as in the example here) or attractive when the charges are unalike.

$$\mathbf{F} = \frac{Q_1 Q_2}{4\pi\varepsilon_0 d^2}\,\mathbf{r} \tag{2.1}$$

where Q_1 and Q_2 are the two charges, d the distance between them, $\mathbf{r}$ is the unit vector directed from Q_1 to Q_2, (as illustrated in Figure 2.1), and ε is the *permittivity* of the space between the charges; this is a quantity that describes the way in which charges relate to one another through space. The *permittivity of free space*, ε_0, has the value 8.85×10^{-12} Farads per meter (F m^{-1}). The presence of matter between the charges increases the permittivity, as we will see in the next section.

While Equation 2.1 is undoubtedly useful, it can only be applied successfully to combinations of two charges. In order to deal with large ensembles of charges, we need to introduce another concept. Since each charge (let us say, Q_1) can be said to exert an influence on all other charges in proportion to the magnitude of this charge and the other charge, we can split Equation 2.1 into the contribution of charge Q_1 — which remains constant, regardless of the other charges present — and the contribution of other charges, which gives us the total value of force. Effectively, Q_1 produces a *force field* that can be plotted across space in the absence of any other charges, and it is the interaction with that force field that produces the electrostatic force. This force field is called the *electric field*, and can be defined such that Equation 2.1 can be rewritten thus:

$$\mathbf{E} = \frac{Q_1}{4\pi\varepsilon_0 d^2}\,\mathbf{r}$$

$$\mathbf{F} = Q_2 \mathbf{E} \tag{2.2}$$

The electric field is a vector quantity, having both magnitude and direction; the magnitude decreases as the square of the distance from the charge, and the direction of the field is either toward or away from the charge according to whether the charge is positive or negative. Figure 2.2 shows the magnitude and direction of the electric field due to a single positive point charge. Note that we often represent electric fields with *field lines* indicating the direction of the field (or rather, the lines along which another charge would move due

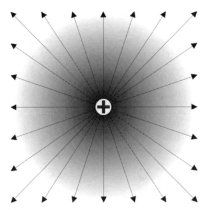

Figure 2.2 A point charge will generate an electric field, such as the one shown above. The electric fields lines travel from positive charges to negative charges (or to infinity if no negative charges are present, as here). The magnitude on the electric field is equivalent to the proximity of the field lines and is indicated here by the depth of tone in the halo surrounding the charge.

to the field); the proximity of the lines to each other gives an indication of the magnitude of the field.

To consider the effect of many charges interacting, we can employ the principle of *superposition*. This states that for a linear system, the total response to a set of stimuli is equal to the combined sum of the responses to each of the stimuli applied separately. In the context of our electric fields, it means that the electric field at any point in a system of multiple charges is equal to the vector addition of all the individual electric fields due to all the charges; for example, Figure 2.3 shows the electric field patterns due to two charges with like and unlike signs at close range.

Another important concept is that of *electrostatic potential*, defined as the work done to move a unit charge between two points with different electric field values. If we consider the force on a unit charge in electric field **E**, from Equation 2.2 the force on the charge will be **E**, and the counterforce required to maintain it in position is −**E**. The work done to move this a small distance dl will be −**E**dl, and so the charge in electrostatic potential can be given by

$$dV = -\mathbf{E}.dl \qquad (2.3)$$

If we integrate Equation 2.3 along the path between two points a and b, then we obtain an expression for the electrostatic potential:

$$V = V_b - V_a - \int \mathbf{E}.dl \qquad (2.4)$$

The electrostatic potential (units: volts) is a scalar quantity and is familiar to those who have studied electronics. The relationship between the electric

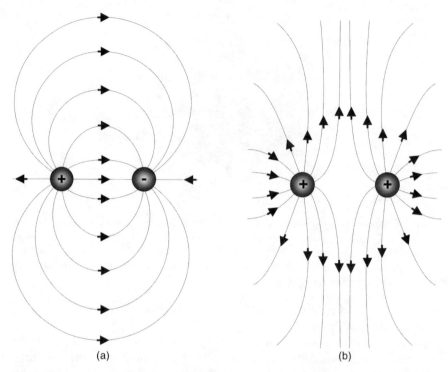

(a)   (b)

Figure 2.3 Electric field lines are diverted by the presence of other charges; here we see how the field lines are (a) bent toward one another when the charges have opposite sign and (b) repelled when the charges have the same sign.

field and the electrostatic potential is perhaps best envisaged as the reverse of Equation 2.4; that is, the electric field is the differential (i.e., the gradient) of the potential. An analogy is that of a gravitational field; if one considers the gravity field generated by the Earth, then the potential is equivalent to the height above the center of the earth, and the electric field is equivalent to the gradient of the surface. A particle (e.g., a ball) placed on the surface of the earth (such as a hill) will roll down the hill in the same manner as a charge being attracted toward a much larger charge in electrostatics, moving between two potentials in the process. When illustrating the potential we often use equipotential lines — these follow the contours of equal voltage and always intersect the field lines due to the same charges at right angles.

Since the electrostatic potential varies in three dimensions, it is necessary to determine the gradient in each direction in order to derive the appropriate vectors. If we consider the movement of a unit charge along a small vector distance dl, then that distance dl can be expressed in terms of three unit vectors:

$$\mathbf{dl} = \mathbf{i}dx + \mathbf{k}dy + \mathbf{k}dz \qquad (2.5)$$

Similarly, we can express the electric field at that point in terms of its three vector components:

$$\mathbf{E} = \mathbf{i}E_x + \mathbf{j}E_y + \mathbf{k}E_z \qquad (2.6)$$

Combining Equations 2.3, 2.5, and 2.6, the potential change along **dl** is

$$dV = -\left(\mathbf{i}E_x + \mathbf{j}E_y + \mathbf{k}E_z\right)\left(\mathbf{i}dx + \mathbf{j}dy + \mathbf{k}dz\right)$$
$$= -\left(E_x dx + E_y dy + E_z dz\right) \qquad (2.7)$$

If we consider the x direction only and rearrange, we find

$$E_x = -\frac{\partial V}{\partial x} \qquad (2.8)$$

and similarly in the other two coordinate directions. Combining all three coordinate components by superposition, we obtain

$$\mathbf{E} = -\left(\mathbf{i}\frac{\partial V}{\partial x} + \mathbf{j}\frac{\partial V}{\partial y} + \mathbf{k}\frac{\partial V}{\partial z}\right) \qquad (2.9)$$

This can be written as

$$\mathbf{E} = -\nabla V \qquad (2.10)$$

where ∇ is the gradient operator, given by the equation

$$\nabla = \left(\mathbf{i}\frac{\partial}{\partial x} + \mathbf{j}\frac{\partial}{\partial y} + \mathbf{k}\frac{\partial}{\partial z}\right) \qquad (2.11)$$

The gradient operator (sometimes called the Del vector operator) is an important concept that we will return to later.

2.3 Gauss's, Laplace's, and Poisson's equations

We now introduce the concept of electric flux. This is a difficult concept to visualize but is best explained by analogy; if the end of a pipe were immersed in water, and more water pumped through the pipe, then the water emerging from the end of the pipe would spread out in all directions; the rate of flow through a defined surface at the end of the pipe is the electric flux through

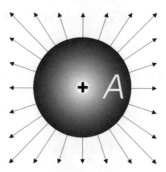

Figure 2.4 Gauss's law determines the flux through a surface enclosing a charge; here a sphere of surface area A encloses a positive charge, and the flux through the surface is related to the field lines passing through that surface.

that surface. If we apply this analogy to electrostatics, the flux of an electric field through an enclosing surface relates the electric field through the surface to the total charge enclosed within the surface. Alternatively, we can describe the flux as being related to the number of field lines through the surface, such as those through the surface of sphere A due to an enclosed positive charge in Figure 2.4. For the charge Q enclosed by a surface A, Gauss's law can be expressed as

$$\int_A \mathbf{E} \cdot dA = \frac{Q}{\varepsilon_0} \qquad (2.12)$$

where dA is a small element of the surface and ε_0 is the permittivity of free space. We can also express Gauss's law in differential form:

$$\nabla \cdot \mathbf{E} = \frac{\rho}{\varepsilon_0} \qquad (2.13)$$

where ρ is the total *charge density* (the number of charges per unit volume) enclosed by the surface. If we substitute Equation 2.10 into the differential form of Gauss's law, then it follows that

$$\nabla^2 V = -\frac{\rho}{\varepsilon_0} \qquad (2.14)$$

This is known as Poisson's equation for the potential. Where the total charge density is zero, the expression reduces to Laplace's equation,

$$\nabla^2 V = 0 \qquad (2.15)$$

indicating a constant electric field.

2.4 Conductance and capacitance

2.4.1 Conductance and conductivity

We can now consider how the above principles apply to physical materials. We are particularly interested in two factors in this instance, those being *conductance* and *capacitance*. These are common terms for describing the electrical properties of materials and will play an important role later on.

When an electrical potential is applied across a material, then free electrons are in the material and these will move toward the positive potential. If the material is also connected to a source of electrons, those will move to fill the gaps left by the vacating electrons so that a continuous flow of electrons is achieved. This process is called *electrical conduction*. The extent to which a flow of electrons (an electrical current) can flow in an electrical field is governed by the material property of conductivity, given the symbol σ and the unit S m^{-1} (Seimens per meter). The lower the conductivity, the less able the material is to conduct, owing to the inability of material to produce free electrons (in a solid), the inability for free charges such as ions to move freely through a liquid, or the lack of ions within a nonconducting liquid. The inverse of conductivity is *resistivity* (units: Ω m, ohm meters). For m species of free conducting particles of concentration n, with mobility μ and charge q, the conductivity is given by the equation:

$$\sigma = \sum_{i=1}^{m} nq\mu \qquad (2.16)$$

For a given potential, there will be a net drift of charged particles in the electric field. We can define the *current density*, J, as the number of charges passing through an area of 1 m^2 per second. For an applied electric field **E**, the current density is given by

$$J = \sigma \mathbf{E} \qquad (2.17)$$

While resistivity and conductivity are parameters that describe the way in which a material interferes with an electrical current, they do not describe the specific extent of the interference for a particular piece of that material. In order to do this, we introduce terms to describe the way in which a piece of material of specific size and shape will conduct electricity. These terms are the commonly known parameters, *resistance* (symbol R), and its inverse, *conductance*. At this point, we must address a discrepancy in convention when discussing symbols. Electronic engineers use the symbol G to represent conductance; chemists use the symbol K. Since the main use of conductance in this book is in the study of surface conductance — a phenomenon studied through the language of colloid chemists — I have elected to use K to represent conductance, but other works do differ on this.

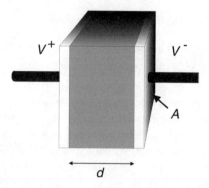

Figure 2.5 A cuboid of material is placed between two conductive plates, which carry voltages V+ and V- across their surfaces. The cross-sectional area of the plates is A; they are a distance d apart.

Conductance relates the current to the applied potential through Ohm's law:

$$V = \frac{I}{K} \tag{2.18}$$

where K is related to σ through the geometry of the conducting volume and I corresponds to the charge per second moving through the material passing through a cross-section of the material per second (the electric current). For a cuboid containing a material of conductivity σ between two equipotential planes, such as conducting plates, of area A and a distance d apart, such as the one shown in Figure 2.5, the conductance is given by

$$K = \frac{\rho A}{d} \tag{2.19}$$

If a cuboid of a second material with different conductance (e.g., K_2) were to be added after the cuboid (of conductance K_1) — the term is in series — then the resultant conductance can be derived from Equation 2.18. Since I will be constant in both conductors (since the flow of charge from one material will replenish charge in the other), we find that

$$\frac{1}{K_T} = \frac{1}{K_1} + \frac{1}{K_2} \tag{2.20}$$

Similarly, if the conductors are side by side, so that both have the same potentials applied across them but can carry different currents, we find

$$K_T = K_1 + K_2 \tag{2.21}$$

Materials that conduct well are referred to as conductors; those that do so poorly are called insulators. However these terms are relative, and in this book we are more likely to use these terms as adjectives, often compounded (e.g., when describing a weakly conducting medium).

2.4.2 Capacitance

Having described materials that have a conductance, we now examine a second parameter, that of capacitance (C, unit Farad). This term comes from the electronic component known as a capacitor, so named because it has a capacity to store charge. Ideally, capacitors have no conductance (or infinite resistance). Instead, the potential across a capacitor is dependent on the ability of the charge on one side of the capacitor to cause a countercharge to appear on the opposing side. This occurs due to a mechanism known as *polarization*.

Polarization is the process of charge redistribution in an electric field, such that positive and negative charges are centered in different places. A capacitor at its simplest consists of two parallel plates similar to the ones shown in Figure 2.5 (we will ignore the additional material between the plates for the moment and will instead consider the gap to be filled with vacuum). If we apply a potential across the plates (let us say +/–5 V), then charge will accumulate on the plates due to the applied charge and associated electric field. The –5 V plate will accumulate electrons across its surface, the +5 V plate will lose them. Since between the plates there are now separate accumulations of positive and negative charge, we have induced a dipole, and the capacitor is polarized. The polarization process takes a finite time to occur (and the charges take a finite time to accumulate), and, when the potential is removed, it takes a finite time for the charges to return to their equilibrium positions; this time is a characteristic of the system and is termed the relaxation time.

The ability of a given capacitor to store charge in this manner is termed its capacitance. The capacitance is related to the charge accumulated on a capacitor's outer faces, Q, and the applied potential, V, by the expression:

$$V = \frac{Q}{C} \qquad (2.22)$$

which we can consider as an analogy to Equation 2.18, but with resistance replaced with capacitance, and static charge Q replacing moving charge I. Indeed, we can carry the analogy further; using the same principles outlined in Section 2.4.1, we can determine the value of series and parallel combinations of capacitors, and indeed the results are similar; for example, capacitors arranged in series will share a common charge along their joining faces, so that from Equation 2.22 we can show that the total capacitance C_T is given by

$$\frac{1}{C_T} = \frac{1}{C_1} + \frac{1}{C_2} \qquad (2.23)$$

We can determine the value of capacitance for two opposing faces at equipotentials, with a potential difference V, by considering the electric field between the plates. Since there are induced charges on the faces at the applied potential faces (let us say, $Q+$ and $Q-$, which are equal and opposite and hence sum to zero), with those charges equally distributed across the flat, parallel surfaces, we can assume that a uniform electric field will be present between the plates. It can be shown from Gauss's law (Equation 2.12) that the electric field between two plates of cross-sectional area A containing a material of permittivity ε_0 (and hence charge density Q/A) is

$$E = \frac{Q/A}{\varepsilon_0} \tag{2.24}$$

Let us say that, as in Figure 2.5, the cuboid has cross-sectional area A and the distance between applied potentials is d. Since the applied voltage is V, the applied electric field E is given by V/d. Combining this with Equations 2.24 and 2.22 and rearranging gives us

$$C = \frac{\varepsilon_0 A}{d} \tag{2.25}$$

We can now consider the effect of putting an insulating material between the plates. Since the material does not conduct charge, it will accumulate countercharge along the faces adjacent to the capacitor places; the negative charge accumulated along the one side of the capacitor will cause an accumulation of positive charge in the adjacent face of the insulator, and the converse will happen on the opposing side. This countercharge will have the effect of altering the electric field between the plates by changing the net amount of charge at either side of the capacitor. If the electric field in the absence of the material is given by Equation 2.24, then with the material present the electric field is reduced to

$$E = \frac{(Q - Q_m)/A}{\varepsilon_0} \tag{2.26}$$

where Q_m is the amount of countercharge accumulated at the surface of the material. We can rewrite this as

$$E = \frac{(Q - Q_m)/A}{\varepsilon_0 Q/(Q - Q_m)} \tag{2.27}$$

Rearranging, we can write

$$\left.\begin{array}{l} E = \dfrac{Q/A}{\varepsilon_0 \varepsilon_r} \\[3mm] \varepsilon_r = \dfrac{Q}{Q - Q_m} \end{array}\right\} \tag{2.28}$$

where ε_r is the *relative permittivity* of the material. Since Q_m cannot exceed Q, ε_r must be greater than or equal to 1. Relative permittivity is dimensionless and carries no units; it effectively indicates the extent to which the material has a greater permittivity than vacuum (which has a relative permittivity of 1). Most dielectric materials (which may be solid, liquid, or gas) have single figure relative permittivities (most gases have $\varepsilon_r \sim 1$); water has an unusually high value of ε_r of 78–80 at room temperature. The mechanism by which the above process occurs — that is, how the dielectric material polarizes — is discussed in Section 2.5.

A note of caution about symbols is required here. Where a single ε appears in this book with a subscript other than 0, it can be taken to mean the *total* permittivity of the material to which the subscript refers, but when comparing the values of permittivity it is more convenient to compare the relative permittivity contribution to the total permittivity value.

The greater the relative permittivity of the material, the greater the effect of charge redistribution and the longer the charges take to dissipate; thus, the greater the effective "stored charge" and the greater the capacitance. For a capacitor containing an insulating material (we refer to materials having the ability to polarize as *dielectric*), the capacitance of the capacitor (previously expressed in Equation 2.25) becomes

$$C = \frac{\varepsilon_0 \varepsilon_r A}{d} \tag{2.29}$$

Note that in both Equations 2.19 and 2.29, the quantity (conductance, capacitance) is related to an intrinsic property of the material (conductivity, permittivity) multiplied by area and divided by distance. The closeness of these two quantities will be important when we come to examine the reaction of materials possessing both qualities later on; thus far we have considered materials possessing only one or the other.

2.4.3 Impedance

One more, closely related, quantity to consider is that of *impedance*, Z. Although a capacitor is in theory nonconducting, the accumulation of charge on one plate and ensuing accumulation of charge on the other plate as a function of the polarization of the material between them means that there is an effect on the flow of current beyond the capacitor. This is particularly true where the applied potential is alternating; the charges accumulate in alternating states of charge polarity, and the potential signal is passed through. The relationship between the voltage and corresponding current through a capacitor under these conditions is given by

$$I = C\frac{dV}{dt} \tag{2.30}$$

When used in an AC circuit, the effect of the capacitor is similar to that of a resistor in that it restricts the current flowing through the circuit for a given value of applied voltage. However, the effect is more complex than resistance, as can be seen from Equation 2.30; applying a sinusoidal input will produce its differential, a cosinusoidal output. This change in phase compared to the effect of a resistor of merely changing the value of the current leads us to the idea that the resistive effect of the capacitor — called the reactance, X — is the *out-of-phase* component of a more generic term, *impedance*, whereas resistance (or conductance) represents the *in-phase* component. All have units of Ω. The impedance of a capacitor with an AC signal of frequency f applied is given by

$$X_c = \frac{1}{2\pi f C} \tag{2.31}$$

where C is the capacitance. When combining resistance and reactance we use the imaginary notation $j = \sqrt{-1}$ to indicate the phase relationship between reactance and resistance impedance

$$Z = R + jXc \tag{2.32}$$

2.5 Polarization and dispersion

2.5.1 Dipoles and polarization

Polarization is the process of alignment of charge on a body such that the positive and negative bound charges move in an electric field, resulting in the centers of charge aligning along the field lines. One example of this is the capacitor described above. Electrically neutral when no electric potential is applied, the capacitor acquires opposing charges on opposite sides when a potential is applied across it. This form of polarization — where materials become polarized, with the formation of one region of positive charge and another of negative (but without any net increase in charge) — is the form most commonly dealt with here. This charge formation is referred to as an *induced* dipole ("two pole"), although as we will discover later, higher order

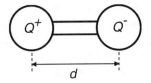

Figure 2.6 An electrostatic dipole is formed by two physically connected (or otherwise associated) charges, one positive, one negative, a distance d apart.

organizations of charge are possible. Figure 2.6 schematically illustrates the key features of an example of a dipole — the magnitudes of the positive and negative charges, and the distance between them.

Electric dipoles occur in many forms and at many levels, from the nano to the macro. Many molecules possess permanent dipoles, where there are fixed regions of positive and negative charge that are not entered at the same point. One example of this is the water molecule that can be seen in Figure 2.7. The water molecule is formed by a single oxygen atom joined to two hydrogen atoms; the hydrogen atoms share their electrons with the oxygen. This means that the oxygen atom gains electrons (negative charges) over its neutral state and becomes negatively charged. The hydrogen atoms lose their electrons and thus gain a net positive charge. Since the hydrogen atoms are arranged on the same side of the oxygen atom, the center of the positive charge is displaced from the negative charge; the positive pole is moved apart from the negative pole and we have a dipole. This is what is known as a *permanent* dipole; there is a permanent, fixed displacement between the positive and negative charges. This is the reason for the high value of relative permittivity for liquid water; since water is composed entirely of dipoles that are free to rotate and align with an electric field, water as a substance is highly polarizable.

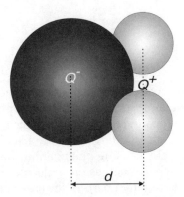

Figure 2.7 A water molecule contains a dipole due to the difference in the locations of the average center of positive charge (between the hydrogen atoms) and the average center of negative charge (on the oxygen atom). This gives water a fairly high relative permittivity value of 78 (at room temperature).

In other materials we do not find permanent dipoles, but molecules that are electrically unpolarized (with positive and negative regions sharing a common average position) when no electric field is applied. When an electric field is applied, the positive and negative charges — the electron cloud surrounding the positively charged nucleus (so-called bound charge, as referred to in Section 2.3) — displace toward the appropriate potential by a tiny amount, of the order of billionths of the diameter of the atom; this means the molecule has become polarized. This is an *induced*, rather than permanent, dipole; the amount of charge and location on the particle are not fixed but are determined by the nature of the electric field causing the polarization process. Although the magnitude of this dipole is small, when applied throughout a medium, very many dipoles acting together produce a significant enough effect to make a capacitor work. A similar form of induced dipole can form where charges can move but are electrostatically attracted to an object in solution, or are contained within an insulating shell but are otherwise free to move; when an electric field is applied, the charges can move across the object to the other side and thus form a dipole. This form occurs in, for example, electrical double layers as described in Chapter 3. The above processes describe dipoles and how they form, but there is another process involved in polarization; the alignment of the dipole *along the electric field lines*. In order for a permanent dipole to align the dipole with the field, a torque is induced to reorient the dipole; this does not occur for induced dipoles, since the dipole is induced along the lines of the electric field.

The extent to which a material will polarize for a given electric field is quantified by two important variables: the polarization per unit volume **P** and the dipole moment **m**; for a material containing, e.g., N dipoles per unit volume, each of which consists of two charges with values Q and $-Q$ a vector distance **d** apart, the polarizability and dipole moment are given by the expressions:

$$\left.\begin{array}{l} \mathbf{J} = N\mathbf{m} \\ \mathbf{m} = Q\mathbf{d} \end{array}\right\} \tag{2.33}$$

We can categorize the processes by which the aforementioned dipoles polarize into three types. The most fundamental type of polarization is electronic polarization, where the (positively charged) atomic nucleus and (negatively charged) electron cloud surrounding it are attracted in opposite directions by the applied electric field, causing polarization. On a slightly larger scale is atomic polarization, where charges on molecules can be displaced in an electric field and move to different parts of the molecule, causing it to become polar. On the largest and slowest scale, the entire molecule can rotate so as to align a permanent dipole along the electric field line; this is orientational polarization. In addition, charges may be induced at the interface between different materials; these can have polarizations of their own, known as Maxwell–Wagner polarizations. The former two processes are so fast as to occur almost instantly; the latter two are sufficiently slow for us to observe them change — or disperse — with frequency.

2.5.2 Complex permittivity

In Section 2.4, we examined ideal conductors and ideal capacitors. However, most materials have elements of both — that is, they have both a capacitance and a conductance. If we return to our cuboid of material between two conducting plates described in Section 2.4 and shown in Figure 2.5, our object had reactance given by

$$X_c = \frac{1}{\omega C} = \frac{d}{\omega \varepsilon_0 \varepsilon_r A} \tag{2.34}$$

where ω is the angular frequency of the electric field and is equal to $2\pi f$. Let us consider that this cuboid of materials also has an inherent resistivity; that is, it is a *lossy* dielectric (a dielectric that also has electric loss). This means that there is an electrical resistance between the plates given by this expression, derived from Equation 2.24,

$$R = \frac{d}{\sigma A} \tag{2.35}$$

Since these are acting between the same plates and have the same potential at either end, we can consider them as being two impedances in parallel; we can then determine the total impedance of the circuit from Equation 2.25, which is given by

$$Z = \frac{R}{1 + RX_c} \tag{2.36}$$

We can actually use this to reexpress the impedance in terms similar to Equation 2.34 by redefining the capacitance — and more specifically the permittivity — to include components from both the conductivity and the permittivity:

$$C = \varepsilon^* \frac{A}{d} \tag{2.37}$$

where the term ε^* (the *complex* permittivity) replaces the permittivity $\varepsilon = \varepsilon_0 \varepsilon_r$ in Equation 2.30:

$$\varepsilon^* = \varepsilon_0 \varepsilon_r - j \frac{\sigma}{\omega} \tag{2.38}$$

The significant factor regarding complex permittivity is that it is frequency dependent (i.e., it contains an ω term). If we consider the behavior of Equation 2.38 at very high frequencies (w $\rightarrow \infty$), the imaginary term tends to zero and ε^* is dominated by the permittivity. At very low frequencies ($\omega \rightarrow 0$), the conductivity term becomes very large and dominates over the permittivity. Therefore, we have two types of behavior — permittivity dominated and conductivity dominated. Between them, there is a transition in the dielectric behavior from one type to another. We call this process a *dielectric dispersion*.

2.5.3   Dispersion and relaxation processes

We now understand the concept of dipoles, and we understand that these can undergo transitions that cause a change in emphasis between different electrical behaviors at different frequencies. There are two prime sources of dielectric dispersion that we need to consider: those that happen *within* materials (Debye relaxations) and those that happen at the *interface* between materials of two different types (Maxwell–Wagner relaxations). We will consider these separately.

2.5.3.1   Debye relaxation

The first type of relaxation process, first described by Peter Debye, is associated with the inability of a dipole to reorient in time with the applied field. The reason for this is that induced dipoles (those that do not exist until an electric field is applied) take a finite time to form; the movement of charges from colocated to separate is not instantaneous. Similarly, as expressed in Section 2.4, the dipoles that form in a capacitor take a finite time to return to their nonpolarized state after the field is withdrawn. Because of this effect of taking time to relax to a nonpolar state, the time for a dipole to form and unform is known as the relaxation time, τ. When a low-frequency AC field is applied, the dipole has plenty of time per cycle to form, collapse, and reform in the opposite polarity as the applied electric field changes polarity. At higher frequencies, however, the dipole does not have time to form every cycle since the field alternates too quickly; the induced dipole does not form in any significant manner, and the transfer of charge by capacitive coupling does not occur. This effect can also occur in permanent (molecular) dipoles, where the electric field is changing in shorter time periods than the dipole can physically take to move; it cannot keep up with the field. This effect is observed for the electrons around the nucleus of an atom, or can happen to ions moving across the surface of an object in a liquid; however, it always acts to reduce the total polarizability of the material.

Some materials will exhibit more than one dispersion because the net polarizability of the particle is composed of more than one polarization mechanism (of the three processes — orientational, atomic, and electronic — described above). In the case of a material experiencing all three types of dispersion, we might expect the dispersion having the lowest frequency to be due to orientational effects (since the physical momentum of molecules is relatively large), followed by the atomic and electronic relaxations. At low frequencies, the polarizability of the material is equal to the sum of the three polarization types; as each polarization phenomenon disperses, the value of the total polarization decrements by the value equivalent to the polarizability of that particular polarization.

Expressing this analytically, we assign the terms χ_{or}, χ_{el}, and χ_a to describe the polarizations due to orientational, electronic, and atomic processes, respectively, where χ refers to the increment in polarization due to that polarization process. This gives polarizabilities

$$\mathbf{P}_{or} = \varepsilon_0 \chi_{or} \mathbf{E}$$
$$\mathbf{P}_{el} = \varepsilon_0 \chi_{el} \mathbf{E} \qquad (2.39)$$
$$\mathbf{P}_a = \varepsilon_0 \chi_a \mathbf{E}$$

The total polarizability at low frequencies is given by

$$\mathbf{P} = \mathbf{P}_{or} + \mathbf{P}_{el} + \mathbf{P}_a \qquad (2.40)$$

while at higher frequencies (above the orientational dispersion), this reduces to an effective value of

$$\mathbf{P}_{eff} = \mathbf{P}_{el} + \mathbf{P}_a \qquad (2.41)$$

Since the dispersion frequencies of the atomic and electronic polarizations are at least of the order of terahertz and upward, they are far beyond the frequencies of the work presented here, allowing us to treat $\mathbf{P}_{eff}$ as a constant and concentrate on the orientation of dipoles instead. It can be shown that the frequency-dependent change in polarization can be given by the expression:

$$\mathbf{P}(\omega) = \mathbf{P}_{or}(\omega) + \mathbf{P}_c$$
$$= \varepsilon_0 \left(\frac{\chi_{or}}{1 + j\omega\tau_{or}} + \chi_{eff} \right) \mathbf{E} \qquad (2.42)$$

where τ_{or} is the relaxation time of the orientational dipole. The first term in brackets corresponds to the Debye relaxation of the orientational dipole.

We can relate Equation 2.42 to the complex permittivity by replacing χ_{eff} with ε_∞, representing the permittivity of the material at very high (effectively infinite with respect to orientational dispersion) frequencies. Since the value of the polarizability χ_{or} is equal to the difference between Equations 2.40 and 2.41, we can define it as being equal to the difference of the permittivities corresponding to the low and high frequency limiting cases. From Equation 2.42 we can derive the complex permittivity by using the relationship

$$\mathbf{P}(\omega) = \varepsilon^* \mathbf{E}$$

$$\varepsilon^* = \varepsilon_\infty + \frac{\varepsilon_s - \varepsilon_\infty}{1 + j\omega\tau_{or}} \tag{2.43}$$

Multiplying this term by its complex conjugate allows us to determine its real and imaginary parts, which can be attributed thus:

$$\varepsilon^* = \varepsilon' - j\varepsilon'' \tag{2.44}$$

where

$$\varepsilon' = \varepsilon_\infty + \frac{\left(\varepsilon_l - \varepsilon_\infty\right)}{1 + \omega^2\tau_{or}^2}$$

$$\varepsilon'' = \frac{\left(\varepsilon_l - \varepsilon_\infty\right)\omega\tau_{or}}{1 + \omega^2\tau_{or}^2} \tag{2.45}$$

These terms are frequency dependent (as evidenced by the presence of an ω term). A plot of ε' (permittivity) and ε'' (dielectric loss) showing a typical dispersion behavior is shown in Figure 2.8. Similarly, we can extend the above analysis to the dispersions due to the other polarization mechanisms, but they do not have relevance here.

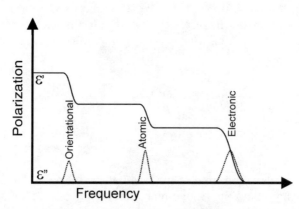

Figure 2.8 The frequency response of a material exhibiting Debye dielectric dispersions due to orientational, atomic, and electronic relaxation processes. The solid line shows the permittivity; the dotted line shows the dielectric loss.

Note in Figure 2.8 that the real part of the graph moves from a higher to a lower value as frequency increases (that is, it is a *monotonically decreasing* function of frequency), whereas the imaginary part is zero at low and high frequencies and rises to a peak through the dispersion process, which occupies a frequency range of over two decades. This is because the imaginary part represents the dielectric loss, which is at its peak when the real part is changing the most. We will observe this behavior again later when we examine the Clausius–Mossotti factor.

2.5.3.2 The Maxwell–Wagner relaxation

While the Debye dispersion described above holds for the reorientation of dipoles within a material, there is a second class of dielectric dispersion that occurs at the interface between two unlike dielectric materials. This dielectric dispersion exhibits a different dispersion from that shown by the materials themselves and can be somewhat lower in frequency. The solution to this type of behavior was developed by James Clerk Maxwell and published in *A Treatise on Electricity and Magnetism* in 1873; the theory was later refined by German scientist K.W. Wagner in 1914. Today the theory is named after both — the Maxwell–Wagner interfacial polarization theory.

The system described by Maxwell considers two dielectric materials, such as two cuboids of the type previously discussed. Both share the same cross-sectional area A, but one has length d_1 and the other d_2 (with their sum being d). Similarly, the two have conductivities σ_1, σ_2 and permittivities ε_1, ε_2, respectively. This is shown schematically in Figure 2.9. We can regard the system as two capacitors in series, for which the total capacitance can be found by using Equation 2.28, except that this time we are assuming that the capacitances possess a complex permittivity such as the one expressed in Equation 2.38. Taking this result and rearranging, it can be shown that for the low-frequency limit ($\omega \to 0$), the effective permittivity is given by

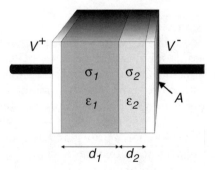

Figure 2.9 Two cuboids of material, each with their own conductivity and permittivity values, are placed in series between two conducting plates; at the interface between these materials, a Maxwell–Wagner polarization process occurs.

$$\varepsilon_{lf} = \frac{\varepsilon_2 d}{d_2} \qquad (2.46)$$

while for the high frequency limit ($\omega \to \infty$), the effective permittivity is given by

$$\varepsilon_\infty = \frac{\varepsilon_1 \varepsilon_2 d}{d_2 \varepsilon_1 + d_1 \varepsilon_2} \qquad (2.47)$$

from which it can be seen that $\varepsilon_{lf} > \varepsilon_\infty$, which indicates that the system exhibits a dielectric dispersion.

By extending this analysis of the two layer system in the terms described in the previous section, it can be shown that the dielectric dispersions of a two layer system can be described in terms of Debye equations (Equation 2.45), but the high and low permittivity terms ε_∞ and ε_{lf} can now be described in terms of the two materials sharing an interface:

$$\varepsilon_{lf} = \frac{d\left(\varepsilon_1 d_1 \sigma_2^2 + \varepsilon_2 d_2 \sigma_1^2\right)}{\left(\sigma_1 d_2 + \sigma_2 d_1\right)^2}$$

$$\varepsilon_\infty = \frac{d\varepsilon_1 \varepsilon_2}{\left(\varepsilon_1 d_2 + \varepsilon_2 d_1\right)} \qquad (2.48)$$

Similarly, we can define the relaxation time for the interfacial polarization thus:

$$\tau = \frac{\varepsilon_0 \left(\varepsilon_1 d_1 + \varepsilon_2 d_2\right)}{\sigma_1 d_2 + \sigma_2 d_1} \qquad (2.49)$$

Note that the relaxation time (and hence, dispersion frequency) is related to the permittivity and conductivity values of both materials. The cause for the effect can be explained as follows: the dielectric material between the two plates has a net total effective impedance, and hence there must be a constant current through the structure. Since the two materials transfer charge in different ways (by conduction or polarization), there exists a discontinuity at the interface between the materials where the mode of charge transfer changes. Where one transfers charge principally by conduction and the other by polarization, it causes an accumulation of charge at the interface (as we would expect charge to accumulate on a capacitor), then there will be a dipole created at the interface.

Since we can treat the interfacial polarization as a type of Debye polarization, it then stands that we can treat it (and the additional polarization it causes) as part of the overall polarizability of the particle as expressed in Equation 2.40, such that the total polarizability is given by the expression

$$\mathbf{P} = \mathbf{P}_{M-W} + \mathbf{P}_{or} + \mathbf{P}_{el} + \mathbf{P}_{a} \tag{2.50}$$

where $\mathbf{P}_{M-W}$ is the polarization due to the Maxwell–Wagner interfacial polarization.

2.6 Dielectric spheres in electric fields

Thus far, we have considered the effects of polarization and dispersion on materials, and mixtures of materials, arranged as cuboids in electric fields; the majority of this book contains studies of particles in suspension. However, our particle is a lossy dielectric material, as is the solution in which it is dissolved, so we merely need to adapt the existing theory to fit this particular structure.

When a polarizable particle is exposed to an electric field, charge builds up at the interface between the surface of the particle and its surroundings; differences in the numbers of positive and negative charges accumulating on the surface mean that the particle is polarized. Since any electrostatic interaction with the particle can be treated as if it were an interaction with the dipole across the particle, by determining the dipole we can determine the behavior of the particle.

If we apply an electric field to a polarizable particle, then charges accumulate at opposing surfaces of the particle along the field vector. However, if the particle contains no excess charge (i.e., it is charge neutral), then, from Laplace's theorem (Equation 2.15) solved for a sphere, there is a uniform electric field *inside* it. There is also a nonuniform electric field generated external to the particle, which we will discuss shortly.

The electric field induced in an ellipsoid exposed to electric field vector **E** applied along axis *a* is given by

$$\mathbf{E}_a = \frac{\varepsilon_0 \mathbf{E}}{\varepsilon_m^* + A_a\left(\varepsilon_p^* - \varepsilon_m^*\right)} \tag{2.51}$$

where A_a is the depolarization factor along the *a* axis and the subscripts *m* and *p* refer to the medium (outside the body) and particle (inside the body), respectively. There are three depolarization factors, A_a, A_b, A_c, one for each axis, which sum to 1; we will deal with these in more detail in Chapter 5. However, since we are dealing with a sphere, the depolarization factors are equal to one another, and thus have the value $1/3$.

The induced polarization **P** per unit volume within the particle is given by the expression

$$\mathbf{P}_a = \left(\varepsilon_p^* - \varepsilon_m^*\right)\mathbf{E}_a \tag{2.52}$$

from which we can determine the dipole moment by multiplying by particle volume. For a sphere of radius r,

$$\mathbf{m}_a = \tfrac{4}{3}\pi r^3 \mathbf{P}_a \tag{2.53}$$

From this, we can define the polarizability, or dipole moment per unit electric field, thus:

$$\alpha_a = \frac{\mathbf{m}_a}{\mathbf{E}} \tag{2.54}$$

Combining Equations 2.51, 2.53, and 2.54, we obtain

$$\alpha_a = \frac{\tfrac{4}{3}\pi r^3 \left(\varepsilon_p^* - \varepsilon_m^*\right)\varepsilon_0}{\varepsilon_0 + \tfrac{1}{3}\left(\varepsilon_p^* - \varepsilon_m^*\right)}$$

$$= 4\pi r^3 \varepsilon_0 \left(\frac{\varepsilon_p^* - \varepsilon_m^*}{\varepsilon_p^* + 2\varepsilon_m^*}\right) \tag{2.55}$$

where the bracketed term is the referred to as the Clausius–Mossotti factor, $K(\omega)$

$$K(\omega) = \left(\frac{\varepsilon_p^* - \varepsilon_m^*}{\varepsilon_p^* + 2\varepsilon_m^*}\right) \tag{2.56}$$

which is dependent on applied frequency (due to the frequency dependence of the complex permittivities), and which can take values between +1 and –0.5. We will study the Clausius–Mossotti factor and the implications of its behavior in more detail in Chapter 4.

We can recombine Equations 2.54 and 2.55 to find the magnitude of the dipole moment of the particle:

$$\mathbf{m} = 4\pi r^3 \varepsilon_0 \left(\frac{\varepsilon_i^* - \varepsilon_o^*}{\varepsilon_i^* + 2\varepsilon_o^*}\right)\mathbf{E} \tag{2.57}$$

This describes the dipole moment for a spherical dielectric particle suspended in a dielectric medium; note that $\mathbf{m}$ is dependent on the dielectric properties of both particle and medium and can change sign (and hence

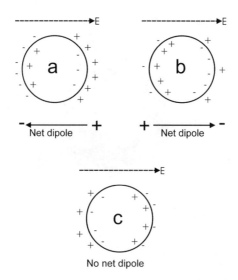

Figure 2.10 A lossy dielectric sphere in a lossy dielectric medium, and exposed to an electric field, will experience interfacial polarization. The amount of charge on each side of the interface will depend on the relative capacitances and conductances of the particle and medium, leading to a net dipole across the particle. (a) Particle more polarizable than medium, (b) particle less polarizable than medium, (c) particle and medium equally polarizable.

polarity) according to the complex permittivities of particle and medium. In order to visualize the processes by which this dipole orientation occurs, let us consider a single particle in an electric field. To demonstrate the physical significance of Equation 2.57, consider three situations: those where the particle is more polarizable than the medium, those where the particle is less polarizable than the medium, and those where they are equal. These are shown in Figure 2.10.

Where the impedance is dominated by conduction in the particle and capacitance in the medium, at the interface between particle and medium there will be a greater amount of charge accumulating on the medium side of the interface (which is acting like a capacitor) than on the particle side (which is acting like a conductor). This situation is shown in Figure 2.10a. As can be seen in the figure, the imbalance between the charges means that across the particle as a whole (including the charge on the interface) there is a dipole oriented opposite the electric field. If the medium is more conductive than the particle, then there will be more charge on the particle side of the interface than on the medium side, and hence the net dipole is oriented in the same direction as the field (Figure 2.10b). Where the complex permittivities of the particle and medium are equal (Figure 2.10c), the net charge is zero and no dipole is present.

This induced dipole will have a perturbatory effect on the electric field in the surrounding medium; by the principle of superposition, the total electric field will be equal to the superposition of the electric field due to a

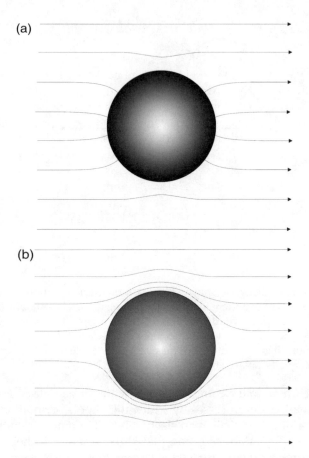

Figure 2.11 Since dipoles consist of positive and negative charges a distance apart, they generate their own electric field; this then warps the electric field that induced the dipole. (a) If the polarized object is conducting, the induced electric field is aligned counter to the external field, and the field is warped toward the object (and intersects the conducting surface at right angles). (b) If the particle is insulating, the dipole is oriented in the same direction as the external field and the field lines warp around the particle.

dipole (shown in Figure 2.3a) with the lines of the applied electric field. Taken together, the net effect is as shown in Figure 2.11. In the case of a particle more polarizable than the medium, the effect is to distort the field lines (previously parallel in this case) toward the surface of the particle, so as to intersect at near right angles. Where the dipole is oriented in the opposite direction, the superposition of field lines means that the electric field passes around the particle. Inside the particle, the electric field is different in these two cases; in the first case (conducting particle) the electric field is low, since the dipole (and its associated electric field) oppose that of the externally applied field. In the capacitive case, the electric field is much higher than the exterior field because the external and internal field are oriented in the same direction.

2.7 Forces in field gradients: dielectrophoresis and electrorotation

We now come to examine the forces on particles in electric fields with nonuniformities of either magnitude or phase. These primarily concern the forces that form the fundamental core of this book, those being dielectrophoresis and (to a lesser degree) electrorotation, plus the effects of inter-induced-dipole attraction and electro-orientation. The reader should note that there are a number of different approaches to deriving the equations for these forces; the one presented here is somewhat less elegant than, for example, the derivation using phasor vector notation used by Jones (see supplementary reading), but uses somewhat simpler mathematics. Those readers who feel more comfortable with vector phasor mathematics are urged to investigate both methods.

2.7.1 Dielectrophoresis

Consider a polarizable particle exposed to an electric field. The applied electric field causes the formation of a dipole within the material and an accumulation of charge at the surface. If the electric field is uniform, then the Coulomb forces on the charges on both sides of the particle are equal and opposite, as are the forces on both sides; therefore, they cancel out and there is no net force on the particle. However, if the field is nonuniform (that is, varying in magnitude across the region occupied by the particle), then the Coulomb forces on either side will not be equal and there will be a net force on the particle. This is called the *dielectrophoretic force*; the action of movement by it is called *dielectrophoresis*.

Let us consider a spherical particle in a nonuniform field such as the one shown in Figure 2.12. When this particle polarizes, it will have centers of positive and negative charges that are equal in magnitude (though opposite in sign) but separated by a distance **d** along vector **r**. Since the electric is nonuniform, the positive and negative charges will experience different electric field strengths, giving rise to a total force on the particle (from Equation 2.2) of

$$\mathbf{F} = Q^+\mathbf{E}(\mathbf{r}+\mathbf{d}) - Q^-\mathbf{E}(\mathbf{r}) \tag{2.58}$$

Where **d** is small relative to the size of the electric field nonuniformity, we can approximate this as

$$\mathbf{E}(\mathbf{r}+\mathbf{d}) = \mathbf{E}(\mathbf{r}) + \mathbf{d}\cdot\nabla\mathbf{E}(\mathbf{r}) \tag{2.59}$$

allowing us to rewrite the force thus:

$$\mathbf{F} = Q\mathbf{d}\cdot\nabla\mathbf{E} \tag{2.60}$$

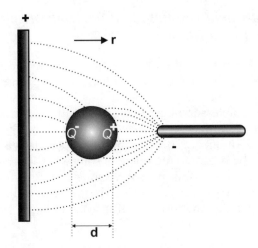

Figure 2.12 Dielectrophoresis occurs when a polarizable particle is suspended in an electric field of nonuniform magnitude, so that the Coulombic forces induced on the charges on each half of the dipole are different.

Since Qd defines the dipole moment, the force can be written thus:

$$\mathbf{F} = (\mathbf{m} \cdot \nabla)\mathbf{E} \qquad (2.61)$$

In order to proceed further, we need to consider the nature of the electric field in a more realistic sense, considering, for example, the fact that it may be changing either in space (having a magnitude gradient) or time (having a phase gradient). For this examination, let us assume that we are examining an AC electric field that varies in three dimensions. This variation can be in the magnitude of the wave (the amplitude of the sine wave varies according to position), phase (the sine wave reaches its maximum at different times according to position), or both. An example of a sine wave changing spatially in phase is a Mexican wave seen in football stadiums; all spectators exhibit the same behavior, but slightly out of phase, resulting in a wave that moves around the stadium even though the participants do not.

We can define an electric field of angular frequency ω according to its position in Cartesian coordinates (at location x, y, z) in terms of its magnitude E_a and phase shift ϕ:

$$
\begin{aligned}
\mathbf{E} &= E_x(t)\mathbf{i} + E_y(t)\mathbf{j} + E_z(t)\mathbf{k} \\
&= E_{x0}(x,y,z)\cos\!\big(\omega t + \phi_x(x,y,z)\big)\mathbf{i} \\
&\quad + E_{y0}(x,y,z)\cos\!\big(\omega t + \phi_y(x,y,z)\big)\mathbf{j} \\
&\quad + E_{z0}(x,y,z)\cos\!\big(\omega t + \phi_z(x,y,z)\big)\mathbf{k}
\end{aligned}
\qquad (2.62)
$$

From Equation 2.57, we know that the induced dipole moment acting on a spherical particle of radius r and relative permittivity ε_m in this electric field can be given by the equation

$$\mathbf{m} = m_x(t)\mathbf{i} + m_y(t)\mathbf{j} + m_z(t)\mathbf{k}$$

$$= 4\pi\varepsilon_0\varepsilon_m r^3$$

$$\cdot \begin{pmatrix} \left(E_{xo}(x,y,z)\operatorname{Re}[K(\omega)]\cos(\omega t + \phi_x(x,y,z)) - \operatorname{Im}[K(\omega)]\sin(\omega t + \phi_x(x,y,z))\right)\mathbf{i} \\ +\left(E_{yo}(x,y,z)\operatorname{Re}[K(\omega)]\cos(\omega t + \phi_y(x,y,z)) - \operatorname{Im}[K(\omega)]\sin(\omega t + \phi_y(x,y,z))\right)\mathbf{j} \\ +\left(E_{zo}(x,y,z)\operatorname{Re}[K(\omega)]\cos(\omega t + \phi_z(x,y,z)) - \operatorname{Im}[K(\omega)]\sin(\omega t + \phi_z(x,y,z))\right)\mathbf{k} \end{pmatrix}$$

$$(2.63)$$

From Equations 2.61–2.63, we can derive the force on this particle due to the interaction between the electric field and the induced dipole (and taking the x, y, z term as given for ease of reading):

$$\mathbf{F}(t) = F_x(t)\mathbf{i} + F_y(t)\mathbf{j} + F_z(t)\mathbf{k}$$

$$= \begin{pmatrix} m_x(t)\dfrac{\partial E_x(t)}{\partial x} + m_y(t)\dfrac{\partial E_x(t)}{\partial y} + m_z(t)\dfrac{\partial E_x(t)}{\partial z} \\ +m_x(t)\dfrac{\partial E_y(t)}{\partial x} + m_y(t)\dfrac{\partial E_y(t)}{\partial y} + m_z(t)\dfrac{\partial E_z(t)}{\partial z} \\ +m_x(t)\dfrac{\partial E_z(t)}{\partial x} + m_y(t)\dfrac{\partial E_z(t)}{\partial y} + m_z(t)\dfrac{\partial E_z(t)}{\partial z} \end{pmatrix} \quad (2.64)$$

$$= \sum_{a=x,y,z}\left(m_x(t)\dfrac{\partial E_x(t)}{\partial a} + m_y(t)\dfrac{\partial E_x(t)}{\partial a} + m_z(t)\dfrac{\partial E_x(t)}{\partial a}\right)$$

Taking the first term, we can expand this using Equation 2.63 to

$$m_x(t)\dfrac{\partial E_x(t)}{\partial a} = 4\pi\varepsilon_0\varepsilon_r r^3\left(\operatorname{Re}[K(\omega)]\cos(\omega t + \phi_x) - \operatorname{Im}[K(\omega)]\sin(\omega t + \phi_x)\right)$$

$$\times\left(\dfrac{\partial E_{x0}}{\partial a}\cos(\omega t + \phi_x) - \dfrac{\partial\phi_x}{\partial a}E_{x0}\sin(\omega t + \phi_x)\right)$$

$$(2.65)$$

and so on for the remaining terms. Note that Re[ ] and Im[ ] refer to the real and imaginary parts of $K(\omega)$, respectively. If we take the average of this term over many cycles, it has the value

$$\left\langle m_x(t)\dfrac{\partial E_x(t)}{\partial a}\right\rangle = 2\pi\varepsilon_0\varepsilon_m r^3\left(\operatorname{Re}[K(\omega)]E_{x0}\dfrac{\partial E_{x0}}{\partial a} + \operatorname{Im}[K(\omega)]E_{x0}^2\dfrac{\partial\phi_x}{\partial a}\right) \quad (2.66)$$

and similarly for the remaining terms. Combining this with the expression for the force gives a time-averaged force value

$$\langle \mathbf{F} \rangle = 2\pi\varepsilon_0\varepsilon_m r^3 \left(\text{Re}[K(\omega)]\nabla \mathbf{E}_{rms}^2 + \text{Im}[K(\omega)]\left(E_{x0}^2\nabla\phi_x + E_{y0}^2\nabla\phi_y + E_{z0}^2\nabla\phi_z\right)\right) \quad (2.67)$$

where E is the root mean square *magnitude* of the electric field, equal to the peak electric field divided by the square root of 2. In conventional dielectrophoresis, we apply the electric field as a single sinusoid, or two sinusoids 180° out of phase, so there is no phase gradient and the second part of the term goes to zero, leaving us with the expression

$$\mathbf{F}_{\text{DEP}} = 2\pi\varepsilon_0\varepsilon_r r^3 \text{Re}[K(\omega)]\nabla E^2 \quad (2.68)$$

This is the general expression for dielectrophoretic force. There are cases where there is a phase gradient through an electric field, and the second part of the force equation (containing the imaginary part of the Clausius–Mossotti factor) is responsible for the induction of traveling-wave dielectrophoresis, which we will return to in Chapter 8. Similarly, we will examine the variations of this equation for nonspherical and multishelled particles in Chapter 5.

One very important feature to note is the fact that the expression for dielectrophoretic force contains the Clausius–Mossotti formula (Equation 2.57), which we know can take either positive or negative values. This has implications for the direction of the force; since $\mathbf{F}_{\text{DEP}}$ is a vector quantity (and thus has a directional component), a change in sign will result in a change in direction. If $\text{Re}[K(\omega)]$ is positive, then the force acts in the direction of the increasing field gradient (since the attractive force on the side where the electric field is strongest gains the greatest attractive force) and hence moves toward the region of the highest electric field. This is called *positive dielectrophoresis*. However, if the value of $\text{Re}[K(\omega)]$ is negative, then the value of the force is negative and the particle is repelled from regions of a high electric field. Since the dipole orients such that like charge signs on electrode and particle are facing, the force is repulsive but is most repulsive on the side where the electric field is strongest and so the particle is pushed down the field gradient. This is referred to as *negative dielectrophoresis*; particles experiencing this force will often be observed to collect in local electric field minima, as we will learn in Chapter 4.

2.7.2   Electrorotation

We now move on to consider the torque on particles in a rotating field. While such fields can be described as varying spatially in phase (and thus having a force component as described above), they can also induce a torque in stationary particles. Consider the particle in a rotating electric field (generated by four electrodes with phase differences between their applied sinusoids)

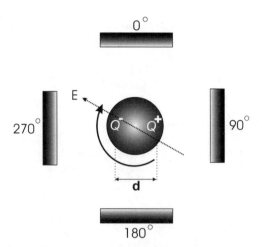

Figure 2.13 Electrorotation occurs when there is a physical displacement between the applied electric field and the induced dipole due to the dipole taking a finite and significant time to respond to the electric field (due to the limit of the relaxation time on dipole formation).

shown in Figure 2.13. The electric field induces a dipole in the particle as before. However, as the field rotates, the dipole must keep up with it; if the dipole lags behind the electric field, then the interaction between charges and field act to induce a torque (rotational force) on the particle. Since the electric field is rotating continuously, the torque is constant and causes the particle to rotate. The torque is at its minimum (zero) when the phase angle between the dipole and the applied field is zero, and it reaches its maximum when the phase angle is ±90°. If the induced dipole moment lags behind the field, then the direction of rotation is with the field and *vice versa* for a moment that leads the field.

The torque exerted by an applied electric field on a dipole can be found by considering the net force acting about the center of a dipole, as before. Considering the dipole in the previous example, the torque is given by

$$\Gamma = \frac{\mathbf{d}}{2} \times Q\mathbf{E} + \frac{-\mathbf{d}}{2} \times (-Q\mathbf{E}) = Q\mathbf{d} \times \mathbf{E} \qquad (2.69)$$

from which the torque Γ on a dipole due to an applied electric field can be found

$$\Gamma = \mathbf{m} \times \mathbf{E} \qquad (2.70)$$

We may proceed as before, using the expressions for $\mathbf{m}$ (Equation 2.63) and $\mathbf{E}$ (Equation 2.62). The cross product for two vectors of the form $(a_x\mathbf{i} + a_y\mathbf{j} + a_z\mathbf{k})$ and $(b_x\mathbf{i} + b_y\mathbf{j} + b_z\mathbf{k})$ is given by

$$\mathbf{m} \times \mathbf{E} = \left(a_y b_z - a_z b_y\right)\mathbf{i} + \left(a_z b_x - a_x b_z\right)\mathbf{j} + \left(a_x b_y - a_y b_x\right)\mathbf{k} \tag{2.71}$$

Taking only the **i** term, then from Equations 2.62 and 2.63,

$$\left(a_y b_z - a_z b_y\right) = 4\pi\varepsilon_0\varepsilon_r r^3$$

$$\left(\begin{array}{l} \left(E_{y0}\,\mathrm{Re}[K(\omega)]\cos(\omega t+\phi_y) - \mathrm{Im}[K(\omega)]\sin(\omega t+\phi_y)\right)\left(E_{z0}\cos(\omega t+\phi_z)\right) \\ -\left(E_{z0}\,\mathrm{Re}[K(\omega)]\cos(\omega t+\phi_z) - \mathrm{Im}[K(\omega)]\sin(\omega t+\phi_z)\right)\left(E_{y0}\cos(\omega t+\phi_y)\right) \end{array} \right)$$

$$= 4\pi\varepsilon_0\varepsilon_r r^3 E_{y0}E_{z0}$$

$$\left(\begin{array}{l} \mathrm{Re}[K(\omega)]\cos(\omega t+\phi_y)\cos(\omega t+\phi_z) - \mathrm{Im}[K(\omega)]\sin(\omega t+\phi_y)\cos(\omega t+\phi_z) \\ -\mathrm{Re}[K(\omega)]\cos(\omega t+\phi_z)\cos(\omega t+\phi_y) - \mathrm{Im}[K(\omega)]\sin(\omega t+\phi_z)\cos(\omega t+\phi_y) \end{array} \right) \tag{2.72}$$

The terms containing the real part of the Clausius–Mossotti factor cancel, leaving

$$\left(a_y b_z - a_z b_y\right) = 4\pi\varepsilon_0\varepsilon_r r^3 E_{y0}E_{z0} \left(\begin{array}{l} \mathrm{Im}[K(\omega)]\sin(\omega t+\phi_z)\cos(\omega t+\phi_y) \\ -\mathrm{Im}[K(\omega)]\sin(\omega t+\phi_y)\cos(\omega t+\phi_z) \end{array} \right) \tag{2.73}$$

We can use the trigonometric identity that sin a cos b – sin b cos a = sin (a – b) to rewrite this expression as

$$\left(a_y b_z - a_z b_y\right) = 4\pi\varepsilon_0\varepsilon_r r^3 E_{y0}E_{z0}\,\mathrm{Im}[K(\omega)]\sin(\phi_z+\phi_y) \tag{2.74}$$

Reassembling this into the original expression gives us

$$\Gamma = \mathbf{m} \times \mathbf{E} = 4\pi\varepsilon_0\varepsilon_r r^3\,\mathrm{Im}[K(\omega)] \left(\begin{array}{l} \sin(\phi_z-\phi_y)E_{z0}E_{y0}\mathbf{i} \\ +\sin(\phi_x-\phi_z)E_{x0}E_{z0}\mathbf{j} \\ +\sin(\phi_y-\phi_x)E_{y0}E_{x0}\mathbf{k} \end{array} \right) \tag{2.75}$$

If we consider a field that is rotating only in the *x-y* plane, with constant electric field (i.e., $E_{x0} = E_{y0}$) then for a circularly rotating field $\phi_x - \phi_y = 90°$. Substituting these factors into the above expression gives the equation

$$\Gamma = \mathbf{m} \times \mathbf{E}$$

$$= 4\pi\varepsilon_0\varepsilon_r r^3 \, \text{Im}\big[K(\omega)\big] \begin{pmatrix} \sin(0)E_{z0}E_{y0}\mathbf{i} \\ +\sin(0)E_{x0}E_{z0}\mathbf{j} \\ +\sin(-90)E_{y0}E_{x0}\mathbf{k} \end{pmatrix} \tag{2.76}$$

$$= 4\pi\varepsilon_0\varepsilon_r r^3 \, \text{Im}\big[K(\omega)\big]\big(-E_{y0}E_{x0}\mathbf{k}\big)$$

$$= -4\pi\varepsilon_0\varepsilon_r r^3 \, \text{Im}\big[K(\omega)\big]E^2$$

for rotation about the $\mathbf{k}$ (vertical) axis. This is the general equation for electrorotational torque on a spherical particle, which (unlike the expression for dielectrophoresis) depends on the imaginary (rather than real) part of the Clausius–Mossotti factor and the square of the field rather than its gradient. Note that the idealized condition (equal magnitudes of E_{x0} and E_{y0}, and a 90° phase shift between them) occurs only under certain conditions and is not generally found anywhere except at the center of electrodes used for generating rotating electric fields (such as the quadrupole electrodes described in Chapter 4). We will return to this issue when discussing electrode design in Chapter 10.

As with dielectrophoresis, there is an important connection between the direction of rotation and the value of $\text{Im}[K(\omega)]$; the direction of rotation is opposite to the direction of the rotating field when $\text{Im}[K(\omega)]$ is positive and in the same direction as the field when $\text{Im}[K(\omega)]$ is negative. These forms of rotation are referred to as *cofield rotation* and *antifield rotation*, respectively. The fact that the direction is perhaps opposite to what one would expect is due to the presence of the minus sign in Equation 2.76. The physical reason for this effect relates to the fact that the displaced charges are repelled by like charges on the electrodes. Note also that since the field can interact with the charges to repel as well as rotate the particle, it can be moved by dielectrophoresis and electrorotation simultaneously, with the force and torque having values proportional to the values of the real and imaginary parts of the Clausius–Mossotti factor, respectively.

2.7.3 Electro-orientation

If we consider a *nonspherical* particle in an electric field, then the dipole moment will be stable only when the particle is aligned with its longest axis along the lines of electric field, as shown in Figure 2.14. When such a particle polarizes, it does so along the longest nondispersed axis. Consider a prolate (cigar shaped) ellipsoid; when this polarizes, the closest the charges can accumulate to the sources of the electric field is along the longest axis of the particle. Furthermore, with the charges at these positions, the particle will experience a torque as the charge centers move to be as close to the high

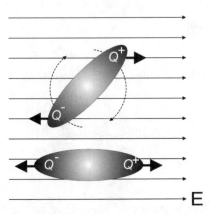

Figure 2.14 Just as dipoles orient with the applied electric field, so polarizable particles experience electro-orientation. Since the dipole is induced, the object will align along the field lines even in AC electric fields. Above the dielectric dispersion, charges cannot move along the particle to form the dipole at the same rate as the field, and the next longest axis will dominate. This is discussed further in Chapter 5.

and low potentials as possible — that is, that the dipole will align along the electric field. If the frequency of the field is high enough, the dipole along this axis will disperse — due to the distance between the charge centers, the dispersion along the longest axis occurs at the lowest frequency — and the next longest nondispersed axis will align with the field instead. We will examine the manipulation of ellipsoidal particles in more detail in Chapters 5 and 7.

2.7.4   Dipole–dipole interactions: pearl chaining

We have established in this section that polarizable particles are attracted to the point of highest electric field. We have also learned in Section 2.6 that such a particle, when in an electric field, will polarize, and this leads to an induced electric field between the two displaced centers of charge. It then follows that when a field is applied, the dipoles formed in particles through-out a volume containing those particles will polarize, and because each then deforms the electric field due to that polarization (in the manner illustrated in Figure 2.11), particles within the region where the two dipole-induced electric fields interact will react to each other (as shown in Figure 2.15). This effect is called mutual dielectrophoresis or dipole–dipole interaction, though it is more commonly referred to as pearl chaining, since its effect is to cause long strings of particles.

The interaction force between two similar particles will be based on their polarizability and the magnitude of the electric field inducing the dipole, but not on whether the particles are experiencing positive or negative dielectro-phoresis. This will change the orientation of the induced dipole, but since like particles will have dipoles induced in similar directions (be they with

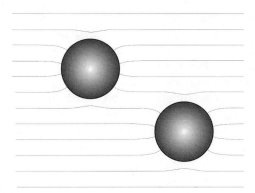

Figure 2.15 When polarizable particles in electric fields come close enough for their induced electric fields to interact, they will experience a dielectrophoretic force to move toward each other. This is called mutual dielectrophoresis, though it is more commonly referred to as pearl chaining because of the chains of particles formed by this phenomenon.

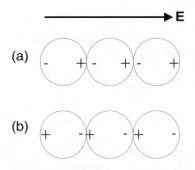

Figure 2.16 The formation of particle chains occurs independently of whether the particles are experiencing (a) positive or (b) negative dielectrophoresis; in both cases, the dipoles align in the same direction to each other, causing Coulombic attraction between charges of opposing signs along the chain.

or against the electric field), they will always attract each other because the dipoles will always arrange in the same direction so that opposing charges will face one another as shown in Figure 2.16.

Since the particles must be aligned such that the dipoles (regardless of direction) are pointed in the same direction and are end to end (since the attraction will be to the point on the surface of each particle where the greatest charge is located), collection by pearl chaining acts along the field lines. Particles of similar type aligning at right angles to this, for example, would repel since similar charge signs are on each side of the particle. However, this position is generally unstable since one particle moving with respect to the other toward or away from the source of the electric field would bring the dipoles into a configuration where they would attract and form chains along the field lines again.

Pearl chaining is usually strongest where the electric field is greatest, that is, at the electrode edge (in the case of an attractive dielectrophoretic force). When a particle arrives at an electrode edge, the field distortion that occurs is often of a sufficient magnitude that incoming particles are attracted to this particle in preference to the electrode edge. This procedure is repeated for subsequent particles, allowing long chains to be formed. In the past, the measurement of chain length has been used as an indicator of the real value of the Clausius–Mossotti factor and is still used in the formation of material structures (Chapter 6) and the formation of conducting nanowires (Chapter 7).

2.7.5   Higher order multipoles

The models of behavior presented thus far are sufficiently accurate to model most of the dielectrophoretic behavior discussed in this book. They are, however, approximations since they consider the behavior of particles under-going dielectrophoresis solely as dipoles. While this approximation holds in the majority of cases and will be used widely in this text, it is important to note that under certain circumstances — in particular, when particles are at a field null, or when the magnitudes of the particle and the electrodes used to generate the electric field are of similar size — higher order terms (called *multipoles*) must be considered. These higher order terms contain many charge centers displaced around the particle and change the way in which the particles relate to the electric field gradient.

The simplest case, called the linear multipole, occurs when the dipole is aligned along the axis of the field gradient, with the electric field being symmetrical about that axis. These terms are straightforward to calculate and provide an extension to the theory thus far presented. We can consider linear multipoles as being the harmonics of the dipole, resulting in arrange-ments of equally spaced positive and negative charge centers along the axis of the dipole, such as are shown in Figure 2.17. Indeed, we can consider the forces on both the monopole (a point charge, as described in Equation 2.2) and the dipole (Equation 2.60) to be the zeroth and first orders of the linear multipole. Working from the diagrams in Figure 2.17, it can be shown that

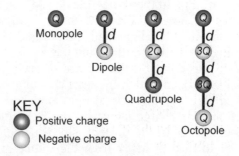

Figure 2.17 The organization of charge in the first four linear multipoles (after Jones).

Table 2.1 The Moments and Forces for the First Four Electric
Multipoles (as shown in Figure 2.17) with the General Term
for an nth-Order Dipole

Multipole term	Pole name	Moment	Force
0	Monopole	Q	$Q\mathbf{E}$
1	Dipole	Qd_1	$Q(d_1 \cdot \nabla)\mathbf{E}$
2	Quadrupole	$2Qd_2^2$	$Q(d_2 \cdot \nabla)^2\mathbf{E}$
3	Octopole	$6Qd_3^3$	$Q(d_3 \cdot \nabla)^3\mathbf{E}$
n	nth-order multipole	$n!Qd_n^n$	$Q(d_n \cdot \nabla)^n\mathbf{E}$

Source: After Jones (see recommended reading).

for the first four dipoles, the moments and forces are given by the expressions
in Table 2.1, with a generalized expression for an nth-order multipole.

Since these higher order components are of much smaller magnitude
than the dipole (so much that we can generally ignore them), the conditions
under which they become apparent are when the dipole does not interact
with the field — that is, it is at a *field null*. The local electric field gradient
can have isolated minima at points in space displaced from the electrodes,
and conditions can arise where, due to symmetry, the electric field has zero
magnitude along an axis. Particles in this situation (such as those in the
center of the quadrupolar electrode arrays described in Chapter 4) still interact with the electric field, even though there is no field along their central
axis. This is because the higher order terms are displaced from that axis. The
first four multipolar terms are shown in Figure 2.18. Particles arranged such
that their central axis is aligned along a field null will therefore interact with
the quadrupolar, octopolar, and higher terms. Since a mathematical treatment of general multipoles involves mathematical processes (such as stress
tensors) that are beyond the scope of this book, those wishing to pursue
them further are referred to the work of Jones (see supplementary reading).

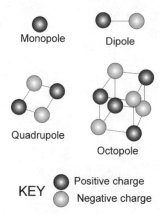

Figure 2.18 The organization of charge in the first four general multipoles (after Jones).

Supplementary reading

Carter, R.G., *Electromagnetism for Electronic Engineers*, Van Nostrand Reinhold, Wokingham, 1986.

Coelho, R., *Physics of Dielectrics for the Engineer*, Elsevier, Amsterdam, 1979.

Daniel, V.V., *Dielectric Relaxation*, Academic Press, London, 1967.

Hasted, J.B., *Aqueous Dielectrics*, Chapman & Hall, London, 1973.

Jones, T.B., *Electromechanics of Particles*, Cambridge University Press, Cambridge, 1995.

Pethig, R., *Dielectric and Electronic Properties of Biological Materials*, Wiley Interscience, Chichester, 1979.

Pohl, H.A., *Dielectrophoresis*, Cambridge University Press, Cambridge, 1978.

chapter three

Colloids and surfaces

3.1 Colloids

This book is concerned with the manipulation of particles that can be described as *colloidal* — that is, the particles can be described as *colloids*. The term was first used by Thomas Graham in 1861 to describe solutions in which the particles diffused slowly but would not sediment under gravity; the term comes from the Greek for "glue." Graham deduced that, when suspended in solution, colloids are affected by Brownian motion to a much greater extent than any other force — their position can vary by many particle diameters in a second due to the effect of random movements of surrounding molecules. This meant the particles must be significantly smaller than 1 μm. However, the slow diffusion rates indicated objects larger than 1 nm.

In general, a loose definition of a colloidal particle is that it should have a maximum size of between 1 nm and 1000 nm; larger particles are sufficiently massive for sedimentation to overcome Brownian motion, whereas smaller particles are governed far more by molecule–molecule interactions and cannot be considered using only classical mechanics. Alternatively, a colloid could be described as a particle that is small enough for electrostatic effects at the surface — the electrical double layer — to dominate the way in which the particle interacts with other particles around it. Particles in the colloidal state can be considered to be too large to act like dissolved molecules, but too small to be considered as bulk matter in solution.

3.2 The electrical double layer

Wherever a charged surface is placed in contact with a fluid, the free charges in the solution will experience a Coulombic force due to the charges present; dissolved ions bearing the same sign as the surface (*coions*) will be repelled from the surface, while ions of opposite charge (*counterions*) will be attracted. This force will act as a function of distance squared from the surface (as described in Chapter 2), leading to an exponential decline in the effect of the surface charge on the medium. At sufficient distance from the surface,

Figure 3.1 A schematic of the electrical double layer surrounding a charged particle (1). The double layer consists of two parties; a layer of charge adsorbed to the surface and having defined thickness, called the Stern layer (2); and an ionic cloud in which the variation in ionic concentration from bulk properties diminishes with distance, known as the diffuse layer (3).

the solution retains its equilibrium concentrations of counterions and coions; we call this the *bulk medium*. Inside that limit, however, the properties of the solution change as the proportion of ions changes. At the interface between the surface and the solution, further structural changes occur where the counterions are sufficiently strongly attracted to the surface to stick or *adsorb* to the surface. These adsorbed charges form a layer within a layer across the surface, with different electrical characteristics from either the bulk medium or the volume where ionic proportions are different.

We call this interface between the surface and the bulk medium the *electrical double layer* (or just *the double layer*), which is shown schematically in Figure 3.1. Since the concentration of charge in the double layer (which is still suspended in solution) varies from that in the bulk medium, we can also expect the conductance of the medium to change near the interface; similarly, since the charge is loosely bound to the surface, we may anticipate a change in the capacitance around the surface. However, the mechanisms of these changes are quite complex and differ for the different parts of the double layer, and we will consider them separately here.

While an electrical double layer exists at any interface between charged surface and electrolytic medium, it is of profound significance when considering colloidal particles; such particles are sufficiently small for the thickness of the double layer to be significant compared to the size of the particle, and may be such that the actual particle may be a relatively small part of the total volume of the particle plus the double layer system.

3.3 The Gouy–Chapman model

First, we will consider the diffuse double layer — the ionic atmosphere that surrounds the particle, first described in the early 20th century by Gouy and

Chapman, working independently. As we have seen in Chapter 2, the presence of charge gives rise to an associated potential, and we can characterize the electrical properties of the surface in terms of its surface charge density u (Coulombs per meter squared) and its electrical potential Φ_0 (volts). It is immersed in a solution containing electrolytes (ions) containing i ion species with a concentration c_i (in molecules per liter) and valency (number of charges per molecule) z_i, and with the medium having a relative permittivity ε_m.

In order to determine the distribution of ions, we use Poisson's equation (Equation 2.14):

$$\varepsilon_m \varepsilon_0 \nabla^2 \Phi = -\rho$$

$$= -e \sum_i z_i c_{ilocal} \tag{3.1}$$

where e is the charge on the electron (1.6×10^{-19}C) and c_{ilocal} is the local ionic concentration for ionic species i. However, while the above equation indicates that the ions should form an ordered distribution governed by the potential gradient, there is also an entropy force (driven by Brownian motion) that acts to distribute the ions in a random, more-or-less homogeneous fashion. The result of the interaction between the imposed order and the entropic forces results in a local charge concentration related to the local electric potential Φ:

$$c_{ilocal} = c_i \exp\left(\frac{-z_i e \Phi}{kT}\right) \tag{3.2}$$

where k is Boltzmann's constant and T is the temperature. Substituting this into Equation 3.1 we find that

$$\varepsilon_m \varepsilon_0 \nabla^2 \Phi = -e \sum_i z_i c_i \exp\left(\frac{-z_i e \Phi}{kT}\right) \tag{3.3}$$

We can interpret this equation for a colloidal particle by approximating to an infinite planar surface, such that variation in concentration only occurs in the direction perpendicular to the surface (which we shall refer to as the z direction). This allows us to reduce the del vector operator to a gradient operator in only one direction, and Equation 3.3 to a differential equation:

$$\frac{d^2\Phi}{dz^2} = \frac{-e}{\varepsilon_m \varepsilon_0} \sum_i z_i c_i \exp\left(\frac{-z_i e \Phi}{kT}\right) \tag{3.4}$$

In order to solve this, we must impose boundary conditions. Since in the bulk solution there can be no net charge (the condition of electroneutrality), we can define it as having no potential gradient and can define the potential as 0 V at large distances from the surface:

$$\left. \begin{array}{l} \Phi = 0 \\ \dfrac{d\phi}{dz} = 0 \end{array} \right\} z \rightarrow \infty \tag{3.5}$$

For our second boundary condition, we must consider the action of accumulated charge within the diffuse layer. Since there is a defined charge accumulation between two equipotential planes (the surface and the bulk), we can consider it to be a capacitor; thus we can work out the potential gradient at the surface with respect to the bulk by considering the stored charge and the permittivity:

$$\left. \dfrac{d\Phi}{dz} = \dfrac{-u}{\varepsilon_m \varepsilon_0} \right\} z = 0 \tag{3.6}$$

With the use of appropriate calculus identities, it can be shown that by imposing these boundary conditions on Equation 3.4 and considering the use of symmetrical electrolytes (those where the valences of the coions and counterions are equal), we can integrate the equation and obtain the following expression for the potential at any distance z from the surface

$$\Phi(z) = \dfrac{2kT}{ze} \ln \left\{ \dfrac{1 + \Gamma \exp(-\kappa z)}{1 - \Gamma \exp(-\kappa z)} \right\} \tag{3.7}$$

where the coefficient Γ is given by

$$\Gamma = \tanh \left(\dfrac{ze\Phi_0}{4kT} \right) \tag{3.8}$$

and we obtain the very important value defining the approximate thickness of the diffuse layer, or *Debye screening length*, given by the value $1/\kappa$:

$$\dfrac{1}{\kappa} = \sqrt{\dfrac{\varepsilon_m \varepsilon_0 kT}{\sum_i (z_i e)^2 c_i}} \tag{3.9}$$

Furthermore, by writing the exponent in Equation 3.4 as a linear series and eliminating terms due to electroneutrality, it can be shown that for particles in weakly conducting media, the equation can be written as

$$\dfrac{d^2\Phi}{dz^2} = \kappa^2 \Phi \tag{3.10}$$

which can be integrated using the boundary conditions outlined above to show that

$$\Phi(z) = \Phi_0 \exp(-\kappa z) \qquad (3.11)$$

From this we can infer that the surface charge density and surface potential are related thus:

$$u = \varepsilon_m \varepsilon_0 \kappa \Phi_0 \qquad (3.12)$$

and the relationship between surface charge density and the surface and bulk potentials leads to an expression for capacitance per unit area:

$$C_d = \frac{u}{\Phi_0} = \frac{\varepsilon_m \varepsilon_0}{1/\kappa} \qquad (3.13)$$

From these expressions we can derive a great deal of qualitative information about the nature and behavior of the electrical double layer. Of primary importance are the distribution of the ions in the double layer, the variation in the electrical potential as a function of distance from the surface, and the thickness of the double layer itself. These are indicated graphically in Figure 3.1. Perhaps the most counterintuitive result is that the double layer thickness, given by Equation 3.9, actually *diminishes* as the concentration of ions in the bulk medium is increased. This is because the number of coions and counterions in bulk solution is greater. The lack of coions in the double layer, as well as the extra counterions, means the surface charge can be countered in a much smaller volume. As distance from the surface is increased, it can be seen from Equation 3.11 that the potential diminishes exponentially across the diffuse layer, achieving electroneutrality on the order of $1/\kappa$. Similarly, we can combine Equations 3.11 and 3.2 to determine the distribution of the coions and counterions as a function of distance from the surface, which both demonstrate a similar exponential change; the coions show an exponential decrease in concentration, the counterions a corresponding increase.

One important aspect of this, arising from Equation 3.12, is that the surface charge density and surface potential are not directly equivalent but are related by the inverse of the Debye screening length, itself proportional to the root of the bulk charge concentration. Therefore, the relationship between these factors varies as a function of medium composition and, when comparing the effects of changing the medium ionic strength, we need to decide which factor should remain fixed (either charge density or potential). Since the majority of the work presented in this book is concerned with particles whose surface charge density is known, Figure 3.2 shows the variation in charge and potential for fixed surface charge density. As can be seen,

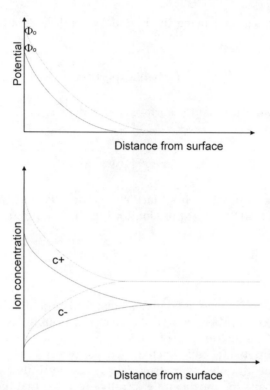

Figure 3.2 The electrical properties of the diffuse layer. The top picture illustrates the variation in electrical potential through the diffuse layer for two cases, those of high (broken line) and low (solid line) salt concentrations in the bulk medium, and for the condition of constant surface charge. As can be seen, the concentration affects both the thickness of the double layer and the value of the surface potential Φ_o. The bottom picture indicates the ion concentration for positive (c+) and negative (c–) ions for the same two cases. For both cases the area between the two divergent lines is equal.

the surface potential varies according to the medium concentration, as does the Debye length (where the concentrations of coions and counterions diverge); note that the area enclosed between the divergent ion concentrations and the surface is the same in both cases.

3.4 The Stern layer

The second aspect of the electrical double layer we will examine is the Stern layer. The diffuse model of countercharge described above assumes that the counterions in the diffuse layer are free to move right up to the interface itself. However, when the counterions come this close to the charged surface, many of them become electrostatically attached or *adsorbed* to the surface. Furthermore, any dipoles that form part of the suspending medium itself — such as water molecules — may also be adsorbed, since the attractive force on the dipole side carrying opposing charge will be greater than the repulsive

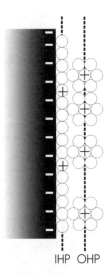

IHP OHP

Figure 3.3 The location of the Helmholtz planes with respect to a negatively charged surface. The inner Helmholtz plane (IHP) consists largely of adsorbed water molecules (empty circles) with some unsolvated ions; the outer Helmholtz plane (OHP) consists of solvated ions.

force on the side carrying like charge, because one side is nearer and thus the Coulombic forces are greater. These molecules form a layer immediately surrounding the surface, the locus of which is called the *inner Helmholtz plane* (IHP). There is then a layer of solvated ions outside the IHP that are also effectively bound to the surface, the locus of which is called the *outer Helmholtz plane* (OHP). These are shown schematically in Figure 3.3. Taken together, the charge structures inside the OHP are described as the Stern layer. The theory was originally proposed in 1924 by Otto Stern (who later won a Nobel prize for work on magnetic moments) as an extension to the Gouy–Chapman theory. Stern's theory considers the ions in solution as objects of finite size rather than point charges, and thus there is a minimum distance to which they can come to the surface, characterized by the IHP.

An important characteristic of the Helmholtz planes is that the charge on the surface and the charge in the IHP form two parallel planes of opposing charge, in the same manner as the capacitors discussed in the previous chapter; we can therefore assume that the potential across the gap between surface and IHP varies linearly, as happens in a capacitor. Similarly, there is a linear variation in potential between the IHP and the OHP.

While it is theoretically possible for the countercharge accumulating across the IHP to cancel out the charge on the surface, this is rarely the case; particularly when one considers that the adsorbed water dipoles do not make a contribution to the countercharge since they are electroneutral. Therefore there is still a net change in the sign of the charge at the surface represented at the OHP, and it is *this* that is "seen" by the ions in the diffuse layer. Similarly, it is the electrical potential at the OHP that marks the equivalent

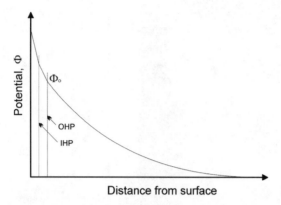

Figure 3.4 The potential across the double layer varies linearly across the gap between the surface and IHP, and again between IHP and OHP, before following an exponential decay through the diffuse layer.

surface potential in the calculations described in the previous section. This is shown schematically in Figure 3.4; as can be seen, the potential is characterized by two linear phases across the Stern layer and an exponential phase across the diffuse layer. In general, we are unable to measure the effects of the IHP, and it is more convenient to consider the effects of the Stern layer as the bound charges of various forms that exist between the OHP and the surface.

It follows that if charge is to be maintained, then the total charge contained in both the Stern (Q_s) and diffuse (Q_d) layers must be equal and opposite to the charge at the surface of the particle Q_p:

$$Q_p - \left(Q_s + Q_d\right) = 0$$
$$Q_p = Q_s + Q_d \tag{3.14}$$

If we consider the effect of a small change in the surface potential Φ_p in terms of the effect on the charge stored in the double layer, the OHP potential Φ_0, and the bulk potential, which we can use as our 0 V reference, then we need to differentiate in terms of the total stored charge. From Equation 3.14 we can obtain

$$\frac{d\Phi_p}{dQ_t} = \frac{d\left(\Phi_p - \Phi_0\right)}{dQ_s} + \frac{d\left(\Phi_0 - 0\right)}{dQ_d} \tag{3.15}$$

Integrating these terms gives us a value for reciprocal capacitance:

$$\frac{1}{C_t} = \frac{1}{C_s} + \frac{1}{C_d} \tag{3.16}$$

which is the expression for the total capacitance for two capacitors in series. This is logical when considering the structure of the interface; a series combination implies that there is only one path from one potential (the bulk) to another (the surface), which requires passing first through one capacitor (the diffuse layer) and then the other (the Stern layer). The total capacitance of the double layer varies with the ion concentration in the bulk, since the capacitance C_d calculated in Equation 3.13 is inversely proportional to the Debye screening length, whereas the Stern layer capacitance is unaffected by changes in the medium. Since the Debye screening length diminishes as ion concentration is increased (Equation 3.9), the value of C_d increases with increasing concentration and the corresponding reciprocal $1/C_d$ decreases, such that at higher ionic concentrations, the total capacitance is dominated by the capacitance of the Stern layer:

$$\frac{1}{C_s} \gg \frac{1}{C_d}$$

$$\frac{1}{C_t} \approx \frac{1}{C_s} ; C_t \approx C_s$$

(3.17)

Similarly, in media of low concentration the value of C_d decreases, until eventually its reciprocal is large enough to dominate over the effect of the Stern layer capacitance, such that $C_t \approx C_d$. If we consider the implications of this in physical terms, then in higher concentrations the majority of stored charge is squeezed into the Stern layer, while at lower concentrations the majority of charge is spread across the much larger diffuse layer. This has a number of implications for the movement of charge around suspended particles undergoing dielectrophoresis, as we will discover later.

3.5   Particles in moving fluids

Having described the arrangement of ions and solvent molecules surrounding a stationary colloidal particle, we now consider what happens when the particle is induced to move relative to the medium.

Fluid moving across the surface of particles at this scale moves in a laminar manner — that is, it is nonturbulent and moves tangentially across the surface of the particle. However, because of electrostatic interactions between medium and surface, and because of the action of viscous drag, there is a layer of molecules immediately adjacent to the particle surface that does not move but remains next to the surface. This is termed the *stagnant* or *boundary* layer, and it is significant in terms of the electrodynamics of the double layer because it contains a significant portion of the double layer charge. This layer is largely independent of surface morphology and exists for both rough and smooth surfaces. It is assigned a thickness d_{ek}.

Since there is a layer beyond which there is no motion of liquid, we can therefore infer that there is a limit that distinguishes the stagnant layer from the free-moving liquid. This limit is known as the *slip plane*. Although it is unlikely that the viscosity effectively undergoes a step change from infinity to a finite value at a specific boundary, the change occupies a sufficiently small thickness for us to consider the model as if that were the case. Since the particle retains its stagnant layer at all times, it effectively acts as though it has a total radius extending to the slip plane and has an electric potential equivalent to the potential at the slip plane. We term this potential the electrokinetic potential or ζ (zeta) potential.

We may ask whether the dynamic model of fluid flow — in which a layer of molecules is immobilized on the surface of the particle due to the action of hydrodynamic interactions resulting in effectively infinite viscosity — corresponds to the similar model of Stern in which molecules are adsorbed on the surface due to electrostatic interactions between dissolved ions, water dipoles, and the charged surface. The answer to this is perhaps unsatisfying but useful; although there does not need to be any direct correlation between the locations of the OHP and the slip plane, it has been demonstrated experimentally that they are approximately equal. Similarly, there is sufficient similarity between the potential at the OHP, Φ_0, and the potential at the slip plane, ζ, for us to consider them approximately equal. Beyond the slip plane, the ionic concentrations fall away as predicted by the Gouy–Chapman model and our previous calculations still apply.

3.6   Colloids in electric fields

So far we have only considered the arrangement of molecules in proximity to a colloid for particles in the absence of an externally applied electric field. However, since this text is concerned with the manipulation of colloidal particles using applied electric fields, it is important for us to determine how the charge accumulated in the double layer will affect the electrical properties of the particle. In Chapter 2 we examined the way in which a particle interacts with an electric field according to whether it is more or less polarizable than the surrounding medium. However, now that we have considered how countercharge accumulates in the atmosphere around the particle, we need to consider how this charge alters the way in which particles polarize.

When an electric field is applied, the charges in the double layer respond by moving toward the appropriate electrode by Coulombic interaction. However, the same charges are also attracted by the particle surface, such that there is a slight net displacement of charge toward the electrode. Since the charged particle and countercharged double layer move in opposite directions under the influence of the electric field, this has the effect of displacing the centers of the charges from being collocated around the center of the particle — that is, the double layer/particle combination becomes polarized. This polarization process occurs in both the Stern and diffuse layers, by slightly different processes; in the Stern layer, the charge is fixed

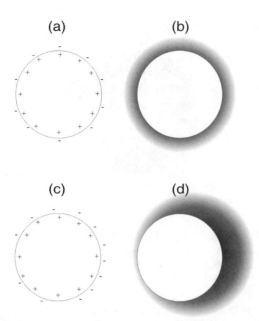

Figure 3.5 When the double layer is at equilibrium, the charge in the Stern layer (a) and diffuse layer (b) is equally distributed in all directions. However, when an electric field is applied, charge is redistributed, with charge in the Stern layer preferentially collecting on one side (c) and the diffuse layer becoming elongated in the direction of the field (d).

within the radius defined by the OHP but can still move in a laminar form along the surface. Because the Stern layer is so thin compared with the radius of the particle, the conduction can be considered to occur in a flat plane across the surface of the particle itself. The conduction process in the diffuse layer is slightly more complex, since the conduction to a polarized state involves the displacement of the ion cloud toward the electrode; the time taken to arrive at this polarized state depends not only on the amount of charge in the diffuse layer, but on how far that charge must move to become polarized. As the medium conductivity is increased, so the diffuse layer contracts and the effective conductivity of the layer increases; this will be significant later. These two polarization processes are shown schematically in Figure 3.5.

As a result of double layer polarization, the particle acquires a dipole moment; this is free to interact with the imposed field and field gradient, such that a particle may experience dielectrophoretic forces due solely to the interaction between the induced dipole and the applied field. The way in which the charge moves across the surface is very important. This is because the process of movement of charge is the process of conduction, and we can envisage the process of electrical conduction in colloids as taking place in two ways: charge can either be transported *through* the particle (as described in Chapter 2), or it may be transported *around* the particle via the double

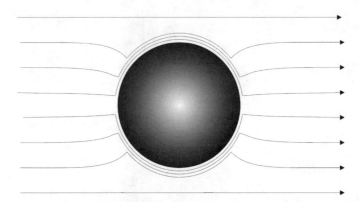

Figure 3.6 Double layer conduction. A particle in an electric field, with a *Du* number greater than 1, will conduct through the double layer, deforming the electric field in the manner shown above. Where *Du* is low, the double layer is insulating and the field pattern is similar to that experienced by insulating particles in electric fields, as shown in Chapter 2.

layer, giving the particle a *surface conductance*. Electrically, this makes no difference; an insulating particle with a highly conductive double layer will be electrically perceived as a conductive particle. This is shown schematically in Figure 3.6.

Stansilav Dukhin introduced a dimensionless number that now bears his name to describe the ratio of surface and bulk conduction. The Dukhin number *Du* is calculated thus:

$$Du = \frac{K_s}{r\sigma_m} \tag{3.18}$$

where K_s is the surface conductance (S), r is the radius of the particle (m), and σ_m is the conductivity of the medium (S m^{-1}). Where *Du* is much less than one, the double layer is effectively insulating and the electric field lines pass around it tangentially, as occurs for generally insulating particles and shown in Figure 2.11b. Where *Du* is high, the field lines meet the double layer at right angles and conduct through it. Even where there is no conduction through the particle itself, there is still effective particle conduction across the surface. In fact, we can go one step further and derive Dukhin numbers for both the Stern and diffuse layers; thus,

$$Du = \frac{K_s}{r\sigma_m} = \frac{K_s^d}{r\sigma_m} + \frac{K_s^i}{r\sigma_m} = Du^d + Du^i \tag{3.19}$$

This relationship will be important for determining the contributions of electrical conduction in the Stern and diffuse layers in the next chapter.

3.7   Electrode polarization and fluid flow

So far we have only considered the effects of electrical double layers at the interface between solution and colloidal surface. However, any charged surface in contact with a liquid acquires an electrical double layer, and so we must also consider these effects. While the influence of charges on the glass surface of capillaries is important for particle manipulation using capillary electrophoresis, the most important electrified interfaces after the ones surrounding the particles are those formed across the electrodes that generate the electric field.

Since the electrodes must supply charge to the surface in order to generate the required electric potential, countercharge accumulates across the electrode surface. However, unlike the fixed charge found on the surface of colloids, the charge on the electrodes is supplied externally and will change according to the magnitude and sign of the surface potential being applied. Furthermore, since the electrodes are also involved in the generation of extremely high electric field strengths, these fields directly couple with the charges in the electrical double layer and are consequently responsible for moving the fluid across the surface. This creates a fluid pumping effect that is ultimately responsible for moving (relatively) large quantities of fluid across the electrode area and particularly across the interelectrode gaps where the electric field is greatest. Unfortunately, this is also where the dielectrophoretic force is greatest, and these forces can act in opposite directions to prevent dielectrophoretic accumulation at electrodes, particularly at lower frequencies.

There are a number of ways in which an electric field can influence the motion of a fluid. At larger scales, the most significant is Joule heating. When a current passes through a medium, the motion of electrons heats the medium. The warmer liquid then rises, causing a displacement of cooler liquid and causing the liquid to circulate. However, at the scales we are concerned with here, the very large electric field strengths used to generate dielectrophoretic forces are applied across very small volumes, so that in low conductivity (and hence highly resistive) media, the actual amount of power applied in the volume between the electrodes is sufficiently small for Joule heating not to be a problem. For example, Ramos et al.[1] determined that the approximate value of the temperature rise ΔT due to an applied root mean square (rms) potential V can be determined thus,

$$\Delta T \approx \frac{\sigma V^2}{k} \qquad (3.20)$$

where σ is the electrical conductivity of the medium and k is the thermal conductivity of the medium. Note that the temperature increase is dependent on the applied voltage rather than the electric field; this is significant in terms of electrode design, since we require that the electric field (and more importantly the electric field gradient) be as large as possible without requiring a

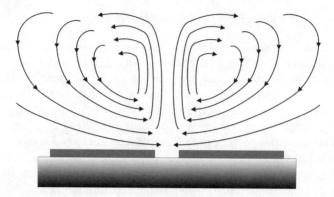

Figure 3.7 Temperature gradients in the medium above the electrode edges (due to electrode heating, particularly by illumination) cause conductivity and permittivity gradients and result in electrothermal fluid flow. Below its frequency f_c, this flow follows a pattern similar to that over the two electrodes shown above (dark gray, in cross-section); above f_c the direction of motion is reversed.

large voltage, which can be achieved by using as small an interelectrode gap as possible. This is particularly important for media of higher conductivity; where a solution conductivity of 10 m S m^{-1} can result in a temperature increase of 1°C in water when a 20 V$_{rms}$ signal is used, a 1 S m^{-1} solution (the conductivity of many physiological media such as blood) in the same electrode array would result in a temperature rise of 100°C, with catastrophic results for biological material within the array.

In addition to the rise in temperature in the solution, the fact that the medium is heated locally gives rise to temperature *gradients* within the medium (that is, the medium temperature is not uniform throughout). Both the conductivity and the permittivity of the medium are proportional to temperature, and the temperature gradient gives rise to conductivity and permittivity gradients, which themselves give rise to forces. The forces due to the two effects are frequency dependent and act counter to one another such that, below a frequency f_c, the force acts to push fluid down the inter-electrode gaps and out across the electrode surfaces, while at frequencies above f_c, the fluid moves across the electrodes to the interelectrode gap and then outward. At frequency f_c there is no fluid motion due to this effect. The direction of motion in parallel, planar microelectrodes is shown schematically in Figure 3.7. The value of f_c is given by the expression

$$f_c = \frac{\sigma}{2\pi\varepsilon} \sqrt{2 \left| \frac{\dfrac{\delta\sigma}{\varepsilon\delta T}}{\dfrac{\delta\varepsilon}{\varepsilon\delta T}} \right|} \tag{3.21}$$

where $\delta\sigma/\delta T$ and $\delta\varepsilon/\delta T$ correspond to the change in conductivity and permittivity due to a change in temperature, respectively. For water, the value of the square root is about three, and for a medium conductivity of 10 m S m^{-1}, we find that the frequency where the force is zero is approximately 7 MHz and scales proportionally to the conductivity. However, the magnitude of the force is related to a number of factors, some surprising; further work by Green and co-workers[2] showed that the magnitude of the fluid flow, which can attain velocities in excess of 100 μm s^{-1}, bears a strong relationship to the intensity of the light used to illuminate the electrode array for particle observation; when the electric field or light intensity are diminished, so is the velocity of the fluid flow. Furthermore, applying a light to only one of an electrode pair causes fluid motion over only the illuminated electrode surface. This has been attributed to the heating of the electrode surface by the illumination source, which greatly exacerbates the temperature (and hence, permittivity and conductivity) gradient across the electrode surface, which combines with the electric field to produce fluid flow.

In addition to this source of fluid motion, there is another that only manifests at low frequencies. This form of fluid flow has been observed for many years and was initially believed to be a third form of dielectrophoresis, adding to the existing forms of positive and negative dielectrophoresis. In experiments using planar electrodes to trap cells, it was observed that at low frequencies, particles that had collected at the electrode edges at higher frequencies would move toward the center of the electrode and collect on its surface. Dubbed an "anomalous dielectrophoretic effect," it remained unexplained until Green and Morgan[3] demonstrated that it was actually due to a combination of positive dielectrophoresis acting to move the particles toward the electrode edges and a fluid flow acting to push them from those edges and onto the electrode array. When this fluid flow is sufficiently strong, the particles take up positions on top of the electrodes where the forces balance, giving the impression of having collected there by dielectrophoretic action alone.

This is dependent on another electrokinetic effect, one that is dependent on the material covered previously in this chapter — that is, the electrical double layer. When the electrode has a voltage applied to it, it acquires an electrical charge of the sign relating to the applied voltage. This then causes the ions within solution to form an electrical double layer across the electrode surface. However, the electric field in electrode arrays designed specifically for the generation of nonuniform electric fields (such as those described here) is such that, although the field lines are perpendicular at the electrode surface, they are tangential through the electrical double layer, so that the vector described by the electric field can be divided into two components: one pointing from the electrode surface and the other pointing along it. The interaction between field line and charge therefore acts not only toward the surface (moving the ions in solution to the surface), it also acts to move the charge along the surface, which by viscous action causes the medium in the vicinity to move along the surface as well. This effect is called *electro-osmosis* and is often used in capillaries as a means of transporting (and sorting) nano-scale particles by a process known as capillary electrophoresis.[4] However,

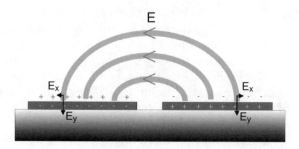

Figure 3.8 When an electric field **E** passes through an electrical double layer at an angle other than 90°, we can consider it as comprising two orthogonal components; one across the double layer and one along it. The one along the surface (E_x in the figure) will induce a force on the charge in the double layer due to Coulombic interaction. Across the electrode surface, this means the fluid will move away from the interelectrode gap; this process is called *electro-osmosis*.

unlike capillary electrophoresis, which operates in DC electric fields, the electro-osmotic fluid flow is, like dielectrophoresis, frequency dependent. It is strong at low frequencies where the double layer has time to form, diminish, and reform with opposing polarity for every cycle of the electric field, but becomes limited at high frequencies where the electrode polarity changes too fast for the double layer to form. The action of the force is then to move fluid along the electrode edges and onto the electrode surface, as shown schematically in Figure 3.8.

Extensive work in the field by Green et al.,[5] González et al.,[6] and Ramos et al.[7] has demonstrated that this force is most particularly active over a range of frequencies between approximately 100 Hz and 10,000 Hz, with the actual frequency response being a complex combination of electric field, electrode geometry, position, and medium conductivity. In fact, the frequency of the peak fluid flow has been observed to change according to the position on the electrode surface. The mathematical treatment of low-frequency electro-osmotic fluid flow is complex and has not been reproduced here; in the work described in much of the remainder of this book, we will consider the manipulation of particles in electric fields of significantly higher frequency. However, since this form of fluid flow can be highly disruptive at frequencies often reaching into the low hundreds of kilohertz, it is important to be aware of both the cause of such fluid flow and, more importantly, its effects.

So far we have discussed the effects of fluid flow when using two parallel electrodes to generate our electric field; however, this type of electrode array is uncommon outside of fluid flow research. More common are electrode arrays such as the quadrupolar and castellated electrodes, which we will discuss in more detail in the next chapter, but which are worth mentioning here in the context of fluid flow in these electrode arrays.

The interdigitated, castellated electrode array was the first planar, microelectrode array developed for dielectrophoresis applications[8] and was

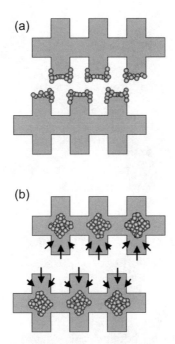

Figure 3.9 The balancing actions of dielectrophoretic force and electro-osmotic fluid flow in a castellated electrode array. Electrodes consist of interdigitated (alternating, interlinked) structures of opposing phases, which have castellated ("castle-like") structures along their sides. In the absence of fluid flow, particles collect along the edges of the castellations nearest the opposing electrode array (a); however, the presence of electro-osmotic fluid flow causes the particles to be pushed onto the electrode surface. Flow from the opposite side of the electrode array means that the forces are in equilibrium when the particles have been pushed into diamond-like structures in the center of the electrode upper surfaces (b).

developed to satisfy the need for electrode arrays with a largely inhomogeneous electric field with well-defined regions of high and low electric field intensity. It was observed in 1988 by Price et al.[8] that, at low frequencies, particles trapped by positive dielectrophoresis moved to form diamond-shaped aggregations on the upper surface of the electrode arrays. Investigations of the electric field across the electrode array surface determined that these diamond-shaped areas corresponded both in location and shape to regions of low electric field strength, and it was thus thought to be due to an unexplained form of *negative* dielectrophoresis. It was not until the studies by Green and Morgan[3] that the motion was shown to be due in fact to fluid flow. When particles collect in this type of electrode array, they do so at the points of highest electric field strength, that is, at the corners and edges of the electrodes nearest to the opposing electrode array, such as are shown in Figure 3.9a. However, as frequency is decreased, fluid flow due to electro-osmosis becomes increasingly prominent; as described above, the location where this is strongest is where the electric field intercepts the

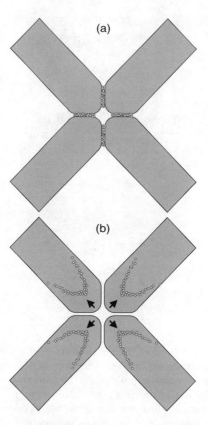

Figure 3.10 The balancing actions of dielectrophoretic force and electro-osmotic fluid flow in a quadrupolar electrode array. Four electrodes meet at a cross and are energized with sinusoidal voltages such that adjacent electrodes have opposing phases and opposing electrodes have like phases. In the absence of fluid flow, particles collect in the gaps between adjacent electrodes by positive dielectrophoresis (a); however, the presence of electro-osmotic fluid flow causes the particles to be pushed onto the electrode surface until they form bow-like structures where the forces are in equilibrium.

double layer at the sharpest angle, which is also at the castellations of the electrodes that are closest together. Therefore, those particles that have collected by positive dielectrophoresis are those that experience the greatest fluid motion, causing them to be swept back onto the electrode surface. As they move further from the electrode edge, the angle of the electric field becomes more orthogonal and the fluid flow diminishes; eventually a neutral point is reached where the two processes are in equilibrium and the particles remain at rest; this is at the center of the array and is responsible for the diamond-shaped collections.

Similar effects are observed in the other common electrode array, the quadrupolar array (Figure 3.10). These were developed by Huang and Pethig[10] for negative dielectrophoresis applications by devising an electrode array that was "uniformly nonuniform." The electric field across the center of the array varied in a mathematically predictable manner (at least in the two dimensions of the electrode plane); the electric distribution follows a polynomial equation, and in fact these electrodes are often referred to as polynomial arrays. Arrays of this type were originally developed by Arnold and Zimmermann for electrorotation studies,[11] and simulations have verified that they are near optimally suited for the purpose.[12] For nanoparticle studies, arrays of this type allow fabrication with very small feature size (interelectrode gaps 500 nm wide have been reported[13]) and again provide well-defined regions of high and low electric field. The fluid effects described above for castellated electrodes are also visible for quadrupolar arrays, though their origin is more clearly visible; particles collect by positive dielectro-phoresis in the gaps between adjacent electrodes, as shown in Figure 4.11a. When low-frequency fluid flow is present, the direction of the flow is ortho-gonal to the electrode edges and pushes the particles on to the electrode array, as before; these form a distinctive bow wave effect on the electrode surface. If the fluid flow is strong enough, the dielectrophoretic force is not strong enough to retain the particles and they are observed to form jets or streams from the center of the electrode array, moving in straight lines along the center of the electrode surfaces.[14]

3.8   Other forces affecting colloidal particles

Although the majority of work in this book describes the use of electric fields to induce a force on particles, there are a number of other forces that also act to move them to a greater or lesser degree. While some of these forces are generally of sufficiently low magnitude to prevent them from causing significant disruption to dielectrophoretic manipulation, others cause signifi-cant, observable changes in motion, and, as such, an awareness of their effects is useful.

3.8.1   Viscous drag

We have already described how viscous drag can cause electro-osmosis across the surface of electrodes and particles under the influence of an electric field, but it is also responsible for retarding the motion of any particle moving through anything other than a vacuum. For example, objects moving through air achieve a terminal velocity (maximum velocity for a given force input) of about 150–200 km h^{-1} in most cases; in water the drag forces are much more significant and, as the object reduces in size, the forces get (relatively) even larger. For colloidal particles, the force is so significant that observing them under a microscope and moving the solution with a pipette gives the impression the particles are embedded in jelly, so little do they move.

Because the viscous drag force is so high, particles on this scale reach terminal velocity in a few nanoseconds. Therefore, the velocity at which they move is directly related to the magnitude of the force propelling them through the medium (objects falling through air achieve constant velocity because the applied force — gravity — is approximately constant) and the viscosity of the medium itself. For a spherical particle of radius r, these factors are related by Stokes's law, which describes the terminal velocity of a particle through a medium with the following expression:

$$V = \frac{F}{6\pi\eta r} \tag{3.22}$$

where η is the viscosity of the medium. This only holds for particles with low Reynolds numbers (less than 0.3); however, the particles and media considered here are such that for all the examples here, the above can be considered correct.

3.8.2   Buoyancy

In the introduction to this chapter, we found that a colloid is defined as a particle that does not precipitate, because Brownian motion overcomes the buoyancy force that would otherwise make particles float or sink. This is a generalization; larger particles such as 250-nm-diameter herpes viruses, which are somewhat denser than water, can sink over time — a fact that allows the measurement of the point at which the sedimentation force and dielectrophoretic force are balanced. The magnitude of this force (dubbed the sedimentation force) is given by

$$\mathbf{F}_{sediment} = v\left(\rho_m - \rho_p\right)\mathbf{g} \tag{3.23}$$

where ρ_p and ρ_m are the densities of the particle and medium, respectively, v is the particle volume, and g is acceleration due to gravity. The force follows a vector toward the direction of gravity, i.e., downward; if the force is negative the force acts upward and the particle will float (in which case, the force may be referred to as the *buoyancy force*). Colloidal particles are of sufficiently small volume that the sedimentation force is small compared with other forces, but when averaged over time the sedimentation effect can be observed, such as when it is used as a method of determining particle properties (Chapter 5), or it can be used as a means of separating particles in field-flow fractionation (Chapter 8).

3.8.3   Brownian motion and diffusion

Far more significant effects are observed due to the action of Brownian motion and diffusion. These are effects that become increasingly significant as the size of the particle is reduced, to the extent that it was suggested by Pohl[15] that such forces would be of sufficiently high magnitude to prevent dielectrophoretic manipulation of colloidal particles from happening at all.

Brownian motion is the result of particles suspended in a liquid being bombarded across their surfaces by the molecules of which the liquid is comprised. If the body is large enough, sufficient numbers of these impacts will happen in almost all directions nearly simultaneously; for small bodies, the net result of these molecular impacts is less likely to sum to zero and there is a net movement induced in the particle.

Although the impacts of Brownian motion may occur with equal chance across the surface of the particle and hence the time-averaged force will be zero, the irregular nature of the particle over time means that there is a net displacement. Studies by Einstein[16] demonstrated that the magnitude of the displacement of the particle from its initial position follows a Gaussian statistical distribution with mean square displacement in one dimension (direction) given by the equation

$$\left| \Delta \bar{d}^2 \right| = 2Dt \tag{3.24}$$

where t is the time and D is the diffusion coefficient. For spherical particles in a liquid, this is given by

$$D = \frac{k_B T}{6\pi\eta r} \tag{3.25}$$

where η is the viscosity of the medium, r is the radius of the particle, T is the temperature (in Kelvin), and k_B is the Boltzmann constant. Combining Equations 3.24 and 3.25, the mean displacement for spherical particles is given by

$$|\Delta d| = \sqrt{\frac{k_B T t}{3\pi\eta r}} \tag{3.26}$$

This is a significant result because we need to overcome the action of Brownian motion when applying dielectrophoresis, as discussed in the next chapter. Note that unlike dielectrophoresis, Brownian motion is not a force; it arises from a succession of force-imparting impacts. These result in an acceleration to terminal velocity in a time period of the order of fractions of a picosecond for particles of the size discussed here, and the time-averaged force is zero, even though there is net displacement over time. In order to compare the effects of different phenomena on particles, we must take different approaches, which will be discussed in the next chapter.

Another force, diffusion, is used to describe the behavior of ensembles of particles in solution; for example, when a drop of one liquid is added to a flask of another liquid, the combined system is strongly inhomogeneous with high concentrations of the additive surrounded by a volume with almost no concentration. After a period of time, this changes and there is an approximately homogeneous distribution of the two liquids (assuming they are soluble in one another). The movement of the ensemble is called diffusion, and we can describe it by a *diffusion force* that moves the particles toward an equilibrium state. However, in the work described here the particles are sufficiently isolated for us to consider each as a specific body independent of all others, and as such we may disregard the diffusion force in our examinations.

3.8.4   Colloidal interaction forces

A significant class of forces that is important in the behavior of colloidal solutions is those between the colloidal particles themselves. The simplest of these to understand in the context of the work so far described is the interaction of the particles and their double layers when they move in close proximity. When two identical particles are close to one another, there is a repulsive force generated due to the Coulombic interaction between the particles and their corresponding Debye atmospheres when they approach to within a distance equal to their combined Debye screening lengths. Another effect, discussed in more detail in Chapter 2, is the induced dipole–induced dipole force (referred to on the quantum level as the London force). When two charged particles interact, any asymmetry in the electric field generated by the locations of distributed charge can lead to the induction of a dipole in one of the particles, which can then induce a dipole in the neighboring particle. Since the second induced dipole will be aligned in the same direction as the first, this leads to an attractive force. Similar forces occur where particles have permanent dipoles (in which case, the force is known as the Keesom force).

The combination of electrostatic, induced dipole, and permanent dipole interactions forms the basis of the van der Waals force, which describes the interaction between colloidal particles on the nanometer scale. Under the conditions of dielectrophoretic manipulation, the influence of van der Waals interactions is sufficiently small for them to merely assist in the processes that are being used; for example, in assisting in the trapping of small colloids to a greater extent than may be possible without them. However, in the presence of dielectrophoresis, electrophoresis, fluid flow, Brownian motion, and so forth, it is virtually impossible to observe the contribution of van der Waals forces at present.

Van der Waals forces play a significant role in the dielectrophoresis of colloidal particles because they govern the extent to which, when two colloidal particles come into contact, they remain separate, free-moving particles. If the electrostatic repulsion force between two particles is low, then they may coagulate; if the forces are larger, then particles may move to very short distances but will not stick. Essentially, van der Waals forces act to counter the electrostatic interactions between particles; where the electrostatic forces between like particles are repulsive, the van der Waals force is attractive. It is this balance that governs the *stability* of dispersed colloidal solutions; if the electrostatic interactions are greater than the van der Waals force, then the particles will remain dispersed even when in close proximity. However, if the van der Waals force is stronger, then the particles will stick together — either *flocculating* (forming a loosely bound cluster that can be broken apart with mechanical agitation) or *coagulating* (forming a permanent structure with the colloids being irrevocably attached to one another). The stability of a solution is therefore dependent on the strength or weakness of the interparticle electrostatic repulsion forces, which can in turn depend on a number of factors, such as medium pH. A well-known example of this is the ability of milk to curdle when a drop of lemon juice is added — the lemon juice lowers the pH of the milk, so that the colloidal fat droplets within coagulate. The stability of a colloidal solution is governed by the DLVO theory (named after its authors, Derjaguin, Landau, Verwey, and Overbeek), which describes the way in which attractive long-range van der Waals forces and repulsive electrostatic forces (generally between the double layers of the two particles) balance. If the attractive forces sufficiently overcome the repulsive forces at short range (for example if a pH change lowers the surface charge, and hence the magnitude of the repulsive force), the particles will coagulate.

Under the vast majority of conditions described here, the force is sufficient for particles not to coagulate, with the notable exception being in the assembly of nanoscopic devices described in Chapter 7.

References

1. Ramos, A., Morgan, H., Greeen, N.G., and Castellanos, A., AC electrokinetics: a review of forces in microelectrode structures, *J. Phys. D: Appl. Phys.*, 31, 2338, 1998.
2. Green, N.G., Ramos, A., Gonzalez, A., Castellanos, A., and Morgan, H., Electric field induced fluid flow on microelectrodes: the effect of illumination, *J. Phys. D: Appl. Phys.*, 22, L13, 2000.
3. Green, N.G. and Morgan, H., Separation of submicrometre particles using a combination of dielectrophoretic and electrohydrodynamic forces, *J. Phys. D: Appl. Phys.*, 31, L25, 1998.
4. Lyklema, J., *Fundamentals of Colloid and Surface Science*, Vols. 1–2, Academic Press, London, 1994.
5. Green, N.G., Ramos, A., Gonzalez, A., Morgan, H., and Castellanos, A., Fluid flow induced by nonuniform AC electric fields in electrolytes on microelectrodes. I. Experimental results, *Phys. Rev. E.*, 61, 4011, 2000.
6. González, A., Ramos, A., Green, N.G., Castellanos, A., and Morgan, H., Fluid flow induced by nonuniform AC electric fields in electrolytes on microelectrodes. II. A linear double-layer analysis, *Phys. Rev. E.*, 61, 4019, 2000.
7. Ramos, A., Morgan, H., Green, N.G., and Castellanos, A., AC electric-field-induced fluid flow in microelectrodes, *J. Coll. Int. Sci.*, 217, 420, 1999.
8. Price, J.A.R., Burt, J.P.H., and Pethig, R., Applications of a new optical technique for measuring the dielectrophoretic behavior of microorganisms, *Biochim. Biophys. Acta*, 964, 221, 1988.

9. Pethig, R., Huang, Y., Wang, X.B., and Burt, J.P.H., Positive and negative dielectrophoretic collection of colloidal particles using interdigitated castellated microelectrodes, *J. Phys. D: Appl. Phys.*, 25, 881, 1992.
10. Huang, Y. and Pethig, R., Electrode design for negative dielectrophoresis applications, *Meas. Sci. Technol.*, 2, 1142, 1991.
11. Arnold, W.M. and Zimmermann, U., Electro-rotation — development of a technique for dielectric measurements on individual cells and particles, *J. Electrostatics*, 21, 151, 1988.
12. Hughes, M.P., Computer-aided analysis of conditions for optimizing practical electrorotation, *Phys. Med. Biol.*, 43, 3639, 1998.
13. Müller, T., Gerardino, A.M., Schnelle, T., Shirley, S.G., Fuhr, G., De Gasperis, G., Leoni, R., and Bordoni, F., High frequency electric field trap for micron and submicron particles, *Il Nuovo Cimento Della Societa Italiana di Fisica D*, 17, 425, 1995.
14. Green, N.G. and Morgan, H., Dielectrophoretic investigations of sub-micrometer latex spheres, *J. Phys. D: Appl. Phys.*, 30, 2626, 1997.
15. Pohl, H.A., *Dielectrophoresis*, Cambridge University Press, Cambridge, 1978.
16. Einstein, A., *Investigations on the Theory of Brownian Movement*, Dover, New York, 1956.

Supplementary reading

1. Bockris, J.O'M. and Reddy, A.K.N., *Modern Electrochemistry*, Vols. 1–2, Macdonald, London, 1970.
2. Evans, D.F. and Wennerström, H., *The Colloidal Domain*, VCH Publishers, Weinheim, Germany, 1994.
3. Lyklema, J., *Fundamentals of Colloid and Surface Science*, Vols. 1–2, Academic Press, London, 1994.

chapter four

Analysis and manipulation of solid particles

4.1 Dielectrophoresis of homogeneous colloids

Dielectrophoretic forces can be induced in a wide range of submicrometer particles from molecules to viruses. However, before we look at how dielectrophoresis might be applied to studying these complex particles, it is wise to consider a simple case to see how the basic principles of dielectrophoresis work at the submicrometer range. This is important since, when the diameter of the particle being manipulated becomes significantly smaller than 1 µm, a number of factors that have relatively little effect on the dielectrophoretic response of larger particles such as cells increase in importance and begin to dominate the response. In order to understand these effects fully, we shall examine the case of a simple spherical particle, consisting of one material only. By examining this, we can develop a model with which we will later test the properties of viruses and proteins.

One of the key experimental tools to understanding the fundamental mechanisms underlying the dielectrophoresis of particles on the nanometer scale is the homogeneous sphere or bead, typically made from polymers such as latex, or occasionally from metals such as palladium. Most common in dielectrophoresis research are latex spheres. These are (as their name suggests) spherical blobs of latex that have been impregnated with fluorescent molecules, enabling the observation of very small particles (sizes as small as 14 nm diameter are available) with a fluorescent microscope. The primary advantage of using latex spheres is that they are very much a known quantity. They are solid and homogeneous (that is, they consist of one material and are consistent throughout). The internal conductivity and permittivity are known, as are the surface properties. Furthermore, there are straightforward chemical methods for changing those surface properties. Since they are spherical, they conform to established models of dielectrophoretic behavior. Finally, they are readily available in a wide variety of sizes and colors. For this reason, latex beads have been used by a number of researchers for investigating fundamental electrokinetic effects in nanoparticle systems.

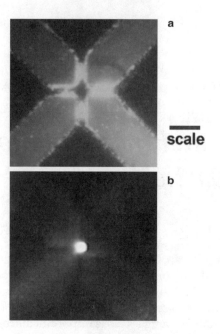

Figure 4.1 Fluorescence photograph of 216-nm-diameter latex beads collecting in an electrode array, exposed to a 10 V_{pk-pk} signal in a dilute KCl solution. (a) Positive dielectrophoresis experienced when the applied signal has frequency 1 MHz; (b) negative dielectrophoresis when the frequency is increased to 10 MHz. The scale bar is 20 μm long.

4.2 Frequency-dependent behavior and the crossover frequency

Now that we have examined our dielectric theory and understood the nature of our colloidal test particles in Chapters 2 and 3, it is time to examine the results of practical experiments and to understand how other factors resulting from the size of the particle and its nature as a colloid affect the way a particle responds to an electric field. A typical experiment might involve suspending a sample of latex beads in a solution of known conductivity (such as ultrapure water with a measured quantity of potassium chloride, KCl), applying the solution to an electrode array written onto a microscope slide, and covering the assembly with a coverslip. The electrode slide is then placed onto a fluorescence microscope (required in order to see particles this small) and the electrodes are connected via attached wires, to a power source (typically a benchtop signal generator, providing perhaps 5 V_{pk-pk} at a frequency between 10 kHz and 10 MHz or more).

When the voltage is applied, the particles are observed to move quickly to the electrodes. Within a few seconds, collections such as those shown in Figure 4.1 are observed; whether the particles collect in the interelectrode "arms" (Figure 4.1a) or in the center of the array (Figure 4.1b) depends on the frequency of the applied voltage; these behaviors occur at low and high frequencies, respectively. At one specific frequency, the force appears to vanish and the particles float freely. Varying the voltage also changes the force, making the particles travel more quickly or slowly to the trap. If the particles are small enough, then the magnitude of Brownian motion is sufficient to require a large voltage be applied in order to ensure the particles remain trapped.

As we have seen, the dielectrophoretic force, $\mathbf{F}_{DEP}$, acting on a homogeneous, isotropic dielectric sphere, is given by

$$\mathbf{F}_{DEP} = 2\pi r^3 \varepsilon_m \, \text{Re}\big[K(\omega)\big] \nabla E^2 \qquad (4.1)$$

where $\text{Re}[K(\omega)]$ is the real part of the Clausius–Mossotti factor, given by

$$K(\omega) = \frac{\varepsilon_p^* - \varepsilon_m^*}{\varepsilon_p^* + 2\varepsilon_m^*} \qquad (4.2)$$

where ε_m^* and ε_p^* are the complex permittivity of the medium and particle, respectively, and $\varepsilon^* = \varepsilon - j(\sigma/\omega)$ with σ the conductivity, ε the permittivity, and ω the angular frequency. The frequency dependence of ε^*, and hence $\text{Re}[K(\omega)]$, implies that the force on the particle also varies with the frequency. The magnitude of $\text{Re}[K(\omega)]$ depends on whether the particle is more or less polarizable than the medium. If $\text{Re}[K(\omega)]$ is positive, then particles move to regions of highest field strength (positive dielectrophoresis); the converse is negative dielectrophoresis, where particles are repelled from these regions.

At frequencies where $\text{Re}[K(\omega)] = 0$, a particle experiences no dielectrophoretic force; since the value of the force changes sign on either side of this frequency, it is commonly referred to as the *crossover frequency*. Crossover frequencies are a product of dielectric dispersions that cause the relative polarizability of the particle to change sign. It is possible to monitor the effects of changing the medium conductivity on the crossover frequency in order to estimate the properties of the particle. In this chapter, where we are only considering the case of homogeneous particles exhibiting a single dielectric dispersion, this is somewhat simplified since there is only a single dielectric dispersion for any given medium conductivity.

Consider the following example. The polarizability of a particular homogeneous sphere will exhibit a single dielectric dispersion such as the one shown in Figure 4.2 when suspended in an aqueous medium of conductivity 1 mS m⁻¹, as calculated using the Clausius–Mossotti factor (Equation 4.2). If the medium conductivity is increased, the polarizability of the particle compared with the medium drops, resulting in the predispersion (positive) side of the curve having a lower value. Eventually the low-frequency polarizability becomes so low that it is below zero at all frequencies; that is, the particle always experiences negative dielectrophoresis. This can be seen in Figure 4.3, where the polarizability is plotted for a range of suspending medium conductivities.

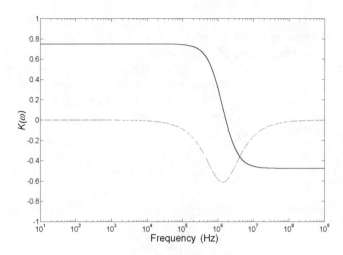

Figure 4.2 The real part (solid line) and imaginary part (dotted line) of the Clausius–Mossotti factor for a solid, homogeneous spherical particle.

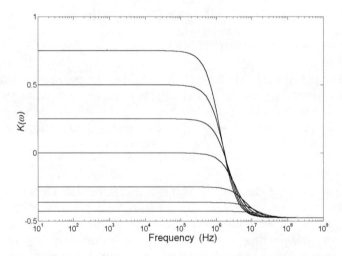

Figure 4.3 The real part of the Clausius–Mossotti factor as a function of frequency for a 216-nm-diameter latex bead, for different values of suspending medium conductivity. The conductivity varies from 0.1 mS m^{-1} (top line) to 500 mS m^{-1} (bottom line). At conductivities above 20 mS m^{-1}, Re[$K(\omega)$] is always negative, that is, the particles always experience negative dielectrophoresis. At lower conductivity, particles cross from positive to negative dielectrophoresis at about 3 MHz.

If we plot the polarizability as a function of both frequency *and* conductivity of the suspending medium, we find a plot such as shown in Figure 4.4a. Ideally, it would be convenient to directly measure the polarizability as a mechanism for determining the dielectric properties of the particle, a method often used for the measurements of cells by determining the rate at which particles collect under positive dielectrophoresis for different frequencies.

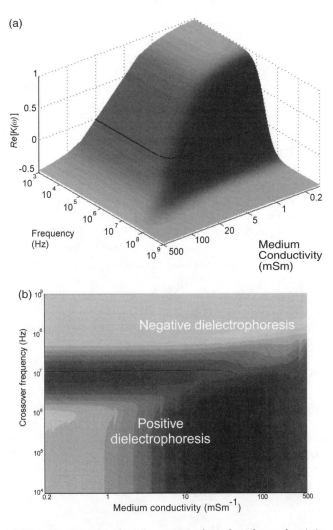

Figure 4.4 (a) The data presented in Figure 4.3 plotted with conductivity on a third axis. The combinations of frequency and conductivity where Re[$K(\omega)$] = 0 form a distinct shape indicated by the black line. Plotting this as a function of conductivity and frequency we obtain figure (b), showing the variation in crossover frequency (where Re[$K(\omega)$] is zero) for various medium conductivities.

However, this is not easy in the case of submicrometer particles where electro-hydrodynamic and Brownian motions can easily disrupt the stable collection of particles. While successful attempts have been made to use a modified collection rate technique to study both latex beads and viruses (discussed in Chapter 5), a far more convenient method of determining dielectric properties is to examine the intercept on the X-Y plane in Figure 4.4a — the plot of frequency against conductivity where the value of Re[$K(\omega)$] is zero — and infer the dielectric properties from that graph. This technique has been used widely to study latex spheres,[1-3] viruses,[4-6] and proteins,[7] as well as larger particles such as cells.[8-10] It is convenient for the measurement of colloids because the zero force frequency can always be seen quite clearly, even in the presence of disruptive fluid flow or Brownian motion. Such a graph — in effect, Figure 4.4a viewed from overhead — is shown more clearly in Figure 4.4b. In reality, data are collected at only a few conductivities and are more likely to be represented as a series of points for different medium conductivities, with a best-fit line being used to determine the most likely data set for the experimental data.

Consider the predicted dielectrophoretic response of a homogeneous latex sphere of diameter 200 nm. We can determine this by inserting values for the conductivity and permittivity of the sphere (conductivity 10^{-6} S m^{-1}, relative permittivity 2.55) and the suspending medium (say, conductivity 10^{-3} S m^{-1}, relative permittivity 78) into Equation 4.2 and calculating values of Re[$K(\omega)$] over a range of frequency. If we consider the mechanism described earlier where the plot of polarizability is gradually reduced, we would expect that the crossover would start at a given value and remain at approximately that value as the conductivity increases, until a threshold is reached where the crossover frequency drops. Above that threshold conductivity, the crossover frequency drops sharply, and above the threshold, the particle exhibits only negative dielectrophoresis; this is the profile shown by the black line in Figure 4.4b. For micrometer-scale homogeneous particles, the observed dielectrophoretic response closely matches the crossover spectrum. However, as the diameter of the particle under study is reduced past 1 μm, this model becomes increasingly inaccurate. The crossover is found to rise with increasing medium conductivity, and above the threshold where the crossover drops rapidly and only negative dielectrophoresis should be seen, the particle still exhibits a crossover but at a lower frequency. The reason for this change in behavior is the increasing effect of the surface charge, and more specifically the electrical double layer.

4.3   Double layer effects

Whereas the anticipated response of our 200 nm diameter latex beads might be expected to have a crossover spectrum such as that shown in Figure 4.4b, the actual experimental response looks like the one shown in Figure 4.5. As can be seen, there are a number of significant differences. In our original model, the response is constant over the lower range of conductivities and always exhibits negative dielectrophoresis at higher conductivities; in reality the crossover frequency exhibits a rise with increasing conductivity, and when it reaches the threshold and the crossover frequency drops, it only does so by about one order of magnitude. Furthermore, when we use the model to determine the permittivity of a latex particle, we find the answers to be quite different from the values we know to be true of bulk latex. For example, in order to find a remotely reasonable fit such as the one shown in Figure 4.5, the conductivity of the particle in the model must be much greater than that

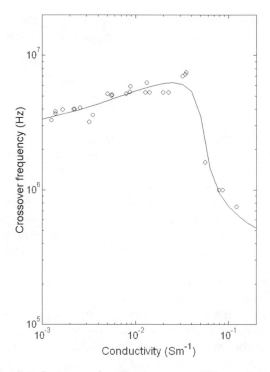

Figure 4.5 The predicted response for a homogeneous, 216-nm-diameter latex sphere (line) and experimental data (dots).

which we know latex to possess. Experiments with latex beads of different sizes show that these effects become increasingly prominent as the size of the particle is decreased. Although there are different causes for these effects, they are all related to the movement of charges in the electrical double layer around the bead; specifically, the movement of charge around the Stern and diffuse layers separately, and the dielectric dispersion experienced by the charges in the double layer (the alpha relaxation). We will examine these separately, to examine how they each affect the dielectric response of the particle.

4.3.1   Charge movement in the double layer

It has been know for some years that the surface charge affects the dielectric response of particles; early studies by Fricke and Curtis in 1936[11] and Schwarz in 1962[12] showed that the net electrical properties of particles could be significantly influenced by surface conduction. Subsequent studies by Arnold et al.[13] and Zimmermann[14] demonstrated that the electrorotation of latex spheres produced anomalously high values of internal conductivity — which in latex spheres should be near zero — which was attributed to the movement of charge around the particle. Solution charges are attracted to the charges on the surface of the particle; when placed in an electric field the charges appear to move in a laminar fashion around it. Arnold and

Zimmermann determined that the component of aggregate particle conductivity σ_p attributed to surface charge movement could be determined using the following equation:

$$\sigma_p = \sigma_{pbulk} + \frac{2K_s}{r} \qquad (4.3)$$

where σ_{pbulk} is the conductivity of the particle interior, K_s is the surface conductance of the particle, and r is the radius of the particle. This formula was used, for example, in the determination of the surface charge of cells infected by malarial parasites by Gascoyne et al.[15] In our study of nanometer-scale particles, this is very useful since it explains why our models of the behavior of our particles indicate a significantly higher conductivity than we know latex to possess. This effect becomes increasingly significant as particle radius is decreased, due to the inverse relationship between radius and the additional conductivity term due to surface conductivity.

For latex spheres, the bulk conductivity is negligible, so the effective conductivity of the particle is dominated by the surface conductance, K_s, where typical values of K_s are of the order of 1 nS. We can extend this further; according to Lyklema,[16] the surface conductance can be calculated directly from the surface charge density, provided the mobility of the ions in the Stern layer are known, using the formula

$$K_s = \sigma^i \mu^i \qquad (4.4)$$

where in this case σ^i represents the charge density that exists on the surface of the particle and μ^i is the mobility of the counterion in the double layer that is usually slightly lower than the value of mobility in the bulk solution.

4.3.2 Charge movement in the Stern and diffuse double layers

When we extend this model to nanoparticles, however, this approximation is too simple. For a cell with a radius that is significantly (by several orders of magnitude) larger than its Debye length, the thickness of the diffuse double layer is so small that charge moving through both components of the double layer can be treated as if occurring in an infinitesimally thin sheet surrounding the particle surface. When the Debye length becomes a more significant fraction of the particle radius — that is, when the particle is a colloid — we can no longer ignore the double layer structure.

The movement of charge though the Stern layer occurs in a layer of finite thickness a few molecules thick and is governed by the surface conductance. There is also a second layer of charge movement in the diffuse double layer. This is different and distinct from charge movement in the Stern layer; where the Stern layer charge is bound to the surface of the particle and moves in thin layer across the surface, charge distributed in the diffuse layer forms an amorphous ionic cloud around the particle as described in Chapter 3. Significantly, the size of this cloud is inversely proportional to the conductivity of the suspending medium — the greater the ionic strength of the medium, the thinner the diffuse double layer is.

As before, we can model the surface conductance effects as contributing to particle conductivity. We can expand this to contain terms due to both the charge movement in the Stern layer and to charge movement in the diffuse part of the double layer.[13] The total surface conductance can then be written as

$$K_s = K_s^i + K_s^d \tag{4.5}$$

where K_s^i and K_s^d are the Stern layer and the diffuse layer conductances, respectively; this then corresponds to a net particle conductivity given by the expression

$$\sigma_p = \sigma_{pbulk} + \frac{2K_s^i}{r} + \frac{2K_s^d}{r} \tag{4.6}$$

Unlike the processes within the Stern layer, charge movement in the diffuse part of the double layer is related to *electro-osmotic transport* rather than straightforward conduction. Electro-osmosis is a process of fluid movement due to an applied potential across a nearby charged surface; the counter-charge accumulates near the surface and then moves in the electric field due to Coulombic attraction. The presence of the surface creates a viscous drag that impedes the motion of the charges. Lyklema[16] gives the following expression for the effective conductance of the diffuse layer containing one ionic species:

$$K_s^d = \frac{\left(4F^2 cz^2 D^d \left(1 + 3m/z^2\right)\right)}{RT\kappa} \left(\cosh\left[\frac{zq\zeta}{2kT}\right] - 1\right) \tag{4.7}$$

where D^d is the ion diffusion coefficient for the ionic species (counterion) in the diffuse layer, z the valence of the counterion, F the Faraday constant, k Boltzmann's constant, R the gas constant, q the charge on the electron, T the temperature, κ the inverse Debye length, c the electrolyte concentration (mol m^{-3}), and ζ the ζ potential. The dimensionless parameter m is given by

$$m = \left(\frac{RT}{F}\right)^2 \frac{2\varepsilon_m}{3\eta D^d} \tag{4.8}$$

where η is the viscosity. A key factor in this expression is the relationship between the surface conductance and the concentration of ions in the bulk medium, which appears twice in this expression. There is a c in the expression itself, and a $c^{1/2}$ in the expression for κ. This gives a net contribution of $c^{1/2}$ to the total diffuse layer conductance. Since the concentration governs the medium conductivity, this expression indicates that since the conductivity of the medium is increased, so the conductivity of the particle will increase but by a lesser degree. This is what we see when the crossover frequency of the particle rises when the medium conductivity is increased; the effective conductivity of the particle is also increased. The remaining values in the equations are more or less constants; the principal unknown variable is the ζ potential. This is known to vary slightly as a function of medium ionic strength but the variation is small, and its mechanism is not fully understood. However, since the concentration of ions is known, determining the diffuse layer conductance allows the direct measurement of ζ potential.

4.3.3 Stern layer conduction and the effects of bulk medium properties

In many circumstances, we can use Equation 4.4 as an approximation for the conductance of the Stern layer. However, although the coionic species does not comprise a significant part of the Stern layer, it was demonstrated by Green and Morgan[2] that latex beads exhibited different behavior in solutions of KCl and KPO$_4$, despite the fact that the counterion was K$^+$ in both cases. The conductance of the Stern layer is in fact dependent on the mobility of the coions in bulk solution, as we can see by examining the factor that relates the bulk and Stern layer conductivities, the Dukhin number. It has been shown[16] that the Dukhin number Du for conductivity in the Stern layer containing one ion species i is given by the equation

$$Du = \frac{u\mu_s^i}{2rz^i Fc^i \mu_m^i} \tag{4.9}$$

where u is the surface charge density of the particle and μ_s^i and μ_m^i are the mobilities of the ion species in the Stern layer and bulk medium, respectively. Combining Equation 4.9 with the definition of the Dukhin number in Equation 3.18 and rearranging for K_s^i, we obtain the following expression:

$$K_s^i = \frac{u\mu_s^i \sigma_m}{2z^i Fc^i \mu_m^i} \tag{4.10}$$

If the electrolyte is symmetrical — that is, the molarity of the counterions is the same as the molarity of the coions — it is possible to replace the conductivity term and concentration c^i with a molar conductivity Λ (expressed in S m^2 mol^{-1}):

$$K_s^i = \frac{u\mu_s^i \Lambda}{2z^i F\mu_m^i} \tag{4.11}$$

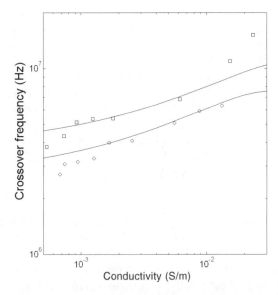

Figure 4.6 A graph showing the crossover frequencies for 216-nm latex beads in KCl solutions (open boxes) and KPO_4 solutions (filled boxes) of similar conductivity.

Values of Λ are constant for solutions of given electrolytes. This is notable since it only produces the same value as Equation 4.4 if the values of the mobilities of the coion and counterion are equal, in which case the value of $\Lambda/2z^{i}F\mu^{i}_{m}$ goes to 1.

For example, Figure 4.6 shows the crossover frequencies for latex beads suspended in solutions of KCl and KPO_4 (a 50:50 mixture of monobasic and dibasic potassium phosphate, with a pH of 7.0) with conductivities between 0.5 mS m^{-1} and 50 mS m^{-1}. The best-fit lines were calculated using values for the relative permittivities of the suspending medium and bead interior of 78 and 2.55, respectively. The best fit to the data gives values of K^{i}_{s} and ζ equal to 0.85 nS and 100 mV for the beads suspended in KCl, and 1.25 nS and 120 mV in KPO_4, respectively. We would expect the dielectrophoretic response to remain the same since the ion species represented in the Stern layer remains the same (the K^{+} ion) for both the KCl and KPO_4 solutions. However, while the diffuse layer characteristic of an increasing effective particle conductivity is similar in both cases, the low medium conductivity response (which is dominated by Stern layer conductance) is different. The molar conductivities of KPO_4 and KCl are 21.2 mS m^{2} mol^{-1} and 14 mS m^{2} mol^{-1}, respectively. We would anticipate from Equation 4.11 that the Stern layer conductivity is approximately 51% greater than that for KCl with all other values remaining the same. This is borne out in the best-fit data, where K^{i}_{s} for the beads in KPO_4 and KCl solutions differs in the ratio 1.47.

Analysis of similar beads[2] using a Coulter counter found that beads of various sizes had surface charge densities ranging between 22 and 40 mC m^{-2}, with 33 mC m^{-1} falling within the tolerance range of all bead sizes; KCl has a molar conductivity Λ of 14 mS m^2 mol^{-1}. KCl is a symmetrical ion, and the bulk mobility of K$^+$ is 7.69×10^{-8} m^2 V^{-1} s^{-1}. Using these values, together with the Stern layer conductance values derived earlier, we can calculate from Equation 4.11 a value for ion mobility within the Stern layer of 3.7×10^{-8} m^2 V^{-1} s^{-1}, or approximately half that of the ion mobility in the bulk suspending medium. This is in keeping with Lyklema's[16] assertion that the mobility of counterions in the Stern layer should be "close to, or somewhat lower than, the corresponding bulk values." If we calculate the value of $K_s^i = u\mu_s^i$ using the above data, we find an estimated value of 3.33×10^{-8} m^2 V^{-1} s^{-1}, i.e., the value of $\Lambda/2z^i F\mu_m^i$ is approximately 0.96, and the approximation holds.

This result is important in that there is an effect on the Stern layer mobility, and hence the surface conductance, due to the coion in solution. For homogeneous colloidal particles such as latex spheres, if the Stern layer mobility is known, dielectrophoretic analysis offers a simple method of determining the surface charge density, ξ potential, and internal permittivity, providing the internal conductivity is known. This is because each quantity affects the dielectrophoretic response in a different way. Three key features are used in best-fit analysis; those being the value of the crossover frequency in low conductivity media, the gradient of the rise in crossover as a function of increasing medium conductivity, and the conductivity at which the cross-over drops by a decade or more. Generally speaking, the internal permittivity causes a scaling change in the frequency only, the surface charge density causes a scaling change in both conductivity and frequency, and the ξ potential affects the slope of the frequency increase with conductivity and the conductivity at which the frequency drops. These unique features allow best-fit techniques to find a unique solution to the fitting of data from homogeneous particles.

4.3.4   Dispersion in the Stern Layer

The above formulae describe the way in which the electrical properties of the particle, as represented in the Clausius–Mossotti equation, are augmented by the movement of charge around the particle. However, in order to describe the low-frequency dispersion visible in high-conductivity media (where no positive dielectrophoresis is expected) it is necessary to add an additional dispersion. Such additional polarizations follow the Debye model of the form $1/(1 + j\omega\tau_e)$, where τ_e is the relaxation frequency of the additional dispersion.[16] In this case the additional term derives from the dielectric dispersion of the charge in the Stern layer. Unlike the diffuse layer, the Stern layer is of fixed size and charge, these being dictated by the surface charge density of the particle. Hence, the frequency of the dielectric dispersion would be expected to be stable over a range of medium conductivities but vary proportionally to the particle radius and the surface conductance.

A good fit to observed data is given when the dispersion has a relaxation time:

$$\tau_e = \frac{\varepsilon_0 \varepsilon_s a}{K_s^i} \qquad (4.12)$$

where ε_s is the relative permittivity of the Stern layer. This is the dispersion frequency of a coupled resistor–capacitor network, where the resistor and capacitor are the surface conductance and Stern layer capacitance. The best fit is provided when ε_s is approximately $14\varepsilon_0$. Since the Stern layer consists of bound ions and water molecules held in specific orientations by electrostatic interactions with the charged particle surface, this is reasonable; values of between 7 and 40 have been suggested in the literature.[16]

If we consider all the above factors — the Clausius–Mossotti factor, surface conduction in the double layer, and Stern layer relaxation, then we can find best-fit lines that correspond well to our data. For example, we can determine the net effects of all these factors on the 216-nm latex beads shown in Figure 4.5. Superposed on the data is a best-fit line derived using the above equations and values $K_s^i = 0.85$ nS, $\zeta = -100$ mV, $\varepsilon_s = 14\varepsilon_0$. The relative permittivities of the particle and medium were 2.55 and 78; the internal conductivity of the beads was considered to be negligible. As can be seen, the model accurately predicts the behavior both below and above the decade transition in crossover frequency at 40–50 mS m^{-1}.

It is interesting to note that the dispersion due to capacitance in the diffuse layer is not in evidence, possibly indicating that its effect is dominated by the capacitance of the Stern layer due to the difference in magnitudes of the permittivities, as discussed in Chapter 3. Diffuse layer dispersions may yet be found at lower frequencies, but since the particles appear to undergo positive dielectrophoresis up to the Stern layer dispersion frequency, this has not yet been observed. That the dispersion is dominated by one double layer component is to be expected, bearing in mind the effect we observed in Chapter 3 where one capacitance is observed to dominate over the other.

4.4   Dielectrophoresis versus fluid flow

As described in Chapter 3, dielectrophoretic force generated by electrodes producing highly nonuniform electric fields of great magnitude is often countered by the force imparted on the medium itself due to electro-osmotic fluid flow. There are two frequency regions where this effect is most noticeable: in the range of frequencies below approximately 10–100 kHz (depending on applied voltage), where the fluid flow is greatest, and at crossover, where the dielectrophoretic force is relatively small due to the low values of Re[$K(\omega)$]. The effect can also be observed where the polarizability is low, and in some cases — particularly where a large area electrode array is used to trap low-polarizability particles — the fluid flow can prevent particles collecting by dielectrophoresis altogether. An example of this is the cover photograph of this book, where plumes of red and green latex beads stream from an electrode array.

In order to study this effect, it is possible to examine the patterns of collection in the electrode plane and in the absence of any of the particles moving above that plane or in the bulk solution. This is performed using evanescent illumination to excite the fluorophores in the beads rather than a normal fluorescence microscope. By illuminating the underside of the slide containing the electrodes using a laser at the angle of total internal reflection, an evanescent field of light is generated. This has wavelength equal to the laser used, but the depth of illumination is only a few hundred nanometers. Thus, only those beads immediately on the surface are illuminated and can be seen along with the edges of the electrodes themselves; the remainder of the beads in solution are invisible.

An example of this effect is shown in Figure 4.7. Latex beads of diameter 512 nm were collected in a quadrupole electrode array by positive dielectro-phoresis at 1 MHz. The frequency was then raised in steps until the onset of negative dielectrophoresis at 2 MHz. As the force diminishes, the balance between dielectrophoresis and fluid flow changes. At these frequencies, the dielectrophoretic force acts to attract the particles toward the array, but the fluid flow acts to push them away. As the frequency is increased, the value of $\text{Re}[K(\omega)]$ decreases, so that the particles are pushed farther from the trap; the fluid flow occurs over a smaller range, and the dielectrophoretic collection occurs in bands stretching across wider interelectrode gaps where pearl-chain-ing effects can enhance the dielectrophoretic collection in regions where the electric field strength is lower. At the crossover frequency itself, there is no longer a force holding the particles together and they follow the contrails of the small fluid flow at the electrode edges, which results in a swirling pattern.

As stated previously, the effect is at its most disruptive at low frequen-cies, where the fluid flow is sufficiently vigorous to push particles collected at electrode edges back onto the upper surfaces of the electrode arrays themselves, which results in collection patterns on top of the electrodes. Similarly, particles collecting under positive dielectrophoresis can be forced away by fluid flow, which causes collection to occur above the points of highest field strength where both force and flow are maximum, resulting in levitated collections of particles above these high field points, where the collected particles are suspended on jets of medium in the same manner as a ball suspended on top of a fountain. Similar collection patterns have been observed at the center of quadrupolar electrode arrays, where the flows generated down the interelectrode gaps meet and form large "fountains" at which large collections of particles have been observed.

While the movement of fluid due to electrohydrodynamic action is usually the enemy of the dielectrophoretic engineer, it can be harnessed. For example, Green and Morgan[2] have used a combination of positive dielectrophoresis, negative dielectrophoresis, and fluid flow to obtain a three-way particle

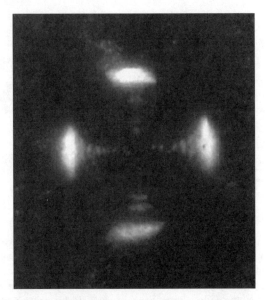

Figure 4.7 The collection patterns of 512-nm-diameter latex beads near the crossover frequency, observed using evanescent illumination. The particles experience positive collection at 1.5 MHz, but fluid flow pushes them out from the array. Particles form bands of pearl chains between the electrodes, which increases the attraction force. The electrode array is similar in configuration and dimension to that seen in Figure 4.1.

separation; since the magnitude of the force experienced by a particle is proportional to its volume, but the viscous drag experienced is proportional only to its radius, particles of different size find equilibrium points at different distances from the point of highest field, forming discrete bands.

4.5 Separating spheres

As well as its usefulness in characterizing the dielectric properties of particles, dielectrophoresis also has important applications as a tool for the separation of heterogeneous mixtures of particles into homogeneous populations. The method underlying the technique is simple: since polarizable particles demonstrate a crossover frequency that is dependent on those particles' dielectric properties, particles with different properties may under specific conditions exhibit different crossover frequencies. As those particles experience positive dielectrophoresis below the crossover frequency and negative dielectrophoresis above, it follows that at a frequency between the two crossover frequencies of the two particle types, one will experience positive dielectrophoresis while the other experiences negative dielectrophoresis. This will result in one group being attracted to regions of high field strength, with the other group being repelled; hence, the two populations are separated. Such separations are typically carried out using electrode arrays with well-defined regions of high and low electric field strength (see Figure 4.8), but this need not be the case.

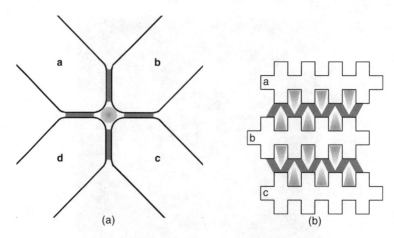

(a)   (b)

Figure 4.8 A schematic of typical electrodes used in dielectrophoresis work. (a) A quadrupole electrode array. The gap between opposing electrodes in the center of the array is typically of the order 10 μm for nanometer-scale work across, but arrays as small as 500 nm have been used. To induce dielectrophoretic motion in particles suspended near the electrode array, electrodes would be energized such that a and c are of the same phase, and b and d are in antiphase to them. (b) A castellated electrode array, where electrodes a and b are in antiphase. More electrodes can be added above and below these two, providing phase alternates between them. Typical dimensions are 5–10 μm along each electrode face.

Dielectrophoretic separation was first demonstrated by Pohl and co-workers in the late 1950s and early 1960s for a number of mineral and polymer suspensiods using coaxial electrode arrangements such as those described in Chapter 10. These particles were attracted from solution to the central electrode and could thus be removed from solution; a review of this early work is presented in Pohl's classic text.[17] Separations generally took place in electrode chambers, perhaps tens of millimeters across, and particles were dissolved in low-permittivity (and hence low-force) solvents, necessitating the use of voltages in the range between hundreds of volts and tens of thousands of volts between the electrodes in order to generate the requisite value for ∇E^2, with 10 kV required to trap particles of 100-μm diameter. Subsequent work by Pohl and Hawk[18] on the separation of yeast cells in water used smaller electrodes (a point and plane arrangement, with 1-mm interelectrode distance) to separate live and dead yeast cells suspended in very low conductivity water solutions, using a 30 V, 2 MHz signal. Subsequent work has demonstrated the separation of blood cells and many other cell types using machined electrode geometries (pin–pin, pin–plane, and isomotive — see again Chapter 10). In the early 1990s, demonstration of separation on microelectrodes was first performed by Gascoyne et al.[19]; the use of microelectrode geometries allowed the use of low voltages, and more inhomogeneous fields allowed separate collection of particles by positive and negative dielectrophoresis. It is the use of microengineered electrodes

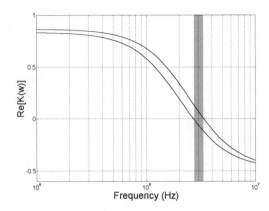

Figure 4.9 The change in $Re[K(\omega)]$ as a function of frequency for latex beads with diameters of 270 nm (leftmost curve) and 216 nm (rightmost curve). As can be seen in the graph, the crossover frequencies (where $Re[K(\omega)] = 0$) are slightly different; between the crossover frequencies (shaded in gray), the particles have different polarities of force and can be separated by dielectrophoresis.

with sufficiently divergent electric fields (with correspondingly high values of ∇E^2 for relatively low applied voltages) that has allowed the separation of particles to be extended to the nanometer scale.

As an example of the separation of homogeneous colloids by dielectrophoresis, let us consider the separation of a mixture of two populations of latex beads, identical except for having different radii. Since the effective conductivity (and hence the polarizability, as expressed in Equation 4.2) of a latex sphere is dependent on double the surface conductances divided by the particle radius (Equation 4.6), it follows that the value of $Re[K(\omega)]$ will be strongly affected by particle radius. This is indeed the case, with larger particles exhibiting lower crossover frequencies than smaller (but otherwise identical) particles. For example, Figure 4.9 shows the dielectrophoretic behaviors of two particles with a difference in radius of 20%, but with all other characteristics the same. As can be seen, in the shaded frequency window the particles exhibit different dielectrophoretic behavior — one experiences positive dielectrophoresis, the other negative dielectrophoresis — and can thus be separated.

Second, particles of identical size and internal composition can be separated according to their surface properties. This was first demonstrated by Green and Morgan in 1997,[20] who reported the separation of 93-nm latex spheres. By using a castellated electrode array with 4-μm feature sizes, the researchers demonstrated that the particles exhibited a narrow range of surface conductances rather than each having an identical value of surface conductance. This caused the population of particles to have crossover frequencies dispersed across a narrow frequency window, and, by applying a frequency in the middle of that range, it was demonstrated that the particles could in fact be separated.

This effect was expanded upon[3,20-23] by actively modifying the surfaces of latex particles to improve the separation and to identify possible biotechnological applications of the technique. The surfaces of some of the beads were chemically modified using EDAC (1-ethyl-3-(3-dimethylaminopropyl) carbodiimide), a reagent used for the chemical coupling of protein to the carboxyl surface of the beads. This caused a significant reduction in the crossover frequency, which was found to equate to a similar reduction in surface conductance from 1.1 nS to 0.55 nS, and ζ potential from –90 mV to –85 mV. The EDAC-activated beads were then mixed with antibodies, and the crossover behavior was measured again. Since the surface of the beads was covered by the antibodies, the crossover spectrum exhibited a further drop in frequency equating to a further drop in K_s^i (to 0.22 nS) and ζ potential (to –80 mV). The crossover frequencies of the IgG-labeled beads were also observed to vary by up to a factor of 2 between different beads as a result of different amounts of antibody coupling between beads. However, the average amount of coverage was determined to be in keeping with near-total coverage when assessed using a protein assay and calculating the maximum coverage for a given number of beads based on the size of a single IgG molecule. These crossover spectra are shown in Figure 4.10. Subsequent work

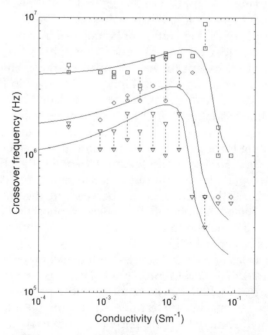

Figure 4.10 Crossover spectra for 216-nm-diameter latex beads (squares), beads whose surfaces have been modified by 1-ethyl-3-(3-dimethylaminopropyl) carbodiimide (EDAC) (circles), and beads that have been EDAC-treated and subsequently coated with a layer of antibodies (triangles). As can be seen, altering the surface has a significant effect on the dielectrophoretic response of the particles. Best fit lines are also shown, and the corresponding values of the surface are described in the text.

demonstrated that by coupling a secondary protein to the primary IgG layer on the bead surface, a much broader range of crossover frequencies was observed corresponding to a range from no secondary antibody attachment to the formation of a complete secondary antibody layer. If one examines the dielectrophoretic responses, it can be seen that for a given medium conductivity, the three particle types exhibit different crossover frequencies; applying a frequency between any two crossover frequencies on the graph for a given conductivity will result in particle separation, with the optimum separation occurring at a frequency roughly equidistant from the two crossovers.

There are a number of potential applications of such a system. First, since the crossover frequency is directly related to the amount of protein attached to the bead surface, it allows the rapid assaying of the amount of protein attached to a sphere, which in turn relates to the amount of protein in the environment, which makes a single sphere a potential biosensor. If the system is calibrated such that the crossover frequency in a particular medium that corresponds to a specific protein coverage is known, then observing the frequencies at which a single bead, or an ensemble, changes dielectrophoretic behavior allows measurement of the protein content in the medium. This could be used for a number of different proteins or other compounds by mixing fluorescent beads of different colors, each with a different surface functionality. By constructing electrodes over a suitable photosensor, systems such as this may form the basis of "lab on a chip" systems being developed as described in Chapter 9.

A second application concerns the fact that very small latex spheres have a large surface area to volume ratio, so that a small volume of beads has a potentially huge surface area. For example, a 1-ml sample containing 1% (by volume) of 200-nm-diameter beads (as used by Hughes and Morgan[23]) has a total surface area of 300 m^2 — which in biosensor terms makes such a sensor exceptionally sensitive. This can, for example, be used to detect very low quantities of target molecules; if a large number of small, activated beads is held near their crossover frequency, then a single molecule attaching to the surface of one bead should change the surface charge of the bead enough to cause that bead to pass the crossover frequency and be detected. A similar system, on the micrometer scale, has been developed at the University of Wales at Bangor as a means of detecting waterborne bacteria.[24] Such separators and detectors as these form the basis of "laboratory on a chip" systems, which we will discuss in more detail later.

4.6 Trapping single particles

One of the many opportunities presented by the manipulation of particles on the nanometer scale is that of manipulating single particles. Such a technology could potentially open up new fields in the study of single-molecule chemistry and molecular biology, and is presently being pursued by a number of workers (e.g., Hughes and Morgan,[25] Watarai et al.,[26] and Washizu et al.[27,28]).

The majority of this research is performed using optical trapping, so-called "laser tweezers," discussed in Section 4.8.

The trapping of single particles is somewhat different and somewhat more difficult to achieve than trapping a larger population of particles. In the latter case, particles need only to *tend* to move toward the trapping region; if some particles leave the trap, this is not considered a problem provided greater numbers of particles are moving into the trap. Furthermore, dielectrophoretic traps — even those constructed on the micron scale — tend to collect large numbers of nanoparticles; a spherical trap with volume 1 μm across could contain over 500 particles of diameter 100 nm.

There are two ways to increase the selectivity of the trapping mechanism. One method is to reduce the size of the electrodes to the order of size of the particle to be trapped. The other is to use a larger trap and follow a regime whereby a single particle is attracted to the electrodes in one operation followed by a second operation to prevent other particles from approaching the trap. Owing to the difficulty of fabrication on the nanoscale, the latter is the method of choice and can be implemented by both positive and negative dielectrophoretic trapping.

Different electrode geometries are required for each trapping method. The basis of electrode design for single particle applications is the dual need to both *attract* a single particle and to *repel* all others. Unlike bulk nanoparticle trapping where the aim is merely to attract particles to a region, it is necessary to both attract a particle to a point and trap it while excluding all other particles from that trap.

4.6.1   Theory of dielectrophoretic trapping

An electric field through a dielectric material containing no free charge (such as those described in Chapter 2) must comply with specific vector rules. The first of these is that the field has zero *curl*, that is, the direction of the field acts outward from a point (as shown in Figure 2.2). A field having curl would spiral out of the point charge; this is obviously not the case here. We can express this mathematically

$$\nabla \times \mathbf{E} = 0$$

Second, if the electric field is applied externally to the dielectric, then in the absence of contained charge the field is not divergent, it is constant everywhere (this expressed in Laplace's law, discussed in Chapter 2). We can mathematically express these statements thus:

$$\nabla \bullet \mathbf{E} = 0$$

Combining these, it was shown by Jones and Bliss[30] that the electric field in free space — that is, away from sources of imposed charge such as electrodes

— cannot contain isolated electric field maxima; the electric field strength is always greatest at electrode surfaces (in our examples). However, the same analysis does allow for any number of isolated *field minima*; these are regions in space from which the magnitude of the electric field increases in all directions. Since the electric field increases in all directions, then a positive dielectrophoretic force would act away from the region in all directions, though at the center of this minimum the electric field gradient (and corre-. sponding force) may be zero (a field *null*). Similarly, a particle experiencing negative dielectrophoresis would experience a force moving it toward the center of the minimum when moved in any direction from that center; it is effectively *trapped*.

One use of this important principle of dielectrophoresis in electric field geometries with maxima at electrodes and isolated minima is the development of *levitators*.[31] The heights to which particles are suspended in electric field nulls by negative dielectrophoresis can be measured, the forces (balancing dielectrophoretic repulsion with gravity) measured, and an estimate of the value of Re[$K(\omega)$] determined. Levitation is also possible by positive dielectrophoresis but is more complex as the applied force must be controlled so as to ensure the particle does not rise up to the electrode. A review of dielectrophoretic levitators and their electrode geometries is presented by Jones;[32] the application of dielectrophoretic levitation to the measurement of virus properties is described in Chapter 5. Dielectrophoretic trapping is similar in principle to dielectrophoretic levitation; however, its purpose is to retain (trap) particles at a particular point in space, either by positive dielectrophoresis (attracting the particle to a point in space from which it cannot escape) or negative dielectrophoresis (enclosing the particle in a dielectrophoretic field minimum from which it cannot escape), rather than levitating them.

4.6.2 Trapping using positive dielectrophoresis

This technique was first demonstrated for nanoscale particles by Alexey Bezryadin and coworkers in 1997.[33,34] Their electrode geometry consisted of two needle-type platinum electrodes facing one another, suspended in free space by etching the silicon substrate beneath the interelectrode gap. The distance between opposing electrode tips was 4 nm. The potential was applied through a high-value (100 MΩ) resistor; a 4.5 V, DC field was used. As we have seen, AC fields are far more common for dielectrophoresis, but this is not a prerequisite.

Colloidal palladium particles of 5-nm diameter were introduced in solution; the particles polarized and were attracted up the field gradient to the electrode tips by positive dielectrophoresis. However, as soon as the first palladium sphere reached the center point between two opposing electrodes, a circuit between the electrodes was made. This resulted in current flowing in the circuit, in which the majority of the supply voltage was dropped across the resistor. With virtually no voltage dropped across the electrodes, the magnitude of the electric field generated in the interelectrode gap was diminished,

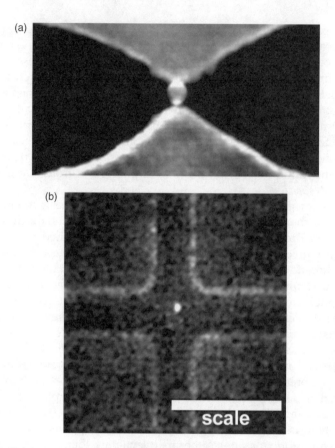

Figure 4.11 Single particles trapped by dielectrophoresis. (a) An electron micrograph of a single 17-nm palladium sphere trapped between two electrodes by positive dielectrophoresis. (b) A fluorescence micrograph of a single 93-nm latex bead suspended above a dielectrophoretic trap by negative dielectrophoresis. The bead is at the tip of the white circle at the center of the array; scale bar, 5 μm.

preventing other colloidal particles from reaching the electrode tips. Once in place, the trapped particle was sufficiently attached to the electrodes for the solution to be removed and the assembly to be observed using a scanning electron microscope, as shown in Figure 4.11a with a slightly larger particle. Similar principles could be applied to the trapping of single fullerene molecules to form single-molecule transistors[35] and have been applied to a range of molecule-scale objects as discussed in Chapters 6 and 7.

4.6.3 Trapping using negative dielectrophoresis

An alternative method for trapping single particles using dielectrophoresis is the use of negative dielectrophoretic trapping. This has advantages over the positive trapping technique described above, where the trapping process is stopped by a conducting particle contacting the electrode structures. For example, many particles that one might wish to trap are of biological origin and therefore nonconducting, preventing the "completed circuit" method of potential removal from being used. Furthermore, since the size of the trapping volume at the center of the arrays is defined not by the electrodes but the geometry of the generated field, the technology required to construct the electrodes is far more readily available. Negative dielectrophoretic trapping of single particles has been achieved by this method using viruses, latex spheres, viral substructures, and macromolecules such as DNA as well as larger structures such as cells.[25,36,37] There are a number of drawbacks to this form of trapping, one being the observation of the particle. Unlike positive trapping where the particle may be detected electrically, particles trapped by negative dielectrophoresis are suspended in the medium at an indeterminate height above the electrode structure. Ultimately, the only means by which such particles can be observed is by fluorescent staining. This is a general problem in the field of single nanoparticle detection, and other methods such as laser tweezers also require the use of fluorescent staining. Another problem is that by the nature of negative dielectrophoresis field cages, the force field extends the trapping volume some distance away from the electrodes, and as such the force is smaller. This does limit the minimum trapping size, and negative dielectrophoresis is less effective than positive dielectrophoresis at trapping the smallest particles.

Electrodes used to trap particles are generally quadrupolar arrays such as the polynomial design shown in Figure 4.8a. This array geometry has the advantage of a well-defined, enclosed field minimum surrounded by regions of high field strength. Ideally, the potential energy minimum would be small enough to contain only one particle. Where this is not the case, particles may first be attracted to the electrode tips by positive dielectrophoresis. When the field frequency is switched to induce negative dielectrophoresis, only those particles on the inward-facing tips of the electrodes will fall into the trap; the others will be repelled into the bulk medium. It is possible to trap single particles this way, though occasionally two or three particles may fall into the potential energy minimum at the center of the trap. Figure 4.11b shows a fluorescent photograph of a 93-nm diameter latex sphere trapped in a polynomial electrode array (as shown in Figure 4.8a) under negative dielectrophoretic forces.[25] Particles have been held in traps such as these for up to 30 minutes, where the limiting factor was the evaporation of the suspending medium.

Single particles trapped by this method move within the confines of the electric field cage under the influence of Brownian motion. During trapping of an object with density greater than that of water, such as a single herpes virus, the particle is levitated in a stable vertical position above the electrodes; Brownian motion is balanced against the weight of the particle. However, particles such as the latex spheres (which have a density approximately equal to that of water) are not constrained in this way because Brownian motion causes constant random movement in the z direction. Such particles may eventually diffuse out of the top of a funnel-type or open trap, in which the field gradient is generated by one set of planar electrodes beneath the trapped particle, though particles can be held for 30 minutes or more. In order to ensure that a particle remains within the trap, a second layer of electrodes must be introduced above the first so that a closed field cage such as those employed by Schnelle and co-workers[37] is created. These have the additional advantage of allowing a degree of three-dimensional positional control by varying the intensity of the field strength at the positional electrodes, as has been demonstrated using a 1-μm-diameter latex bead.[37] Alternatively, a planar (two-dimensional) electrode array can be sufficient provided that any coverslip used to contain the solution above the electrode is sufficiently close to the electrode plane and that any field trap constraining the particle in the *x-y* plane extends the full height of the solution, creating a force field of approximately cylindrical aspect.

4.7   Limitations on minimum particle trapping size

In order to stably trap submicron particles, the dielectrophoretic force acting to move the particle into the center of the trap must exceed the action of Brownian motion on the particle, which, if large enough, will cause a particle to escape. In the case of positive dielectrophoresis, the applied force acts toward a single point (that of greatest field strength), and the force attracting it to that point must exceed the action of Brownian motion. For particles trapped in planar electrode arrays by negative dielectrophoresis, the particle is surrounded by a dielectrophoretic force field barrier through which it must not pass.

The trapping of particles using positive dielectrophoresis is the simplest case to analyze; particles are attracted to the point of highest electric fields strength, rising up the field gradient. Once at that point, the particle remains unless Brownian motion displaces it a sufficient distance that the field is unable to bring it back. The nature of positive dielectrophoresis is such that, given a long enough period of time (be it seconds, or years), any particle polarized such that it experiences positive dielectrophoresis will ultimately fall into the trap;[38] as the electric field gradient extends to infinity, there will be an underlying average motion that will over time cause a displacement toward the high field trap, though the time taken for such motion to occur may be such a large magnitude as to be effectively unobservable.

The concentration of particles from solution by negative dielectrophoresis is slightly different in concept from the above case. Particles are trapped in regions where the electric field strength is very low, and are prevented from escaping by a surrounding force field wall that encloses the particles (although it may be in the from of an open topped funnel). However, since in both the above cases the particle must achieve the same effect — the overcoming of a dielectrophoretic energy barrier that forces the particle into the trap — the mathematical treatment of both cases is similar.

Given this approach to the trapping of particles by overcoming the action of Brownian motion through the application of a quantifiable force, it is possible to determine what the relationship is between the magnitude of the electric field applied by the electrodes and the smallest particle that may be trapped by such electrodes. This approach was developed by Smith et al.[39] for determining the smallest particle that may be trapped by laser tweezers, a technique similar to positive dielectrophoresis. However, it is equally applied to negative dielectrophoresis.

Consider the force on a particle of radius r suspended in a nonuniform electric field, experiencing a trapping dielectrophoretic force $\mathbf{F}_{DEP}$. For times longer than the order of 10^{-8} seconds, from Stokes's law, the particle attains a terminal velocity v according to the equation

$$v = \frac{\mathbf{F}_{DEP}}{6\pi\eta r} \tag{4.13}$$

where η is the viscosity of the medium. Considering a small region of thickness Δd over which the force (i.e., the field gradient) is constant, then the time t_{DEP} taken for the particle to traverse this region is given by

$$t_{DEP} = \frac{6\pi\eta r\Delta d}{\mathbf{F}_{DEP}} \tag{4.14}$$

In addition to the dielectrophoretic force, there is the effect of Brownian motion that acts to move the particle in random directions. From Einstein's equation,[40] the mean time $\langle t_B \rangle$ taken for a particle to move a distance Δd in one dimension can be derived and is given by

$$\langle \tau_B \rangle = \frac{3\pi\eta r(\Delta d)^2}{kT} \tag{4.15}$$

where k is Boltzmann's constant and T is the temperature. For stable trapping to occur, the time for the particle to fall into the trap (from the edge) should be significantly less than the time taken for the particle to escape from it. If we wish the trapped particle to never escape from the field trap, then the dielectrophoretic force must be significantly greater than the Brownian, thermal force (a factor of ×10 was suggested by Smith et al.,[39] though this is arbitrary) by which t_{DEP} is smaller than $\langle t_B \rangle$, then from Equations 4.14 and 4.15, the conditions are

$$\frac{6\pi\eta r\Delta d}{\mathbf{F}_{DEP}} < \frac{1}{10}\left(\frac{3\pi\eta r(\Delta d)^2}{kT}\right) \tag{4.16}$$

and

$$\Delta d > \frac{20kT}{\mathbf{F}_{DEP}} \tag{4.17}$$

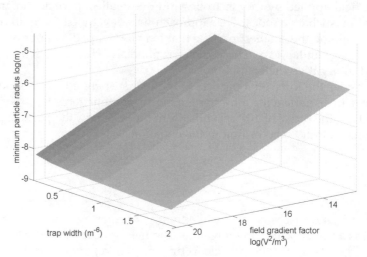

Figure 4.12 A graph showing the variation in the minimum radius of particles that can be trapped, according to Equation 4.19.

We can therefore determine the approximate minimum particle radius for successful trapping by substituting Equation 4.1 into 4.16, giving:

$$r > \sqrt[3]{\frac{10kT}{\pi \varepsilon_m \Delta d \, \mathrm{Re}[K(\omega)] \nabla E^2}} \qquad (4.18)$$

It is evident that the factors that contribute to the critical particle radius for stable trapping are the electric field gradient and the distance across which this gradient exists, i.e.,

$$r \propto \frac{1}{\sqrt[3]{\Delta d \nabla E^2}} \qquad (4.19)$$

The variation of the minimum particle radius for stable trapping can be calculated from Equation 4.18 as a function of field gradient ∇E^2 and trap width Δd. At a temperature of 300 K, and with $\varepsilon_m = 78\varepsilon_0$ and $\mathrm{Re}[K(\omega)] = 1$, this variation is shown in Figure 4.12. In cases where ∇E^2 varies as a function of distance, the trapping efficiency is given as the maximum value of the function $\Delta d \nabla E^2$ for the particular trap. In order to determine what this might mean in terms of a given electrode geometry, it is necessary to simulate the electric field gradient around that geometry using one of the techniques described in Chapter 10. The most accurate method of deriving the trapping force is to then integrate the force across all possible escape paths, but an alternative, rule of thumb method is to measure the shortest escape distance across which the force field has a given value. For example, a numerical

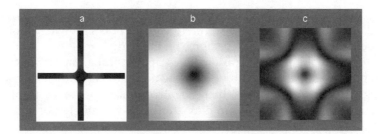

Figure 4.13 A simulation across the central area of the (a) electrode array, (b) electric field, and (c) ∇E^2 for an electrode array similar to that shown in Figure 4.7a. The electrode array had gaps of 2 μm between adjacent electrodes and 5 μm across the center of the array, and the simulation shows the field 3 μm above the electrodes. As can be seen, the field strength is greatest above the interelectrode arms of the array; there is a field null at the center of the array, enclosed on all sides by a region of high electric field.

model based on the Moments method[41] was used to calculate ∇E^2 around the polynomial electrode array. Figure 4.13 shows a three-dimensional plot of E and $|\nabla E^2|$ across the center of the electrode at a height of 7 μm above the electrode array shown in Figure 4.13a, this being the approximate height at which particles trapped in arrays of this geometry and size are observed to be levitated by negative dielectrophoresis. The trap efficiency is governed by the smallest distance that a trapped particle has to travel in order to escape from the trap. In the case of a particle trapped by positive dielectro-phoresis, it is principally governed by the magnitude of the electric field. Since the field diminishes as an inverse square due to Coulomb's law, the field barrier is well established. For particles trapped by negative dielectro-phoresis, Δd is governed by the magnitude of the field barrier that encloses the particle. As can be seen in Figure 4.13, while E is as one might expect (with high field regions in the interelectrode gaps and a field null at the center), the force pattern is more complex, with a force barrier surrounding the field null. Repulsion by this barrier prevents particles trapped by nega-tive dielectrophoresis from escaping. Analyzing these data for both force magnitude and the thickness of the field barrier (i.e., Δd) for an applied voltage of 5 V_{pk-pk}, we find that $\Delta d \nabla E^2$ reaches a broad maximum when Δd is in the range 2 to 4 μm and the corresponding values of $|\nabla E^2|$ are in the range 1 to 2×10^{17} V^2 m^{-3}. If we consider that this is the magnitude of the force barrier that must be overcome in order for the particle to escape, we can determine that the minimum radius of particle for stable trapping (i.e., without any particles escaping from the trap) is approximately 30 nm.

In order to trap large numbers of particles, the DEP force must overcome diffusion, but not necessarily by a factor as great as 10. For example, this model does not predict the stable trapping of proteins by negative dielectrophoresis, or of 14-nm diameter beads in submicron electric field cages as reported by Müller et al.[42]; the minimum particle diameter predicted under the condition $F_{DEP} = 10F_{BROWNIAN}$ would be approximately 66 nm for electrodes with a 500 nm interelectrode gap. However, if the condition is merely $F_{DEP} = F_{BROWNIAN}$, then the DEP force on particles will be *on average* greater than Brownian motion. In that case there will always be a *net* force on the particle mass toward the trap, even if individual particles occasionally escape from the trap. If we consider that trapping will be assisted by fluid flow and particle–particle interactions creating van der Waals forces, then negative dielectrophoretic trapping of populations of proteins or 14-nm diameter beads is entirely possible.

4.8   Dielectrophoresis and laser trapping

Dielectrophoresis is not the only method available for the manipulation of nanometer-scale particles in solution. The principal alternative technique is optical trapping by the use of so-called laser tweezers. Since we have now amassed sufficient knowledge of how dielectrophoresis can be used for trapping and separating particles, we are in a position to compare the two methods and their relative merits and demerits.

Laser trapping makes use of so-called optical pressure to induce force in an optical gradient, as reported by Arthur Ashkin of Bell Labs in 1986.[43] If a transparent particle is exposed to a focused beam of light such that the particle experiences a gradient in the intensity of light, then the particle experiences a net force toward the direction of the increasing gradient. This comprises two components, a *scattering* force and a *gradient* force. The magnitude of the gradient force on a spherical particle is given by the equation[44]

$$\mathbf{F}_{optical} = -\frac{n_m^2 r^3}{2} \left(\frac{n_p^2 - n_m^2}{n_p^2 + 2n_m^2} \right) \nabla E^2 \tag{4.20}$$

where n_p is the refractive index of the medium, n_m is the refractive index of the particle, and $\mathbf{E}$ represents the electric field generated within the laser (light being another form of electromagnetic radiation). Comparisons can be made between this expression and Equation 4.1, with the light beam providing the appropriate electric field. Also note the presence of the r^3 term in both expressions and the similarity of the bracketed term to the Clausius–Mossotti factor. However, against this we must set the additional radiation (or scatter) force terms, which do not have a significant dielectrophoretic analog.

In order to generate field gradients of sufficient magnitude, it is necessary to use a focused laser beam, with the particles attracted to the focal point. This technique has been used to manipulate a range of biological nanoparticles, from cells to proteins.[45] There even exists an equivalent of electrorotation, wherein rotating modes in the laser are used to induce a rotational torque in the particle. These "laser spanners"[46] have been suggested as a possible mechanism for driving nanoscopic cams and gears fabricated from carbon nanotubes.[47]

Given that these techniques exist and have gained a widespread reputation, what place is there for dielectrophoresis? Its advantage over laser tweezing lies in the simplicity of operation. Laser tweezers require the use of one or more powerful and expensive laser and complex optics in order to trap particles of the dimensions we refer to in this book. While it is straightforward to trap particles in solution by either method, it is important in optical trapping that there be a clear line of sight between the laser and the particle, for obvious reasons. On the other hand, with dielectrophoresis, the trapping is performed by small and relatively inexpensive electrode structures, generating electric fields using equipment that can usually be found in any radio repair shop. Provided the electrodes have power lines to the outside world, there does not need to be a line of sight to the electrode chamber, which means that fulfilling the aforementioned application of rotating nanoscale gears would be considerably easier by this method. Dielectrophoresis is much more efficient than laser trapping, since both the generation of the laser beam and the dispersal of laser energy in the medium are wasteful of energy. The separation of heterogeneous mixtures into two trapped populations is far more difficult to realize with laser trapping because of the difficulties of creating a closed trap in which particles are repelled. The two-dimensional nature of dielectrophoresis electrodes and the wide range of geometries that can be fabricated according to the intended application give the technique the edge. Finally, unlike tweezers, objects manipulated by dielectrophoresis are not required to be optically transparent.

That said, it is important to consider that there are other advantages intrinsic to laser trapping. Greatest of these is that since the particles are attracted to the focal point of the beam, the location of which is dictated by the geometry and position of the focusing objective lens, it is possible to achieve full three-dimensional directed motion of a trapped particle by moving the objective lens. In this respect, dielectrophoresis fares poorly since it is only capable of providing limited control in the vertical plane when particles are trapped by negative dielectrophoresis unless complex three-dimensional electrode structures are used. Furthermore, laser tweezers offer the ability to be switched to laser *scissors*, where a high-intensity beam is used to burn the contents of the focal point, making a useful tool for nanoconstruction.

Ultimately, both techniques have strengths and weaknesses and, as such, are complementary; indeed, both techniques have been used simultaneously. For example, a laser tweezer system has been used to hold a single bacterium in place, while negative dielectrophoresis was used to repel other cells from the area[48]; another example is the use of laser tweezers to isolate single sells from a trapped ensemble in a laboratory on a chip system (described in more detail in Chapter 9).[49] More commonly, laser tweezers have been used to maintain a steady position of cells within a rotating electric field during electrorotation experiments in order to ensure that the cells do not move within the chamber (thereby experiencing different electric field strengths and thus torques).[50] Finally, the electric field of the

laser can be used to generate a local electric field and hence induce dielectrophoresis; work by Mizuno et al.[51] demonstrated that a laser, focused through
a slit across which an electric field was applied, could be used for the sorting
of yeast cells.

References

1. Green, N.G. and Morgan, H., Dielectrophoretic investigations of sub-
    micrometre latex spheres, *J. Phys. D: Appl. Phys.*, 30, 2626, 1997.
2. Green, N.G. and Morgan, H., Dielectrophoresis of submicrometre latex
    spheres.1. Experimental results, *J. Phys. Chem.*, 103, 41, 1999.
3. Hughes, M.P., Morgan, H., and Flynn, M.F., Surface conductance in the diffuse
    double-layer observed by dielectrophoresis of latex nanospheres, *J. Coll. Int.
    Sci.*, 220, 454, 1999.
4. Hughes, M.P., Morgan, H., and Rixon, F.J., Dielectrophoretic manipulation
    and characterisation of herpes simplex virus-1 capsids, *Eur. Biophys. J.*, 30,
    268, 2001.
5. Hughes, M.P., Morgan, H., and Rixon, F.J., Measurements of the properties of
    Herpes Simplex Virus Type 1 virions with dielectrophoresis, *Biochim. Biophys.
    Acta*, 1571, 1, 2002.
6. Morgan, H. and Green, N.G., Dielectrophoretic manipulation of rod-shaped
    viral particles, *J. Electrostatics*, 42, 279, 1997.
7. Hughes, M.P. and Morgan, H., Dielectrophoretic manipulation of protein
    molecules in solution, 1st European Workshop on Electrokinetics and Electro-
    hydrodynamics in Microsystems, Glasgow, 2001.
8. Gascoyne, P.R.C., Pethig, R., Burt, J.P.H., and Becker, F.F., Membrane changes
    accompanying the induced differentiation of Friend murine erythroleukemia
    cells studied by dielectrophoresis, *Biochim. Biophys. Acta*, 1149, 119, 1993.
9. Huang, Y., Wang, X.B., Becker, F.F., and Gascoyne, P.R., Membrane changes
    associated with the temperature-sensitive P85gag-mos-dependent transfor-
    mation of rat kidney cells as determined by dielectrophoresis and electro-
    rotation, *Biochim. Biophys. Acta*, 1282, 76, 1996.
10. Gascoyne, P.R.C., Noshari, J., Becker, F.F., and Pethig, R., Use of dielectro-
    phoretic collection spectra for characterizing differences between normal and
    cancerous cells, *IEEE Trans. Ind. Appl.*, 30, 829, 1994.
11. Fricke, H. and Curtis, H.J., The determination of surface conductance from
    measurements on suspensions of spherical particles, *J. Phys. Chem.*, 40, 715,
    1936.
12. Schwarz, G., A theory of the low-frequency dielectric dispersion of colloidal
    particles in electrolyte solutions, *J. Phys. Chem.*, 66, 2636, 1962.
13. Arnold, W.M., Schwan, H.P., and Zimmermann, U., Surface conductance and
    other properties of latex particles measured by electrorotation, *J. Phys. Chem.*,
    91, 5093, 1987.
14. Arnold, W.M. and Zimmermann, U., Electro-rotation — development of a
    technique for dielectric measurements on individual cells and particles,
    J. Electrostatics, 21, 151, 1988.
15. Gascoyne, P.R.C., Pethig, R., Satayavivad, J., Becker, F.F., and Ruchirawat, M.,
    Dielectrophoretic detection of changes in erythrocyte membranes following
    malarial infection, *Biochim. Biophys. Acta*, 1323, 240, 1997.

16. Lyklema, J., *Fundamentals of Interface and Colloid Science,* Academic Press, London, 1995.
17. Pohl, H.A., *Dielectrophoresis,* Cambridge University Press, Cambridge, 1978.
18. Pohl, H.A. and Hawk, I., Separation of living and dead cells by dielectrophoresis, *Science,* 152, 647, 1966.
19. Gascoyne, P.R.C., Huang, Y., Pethig, R., Vykoukal, J., and Becker, F.F., Dielectrophoretic separation of mammalian cells studied by computerized image analysis, *Meas. Sci. Technol.,* 3, 439, 1992.
20. Green, N.G. and Morgan, H., Dielectrophoretic separation of nano-particles *J. Phys. D: Appl. Phys.* 30, L41, 1997.
21. Morgan, H., Hughes, M.P., and Green, N.G., Separation of sub-micron particles by dielectrophoresis, *Biophys. J.,* 77, 516, 1999.
22. Hughes, M.P., Flynn, M.F., and Morgan, H., Dielectrophoretic measurements of sub-micrometre latex particles following surface modification, *Institute of Physics Conference Series,* 163, 81, 1999.
23. Hughes, M.P. and Morgan, H., Dielectrophoretic manipulation and separation of surface-modified latex microspheres, *Anal. Chem.,* 71, 3441, 1999.
24. Burt, J.P.H., Pethig, R., and Talary, M.S., Microelectrode devices for manipulating and analysing bioparticles, *Trans. Inst. Meas. Control,* 20, 82, 1998.
25. Hughes, M.P. and Morgan, H., Dielectrophoretic manipulation of single sub-micron scale bioparticles, *J. Phys. D: Appl. Phys.,* 31, 2205, 1998.
26. Watarai, H., Sakamoto, T., and Tsukahara, S., *In situ* measurement of dielectrophoretic mobility of single polystyrene microspheres, *Langmuir,* 13, 2417, 1997.
27. Washizu, M., Suzuki, S., Kurosawa, O., Nishizaka, T., and Shinohara, T., Molecular dielectrophoresis of biopolymers, *IEEE Trans. Ind. Appl.,* 30, 835, 1994.
28. Washizu, M., Shikida, M., Aizawa, S.I., and Hotani, H., Orientation and transformation of flagella in electrostatic field, *IEEE Trans. Ind. Appl.,* 28, 1194, 1992.
29. Chiu, D.T. and Zare, R.N., Optical detection and manipulation of single molecules in room-temperature solutions, *Chem. A Eur. J.,* 3, 335, 1997.
30. Jones, T.B. and Bliss, G.W., Bubble dielectrophoresis, *J. Appl. Phys.,* 48, 1412, 1977.
31. Lin, I.J. and Jones, T.B., General conditions for dielectrophoretic and magnetohydrostatic levitation, *J. Electrostatics,* 15, 53, 1984.
32. Jones, T.B., *Electromechanics of Particles,* Cambridge University Press, Cambridge, 1995.
33. Bezryadin, A. and Dekker, C., Nanofabrication of electrodes with sub-5 nm spacing for transport experiments on single molecules and metal clusters, *J. Vac. Sci. Technol. B.,* 15, 793, 1997.
34. Bezryadin, A., Dekker, C., and Schmid, G., Electrostatic trapping of single conducting nanoparticles between nanoelectrodes, *Appl. Phys. Lett.,* 71, 1273, 1997.
35. Park, H., Park, J., Lim, A.K.L., Anderson, E.H., Alivisatos, A.P., and McEuen, P.L., Nanomechanical oscillations in a single-C60 transistor, *Nature,* 407, 57, 2000.
36. Schnelle, T., Hagedorn, R., Fuhr, G., Fiedler, S., and Müller, T., Three-dimensional electric field traps for manipulation of cells — calculation and experimental verification, *Biochim. Biophys. Acta,* 1157, 127, 1993.
37. Schnelle, T., Müller, T., and Fuhr, G., Trapping in AC octode field cages, *J. Electrostatics,* 50, 17, 2000.

38. Ramos, A., Morgan, H., Green, N.G., and Castellanos, A., AC Electrokinetics: a review of forces in microelectrode structures, *J. Phys. D: Appl. Phys.*, 31, 2338, 1998.
39. Smith, P.W., Ashkin, A., and Tomlinson, W.J., Four-wave mixing in an artificial Kerr medium, *Opt. Lett.*, 6, 284, 1981.
40. Einstein, A., *Investigations on the Theory of Brownian Movement*, Dover, New York, 1956.
41. Birtles, A.B., Mayo, B.J., and Bennett, A.W., Computer technique for solving three-dimensional electron-optics and capacitance problems, *Proc. IEE*, 120, 213, 1973.
42. Müller, T., Gerardino, A., Schnelle, T., Shirley, S.G., Bordoni, F., DeGasperis, G., Leoni, R., and Fuhr, G., Trapping of micrometre and sub-micrometre particles by high-frequency electric fields and hydrodynamic forces, *J. Phys. D: Appl. Phys.*, 29, 340, 1996.
43. Ashkin, A., Dziedzic, J.M., Bjorkholm, J.E., and Chu, S., Observation of a single-beam gradient force optical trap for dielectric particles, *Opt. Lett.* 11, 288, 1986.
44. Smith, P.W., Ashkin, A., and Tomlinson, W.J., Four-wave mixing in an artificial Kerr medium, *Optics Lett.*, 6, 284, 1981.
45. Greulich, K.O., Pilarczyk, G., Hoffmann, A., Meyer Zu Hörste, G., Schafer, B., Uhl, V., and Monajembashi, S., Micromanipulation by laser microbeam and optical tweezers: from plant cells to single molecules, *J. Microscopy*, 198, 182, 2000.
46. Simpson, N.B., Dholakia, K., Allen, L., and Padgett, M.J., Mechanical equivalence of spin and orbital angular momentum of light: an optical spanner, *Opt. Lett.*, 22, 52, 1997.
47. Srivastava, D., A phenomenological model of the rotation dynamics of carbon nanotube gears with laser electric fields, *Nanotechnology*, 8, 186, 1997.
48. Arai, F., Ogawa, M., and Fukuda, T., Selective manipulation of a microbe in a microchannel using a teleoperated laser scanning manipulator and dielectrophoresis, *Adv. Robotics*, 13, 343, 1999.
49. Arai, F., Ichikawa, A., Ogawa, M., Fukuda, T., Horio, K., and Itoigawa, K., High-speed separation systems of randomly suspended single living cells by laser trap and dielectrophoresis, *Electrophoresis*, 22, 283, 2001.
50. De Gasperis, G., Wang, X.B., Yang, J., Becker, F.F., and Gascoyne, P.R., Automated electrorotation: dielectric characterization of living cells by real-time motion estimation, *Meas. Sci. Technol.* 9, 518, 1998.
51. Mizuno, A., Imamura, M., and Hosoi, K., Manipulation of single fine particles in liquid by electrical force in combination with optical pressure, *IEEE Trans. Ind. Appl.*, 27, 140, 1991.

chapter five

Dielectrophoresis of complex bioparticles

5.1 Manipulating viruses

In the previous chapter, we established that dielectrophoresis can be used to determine the dielectric properties of homogeneous particles with different diameters and surface functionalities. We can extend the technique to allow it to be used as an investigative tool for nanobiology; that is, it can be used to probe into the heart of nanoscale bioparticles and reveal their inner workings. Such applications have been demonstrated with cells for over 30 years,[1] but it is only recently that, along with developments in nanomanipulation by electrokinetic methods in general, the study of complex nanoparticles has become possible.

Of all applications for nanometer-scale AC electrokinetics, the study of viruses is perhaps the most similar to the studies of cellular properties that form a significant part of dielectrophoretic research over the last 2 decades. The physical structure of some virus particles resembles that of cells, with DNA-containing cores surrounded by lipid membranes. However, while electrorotation has become the *de facto* method of choice for cell analysis, it is not possible to gain accurate measurements of such factors as rotation rates of virus particles with conventional equipment, simply because viruses are so small it is virtually impossible to observe rotational motion with present-day optical equipment. Instead, we must rely on observation of dielectrophoretic motion to determine the properties of the virus. By extending our simple model of dielectric response to cover particles made of many concentric shells, we have a tool that will enable us to analyze the electrical properties of viruses.

Furthermore, dielectrophoresis offers great potential for viral particle *detection*. There is a need for portable equipment to be developed in order to detect pathogenic bioparticles such as viruses or bacteria. Such a device would offer great benefits for diagnostics at the point of care, for the diagnosis of veterinary diseases in the field, for the determination of water quality, and for the detection of biological weapons.

Many methods exist for the detection of bioparticles such as bacteria and viruses using biochemical methods, but these generally require the use of complex and expensive laboratory equipment. Furthermore, many methods of particle detection require the particles to contact a surface where detector biochemicals are used to identify them; however, since many biological nanoparticles are neutrally buoyant (or are so small as to be affected by Brownian motion far more than gravity), it is difficult to move particles to such a surface. The ability to attract such particles indicates the advantages of dielectrophoresis in this area, particularly in laboratory-on-a-chip systems as described in later chapters.

In this chapter, we will concentrate on the dielectrophoretic detection of virus particles; these are structurally the most complex particles on this scale, on a size scale below cells (mostly between 5 and 10 μm across) and bacteria (of the order of 1 μm across). We will use this as a basis for examining other aspects of electrokinetic theory, in particular the behavior of ellipsoidal particles and of multishelled particles, and examine how we can use this as a method for particle analysis.

5.2   Anatomy of viruses

The term virus comes from the Latin for "poison." The original distinction between bacteria and viruses as agents of disease in plants, animals, and humans was made on the basis that light microscopy could be used to view bacteria, and a fine filter could be used to remove them from a liquid sample, but no such filter existed for viruses, and samples examined by microscope appeared to show nothing. This is because viruses are fundamentally different from the essentially cellular nature of bacteria, which are generally about 1 μm in diameter. Viruses are considerably smaller (between 5 nm and 250 nm in diameter) and in general too small to see with a light microscope, requiring either fluorescent staining or a non–light based microscopic technique such as electron microscopy or atomic force microscopy.

Viruses are unique in the world of living organisms, in that there is debate over whether or not a virus actually *is* a living organism. Viruses do not metabolize food, nor move; in many respects, they resemble inert protein structures containing DNA. They are not even directly able to reproduce. Viral reproduction occurs by a virus particle gaining entry to a cell, releasing its DNA, and "reprogramming" the cellular processes so as to cause the cell to start producing new virus particles. These are released from the cell either one at a time or in a burst of thousands of viruses, an act that also destroys the cell. The time from infection to destruction, with thousands of replicas having been produced, is usually a matter of hours.

Viruses may be classified in a number of ways. One is according to the target host, with three main types — animal, plant, and bacterial. Another is according to whether the virus contains single-stranded DNA (e.g., parvoviruses), double-stranded DNA (e.g., herpesviruses), or single- or double-stranded RNA (e.g., polioviruses and reoviruses).

A third way of classifying viruses — and the most useful for our studies — is by morphology. All virus particles (or *virions*) contain a DNA or RNA payload encapsulated in a protein shell, called a *capsid*, which is assembled from protein subunits called *capsomeres*. These capsids may be icosahedral — that is, multifaceted but approximately spherical in structure — or helical, where the capsomeres wrap around the DNA or RNA core to produce a long thin cylinder. Species of helical viruses consisting of only RNA and protein subunits almost exclusively infect plants. However, a number of plant and animal viruses consist only of capsids and payload; many of the best-known examples of these are the picornaviridae, whose family includes hepatitis A and the common cold. The capsid may have protein *spikes* protruding from it for cell attachment and entry.

In addition to the payload and the capsid, many viruses have an extra lipid membrane (or *envelope*) similar to that which encloses cells. These are referred to as *enveloped* viruses, whereas those without membranes have *naked capsids*. In the vast majority of cases, there is also a layer of protein gel between the capsid and envelope called the *tegument*. Most such viruses also have *glycoprotein* molecules protruding from the envelope; these are molecules that interact with the surface of the host cell and aid in the infection process. Again, a number of common human viruses share this structure; examples include various herpes viruses (including those responsible for cold sores and chicken pox), and orthomyxovirus (which causes influenza). The capsid structure in enveloped viruses may either be icosahedral or helical; in the latter case the helix may be wrapped into a ball, such as in the case of influenza.

There are exceptions to the above division of spherical and helical structures; for example, bacteriophage T-4 consists of a capsid plus additional structures that can act like a hypodermic syringe so that when the base of the virus comes into contact with the surface of an *E. coli* virus, the needle penetrates the cell wall and propels the viral DNA into the cytoplasm. Viruses with head-and-tail structures such as this exclusively infect bacteria. Another unusual example of virus structure is the *vaccina* virus, which is enveloped but contains no defined capsid structure. A schematic diagram showing a range of virus structures and relative sizes is shown in Figure 5.1; details are given in Table 5.1, where the shapes and sizes of a number of different viruses are listed.

5.3   The multishell model

Thus far we have examined the application of the Clausius–Mossotti relationship as a means of determining the dielectrophoretic response of a solid, homogeneous sphere. However, viruses are not solid, homogeneous spheres; while virus shape is often approximately spherical (enough for the model to give reasonable results), it is never homogeneous throughout, instead it consists of a number of shells surrounding a central core. In the simplest case, a virus might consist of a protein case enclosing a central space wherein

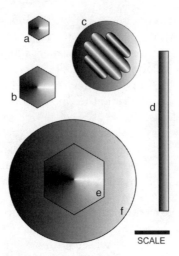

Figure 5.1 A schematic diagram showing the shape and approximate relative sizes of some viruses: (a) poliovirus; (b) polyoma virus; (c) influenza virus; (d) tobacco mosaic virus; (e) herpes simplex capsid; (f) herpes simplex virion. Scale bar: 100 nm

Table 5.1 The Geometries and Sizes of Some Well-Known Viruses

Virus	Family	Approx. Size (nm)	Geometry	Enveloped?
Tobacco mosaic	Tobamoviridae	270×18	Helical	No
Rhinovirus (common cold)	Picornaviridae	18–38	Icosahedral	No
Papilloma (warts)	Papovaviridae	40–57	Icosahedral	No
Influenza	Orthomyxoviridae	100	Helical, spherical	Yes
Rabies	Rhabdoviridae	180×70	Helical, bullet-shaped	Yes
Smallpox	Poxviridae	200–300	Complex, brick-shaped	Yes
Herpes simplex type 1	Herpesviridae	200–250	Icosohedral	Yes

the viral DNA lies; a more complex virus such as herpes simplex encloses that protein shell in a thick protein gel, which is in turn surrounded by a lipid membrane similar to that which encloses a cell. Viruses such as tobacco mosaic virus are not at all spherical, but are long and cylindrical; we will deal with them later in the chapter.

It is possible to adapt the existing model of interfacial dielectric behavior developed in Chapter 2 to account for more complex particle structure. Developed by Irimajiri et al.,[2] it works by considering each layer as a homogeneous particle suspended in a medium, where that medium is in fact the layer surrounding it. So, starting from the core we can determine

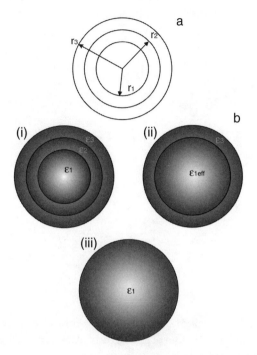

Figure 5.2 (a) A sphere comprises a core and inner and outer shells, with radii $r1$, $r2$, and $r3$, respectively. (b) These three layers have permittivity ε_1^*, ε_2^*, and ε_3^*. We can find the total permittivity of the particle by successively combining the two innermost layers to find the effective combined permittivity.

the dispersion at the interface between the core and the layer surrounding it, which we will call shell 1. This combined dielectric response is then treated as a particle suspended in shell 2, and a second dispersion due to that interface is determined. Then a third dispersion is determined due to the interface between shells 2 and 3, and so on. In this way, the dielectric properties of all the shells combine to give the total dielectric response for the entire particle. This is illustrated schematically in Figure 5.2.

In order to examine this mathematically, let us consider a spherical particle with N shells surrounding a central core. To each layer we assign an outer radius r_i with r_1 being the radius of the core and r_{N+1} being the radius of the outer shell (and therefore the radius of the entire particle). Similarly each layer has its own complex permittivity given by

$$\varepsilon_i^* = \varepsilon_i - j\frac{\sigma_i}{\omega} \tag{5.1}$$

where i has values from 1 to $N + 1$. In order to determine the effective properties of the whole particle, we first replace the core and the first shell surrounding it with a single, homogeneous core. This new core has a radius a_2 and an effective permittivity given by

$$\varepsilon_{1eff}^* = \varepsilon_2^* \frac{\left(\frac{r_2}{r_1}\right)^3 + 2\frac{\varepsilon_1^* - \varepsilon_2^*}{\varepsilon_1^* + 2\varepsilon_2^*}}{\left(\frac{r_2}{r_1}\right)^3 - \frac{\varepsilon_1^* - \varepsilon_2^*}{\varepsilon_1^* + 2\varepsilon_2^*}} \qquad (5.2)$$

We now have a core with $N - 1$ shells. We then proceed by repeating the above calculation, but combining the new core with the second shell, thus,

$$\varepsilon_{2eff}^* = \varepsilon_3^* \frac{\left(\frac{r_3}{r_2}\right)^3 + 2\frac{\varepsilon_{1eff}^* - \varepsilon_3^*}{\varepsilon_{1eff}^* + 2\varepsilon_3^*}}{\left(\frac{r_3}{r_2}\right)^3 - \frac{\varepsilon_{1eff}^* - \varepsilon_3^*}{\varepsilon_{1eff}^* + 2\varepsilon_3^*}} \qquad (5.3)$$

If this procedure is repeated a further $N - 2$ times, then the final step will replace the final shell and the particle will be replaced by a single homogeneous particle with effective permittivity ε_{Peff}^* given by

$$\varepsilon_{Peff}^* = \varepsilon_{Neff}^* \frac{\left(\frac{r_{N+1}}{r_N}\right)^3 + 2\frac{\varepsilon_{N-1eff}^* - \varepsilon_{N+1}^*}{\varepsilon_{N-1eff}^* + 2\varepsilon_{N+1}^*}}{\left(\frac{r_{N+1}}{r_N}\right)^3 - \frac{\varepsilon_{N-1eff}^* - \varepsilon_{N+1}^*}{\varepsilon_{N-1eff}^* + 2\varepsilon_{N+1}^*}} \qquad (5.4)$$

This value provides an expression for the combined effective permittivity of the particle at any given frequency ω. It can also be combined with the complex permittivity of the medium to calculate the Clausius–Mossotti factor, as demonstrated by Huang et al.[3] for yeast cells. This approach of successively replacing shells has become the standard approach in determining the complex permittivities of many concentric shells; it is easily implemented in software, as described in Chapter 11.

5.4   Methods of measuring dielectrophoretic response

5.4.1   Experimental considerations

In order to derive dielectric data that may help us to understand the physical processes within a virus, we need to test the virus's dielectrophoretic response in some way. There are a number of methods that can be used, which vary in degrees of complexity, and which provide varying amounts of information. In each case we use the force — either by measuring it or determining which frequency causes it to cross over — as the basis for deriving our information using the equations for the dielectrophoretic force and the Clausius–Mossotti factor.

The most widely used method of studying virus particles is the same as that used for latex beads; the crossover method, as discussed in the last chapter. Other methods include collection rate measurement and levitation analysis. These techniques are generally performed using fluorescence microscopy to observe the virus particles, but recent developments by Jan Gimsa[4] have demonstrated that a light-scattering technique can be used for direct virus observation without staining. In all of these methods, the numerical result is obtained by finding the best-fit match between the data and the variables of permittivity and conductivity of the membrane, viral interior, and medium. However, with so many variables, there is rarely a unique solution to a given data set, particularly for more complex viruses and simpler data collection techniques.

Viruses are so small, we cannot observe them optically at all. In order to observe them, we must employ fluorescence microscopy. Virus particles are stained using fluorescent dyes, the manner of which will depend on the morphology of the virus. If the virus is surrounded by a lipid membrane (like a cell), then dyes can be purchased that penetrate between the membrane *leaflets* (the inner and outer layer of the lipid bilayer), without entering the interior of the virus itself. Dyes such as NBD-dihexadecylamine have been used, which fluoresce at similar wavelengths to the common fluorescent dye fluorescin. Studies of the infectivity of herpes simplex virus before and after labeling demonstrated that labeling techniques such as this do not affect the infectivity of the virus, although it is not known if it affects the dielectric properties of the virus at all. Comparison of derived data for lipid envelopes surrounding stained viruses and cells indicates the change, if any exists, is minimal, though changes are more likely to occur in stained naked viruses, where the dye attaches to charged sites on the viral surface, thereby changing the surface density and hence surface conductance.

If the virus does not contain a lipid envelope as its outer layer, then the structure consists of a bare protein outer shell and labeling is performed directly onto this. It is possible to directly attach common fluoroprobe molecules such as rhodamine to the amino acids of the protein shell, such as lysine. Protocols such as this have been used for labeling herpes capsids and tobacco mosaic viruses with a good deal of success.

Attempts have been made to examine the enveloped virus influenza using light scattering techniques (described below). However, this approach is unable to discriminate between individual (monodisperse) viruses and those that are clumped; models derived from clumped viruses are far more difficult to construct, since they must consider the radius of the clumps and distribution of the virus matter within the clump. Since, as we have seen, force is proportional to the volume of the particle, these measurements are quite important to the model; furthermore, since the multishell model is based purely on the properties of concentric shells, only an approximate, ensemble value for the clump as a homogeneous sphere can be obtained reliably.

Finally, it is worth mentioning that, despite the arguments that viruses are not truly living organisms, they are nevertheless very fragile biological entities and, as such, need to be treated with great respect when handled. Aside from the important consideration that many viruses are harmful to humans (and should be treated with the due respect warranted by their potential for harm), it is also important to consider the effect of treatment and handling on the viruses themselves. Viruses that have a membrane envelope are most sensitive to environmental conditions, which must, for example, be maintained in an iso-osmotic medium; both enveloped and nonenveloped viruses are sensitive to pH. Since both of these factors are affected by the concentration and type of ions in solution, careful selection of solute is important. Generally, such experiments take place in sugar solutions such as mannitol or sucrose, containing salts such as KCl or NaCl that do not affect the medium pH. However, as described at the end of this chapter, it is possible that changing the medium, even with chemicals that one does not expect to have any effect on the dielectrophoretic response of the virus, can in fact cause a significant change.

5.4.2   Crossover measurements

As with latex beads, crossover measurements provide a rapid, efficient, and accurate means of determining dielectrophoretic response. It has been applied to the observation and analysis of herpes simplex virus,[6] tobacco mosaic virus,[7,8] Sendai virus and influenza virus,[9,10] as well as Herpes capsids.[11]

In many respects viruses are easier to observe by this method than latex beads, since they tend to be more dense than the suspending medium and therefore are more slow moving, remaining near dielectrophoretic traps after the field has been removed; they are also less susceptible to the effects of electrohydrodynamic fluid flow and Brownian motion than beads for the same reason. The principal drawback with viruses is that they tend to adhere to electrodes, glass slides, and each other. Since determination of crossover is most easily achieved by filling a trap by positive dielectrophoresis and then finding the frequency where particles are repelled or released (due to the crossover frequency being reached, so that the particles experience no force), such measurements are made more complex when the particles being observed remain on the electrode array due to adhesion with the surface. The worst examples of this effect bind to the electrodes with such vigor that only dissolving them in concentrated alkali (such as caustic soda) will work, a process that can slow an experiment down when crossovers are required to be taken at a number of different concentrations of medium salts as part of a crossover spectrum. Similarly, viruses have a tendency to stick to one another; such clusters must be excluded from experimental observations, since they no longer conform to the model described earlier as consisting solely of concentric shells.

5.4.3 Collection rate measurements

Collection rate measurement has been used as a means of determining frequency-dependent polarizability using dielectrophoresis since the 1960s.[1] It is based on the assumption that the rate at which particles collect at an electrode is directly proportional to the magnitude of the force applied to those particles. Consequently, by varying the force generated, we proportionally affect the rate at which particles collect. Since the force on the particles is related to the frequency of the field via the Clausius–Mossotti factor, we can investigate the frequencies of dispersions other than the one that causes the polarizability to change sign by measuring the change in the rate at which particles accumulate at different frequencies. In fact, we can also investigate particles that exhibit no crossover, only experiencing positive dielectrophoresis over the range of frequencies the experimental equipment is capable of generating. However, the collection rate is a relative indicator of force, rather than providing a direct measurement of $Re[K(\omega)]$; this is due to the fact that ultimately the collection rate is dependent on other factors such as original particle concentration.

An example of data provided by collection rate measurement is shown in Figure 5.3. In this experiment, measurements were performed by video-capturing images of the electrode array at five 30-second intervals after the application of 5 V_{pk-pk} 500 kHz signal. The fluorescence excitation light was blanked off between measurements to avoid photobleaching. The captured images were converted into bitmaps and an analysis was performed on the

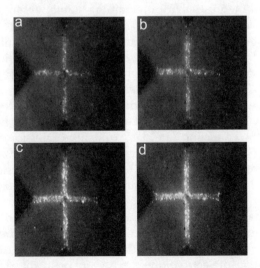

Figure 5.3 A sequence of three fluorescence photographs showing the collection of herpes simplex virus, type 1 particles in the interelectrode gaps of an electrode under positive dielectrophoresis. The photographs were taken (a) 30, (b) 60, (c) 90, and (d) 120 seconds after application of a 500 kHz, 5 V_{pk-pk} sinusoidal signal. Images were taken with an image-intensifying camera. A plot of the variation in rate of increase of particles at the electrodes is shown in Figure 5.4.

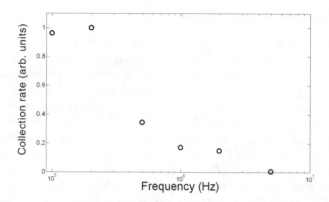

Figure 5.4 The variation in collection rate of viruses at an electrode array, as determined using data such as that shown in Figure 5.3. The collection rate data demonstrate a dielectric dispersion at approximately 4 MHz, as well as a significant rise in collection at low frequencies (below 100 kHz) where collection is assisted by fluid flow as described in Chapter 3.

rate of particle collection, based on the light intensity and total frame area illuminated (i.e., containing a fluorescing virus). The overall rate of particle collection was determined by comparing the light intensity in successive frames. The mean normalized particle collection rate in the range 100 kHz–10 MHz is shown in Figure 5.4; it was found that measurements at frequencies below 100 kHz were disrupted due to interference by fluid convection effects, which prevented the viruses from remaining stable within the trap.

Analysis of the collection rate of particles in the high field regions of the electrodes indicates that the particles exhibit two dispersions — a low-frequency dispersion at 200 kHz and a higher-frequency dispersion at 2 MHz leading to the DEP crossover at 4.5 MHz. For frequencies above approximately 700 kHz, the rate measurement is in agreement with the simple interfacial polarization model with the crossover frequency at 4–5 MHz. Below these frequencies it appears that the particle experiences an increase in the collection rate, i.e., a positive DEP force that increases with decreasing frequency. This lower frequency dispersion seen in the collection rate (at 200 kHz) may be related to the dispersion of the Stern layer, as it occurs at a frequency similar to the crossover frequency of particles in high-conductivity media.

The principal drawback with the measurement of dielectrophoretic collection rates for nanometer-scale particles is that planar electrode arrays can only be used for determining the rate of positive dielectrophoretic collection; negative dielectrophoresis tends to push the particles into the suspending medium, out of the focal range of the observation equipment used. This problem has been overcome for cells[12] through the use of light absorbance measurements. In such systems, the solution containing the particles is sandwiched between two electrode arrays. A laser travels through the center, and its intensity is diminished by the presence of particles in its path. Positive dielectrophoresis pulls the particles from solution onto the electrodes, causing the solution to become clearer; negative dielectrophoresis pushes particles into the path of the beam, decreasing the intensity of the beam at the output. Measurement of the rate of increase or decrease of the absorbance of the solution allows determination of collection rate for both positive and negative dielectrophoresis. However, this technique has not yet been demonstrated for submicrometer particles, where the particles are sufficiently small not to interfere with the beam at optical wavelengths.

5.4.4   Phase analysis light scattering techniques

In order to redress the problem of light not detecting particles of such small dimensions, a new approach developed by Jan Gimsa[4] has overcome the problems inherent in the enhancement of the light absorbance measurement method of detecting $Re[K(\omega)]$. Named phase analysis light scattering (PALS), it is a development of a system originally devised by Kaler and colleagues[5] for observation of cell movement in isomotive electrode arrays (as discussed in Chapter 10). The system employs a technique called Doppler anemometry, wherein two laser beams intersect in a measurement volume, creating a series of fringes. As particles move between fringes, the light intensity varies and, by measuring the time taken for the particles to travel between fringes a known distance apart, the velocity of the particle can be measured directly.

This system can be enhanced by introducing a slight wavelength difference between the interfering laser beams, so that the interference pattern is observed to move. By observing the phase-change of the scattered light with respect to the velocity of the moving interference pattern using a photomultiplier, a direct measurement of velocity can be obtained. A schematic of the setup for PALS analysis is shown in Figure 5.5a. As a technique it offers many advantages, including the direct measurement of particle velocity (rather than inferring it from collection rates) and thus $Re[K(\omega)]$. It does not require that the virus be stained, and it has been demonstrated to work for particles as small as 100-nm-diameter influenza viruses, with unpublished data indicating that the technique can operate with particles as small as 10 nm diameter.

Gimsa and co-workers have also developed an equivalent technique for electrorotation, offering the only method of investigating the electrorotation spectrum of nanometer-scale particles. Dubbed electrorotation light scattering (ELRS), the technique is similar to PALS but, rather than employing two laser beams, a single beam passes through the solution contained within electrorotation electrodes (as shown in Figure 5.5b). The change in phase angle introduced by the beam passing through rotating particles is measured, and the rotation velocity inferred from this, enabling the electrorotation spectrum to be determined. Again, this has been demonstrated to be effective in measuring influenza viruses, producing results that match well with the dielectrophoretic results observed by PALS. These techniques have great

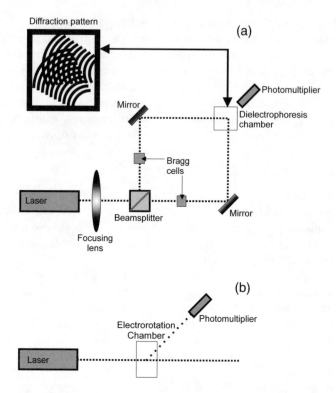

Figure 5.5 (a) Phase analysis light scattering (PALS) allows experimental assessment of particle velocity by examining the movement of particles through a moving diffraction pattern. (b) A similar, though simpler, concept measures phase changes in laser light (so-called electrorotation light scattering, ERLS) to determine rates of electrorotation. Both have been shown to be effective for studying particles below 100 nm without the requirement for fluorescent staining.

potential and may become the standard method by which nanometer-scale bioparticles are measured in the future.

5.4.5 Measurement of levitation height

This is a rapid method of analyzing the dielectric properties of *single* particles. It relies solely on negative dielectrophoresis, in a manner similar to the connection between positive dielectrophoresis and collection rate observation. Demonstrated to be highly effective for cells and other particles on the micrometer scale, the method is less effective for virus analysis because Brownian motion acts to make the accurate determination of levitation height difficult but still has applications where there is a requirement for the determination of the properties of single particles (e.g., Jones and Bliss[13] and Kaler and Jones[14]).

Consider a particle suspended in a negative dielectrophoretic field trap such as that described in the previous chapter for the study of single beads. As well as the negative dielectrophoretic force containing it at the center of the trap, the particle also experiences some levitating force due to the action of negative dielectrophoresis to push the particle above the array. This levitating force is balanced against a known quantity: the negative buoyancy of the particle as it sinks in the suspending medium. By determining the force applied at that point, it is possible to calculate the properties of the particle. Where the particle has a lower density than the suspending medium and tends to float upward, measurements can still be taken by inverting the electrode array, so that the electrodes are at the top of the chamber and pushing particles downward.

When the forces are in equilibrium, the particle remains at a stable position so that on average the total force in the z-direction is zero, i.e.,

$$\mathbf{F}_{TOTAL,z} = \mathbf{F}_{DEP} + \mathbf{F}_{BUOYANCY} = 0 \tag{5.5}$$

i.e.,

$$\mathrm{Re}[K(\omega)]2\pi r^3 \varepsilon_m \nabla E^2 = \frac{4}{3}\pi r^3 (\rho_p - \rho_m)g \tag{5.6}$$

or

$$\mathrm{Re}[K(\omega)]\nabla E^2 = \frac{2(\rho_p - \rho_m)g}{3\varepsilon_m} \tag{5.7}$$

where ρ_p and ρ_m are the densities of the particle and medium, respectively, and g is acceleration under gravity. By simulating the electric field in order to determine the value of ∇E^2, it is possible to use the other known values to determine the value of $\mathrm{Re}[K(\omega)]$ for that applied frequency. Measurement of the particle height for a range of frequencies, or the range of voltage required to ensure levitation at a constant height for different frequencies, allows the determination of the spectrum of $\mathrm{Re}[K(\omega)]$ and hence the dielectric properties where the particle experiences negative dielectrophoresis. The latter procedure, maintaining constant particle height and varying the voltage in order to determine $\mathrm{Re}[K(\omega)]$, has the advantage of allowing electronic feedback to improve measurement accuracy and convenience. Note also that the procedure can be reversed for determining the positive part of the dielectrophoretic spectrum by reversing the procedure; using positive dielectrophoresis to attract a sinking particle upward.

For example, a herpes simplex virus is levitated to a height of 7 μm when a 5 V pk, 10 MHz signal is applied using the quadrupole electrode array

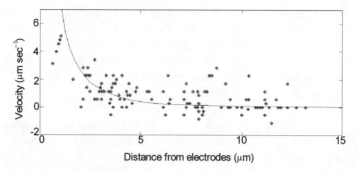

Figure 5.6 The velocity of 282-nm colloidal particles in an electrode array, as a function of the distance from the edge of the nearest electrode. Although the action of Brownian motion causes the movement of particles to appear very noisy, it is possible to statistically unravel the velocity from the applied force gradient (determined by simulation and shown as a line).

described in the previous chapter. The simulation of the electric field around the electrode array, shown in Figure 4.9, gives a value of field gradient ∇E^2 of approximately 8×10^{13} V^2 m^{-3} at this height. Although the electric field gradient at the very center of the array has zero field gradient in the z direction, the displacement of the virus across the enclosed trapping volume means that the virus most probably experiences dipolar, rather than quadrupolar, force interactions. Using Equation 5.7 and the density of HSV-1, $\rho_p = 1.4$ g cm^{-3}, the model indicates that Re$[K(\omega)] = -0.04$ at 10 MHz. This experiment can then be repeated for other applied frequencies, allowing the value of Re$[K(\omega)]$ to be determined for the negative part of the spectrum.

5.4.6   Particle velocity measurement

A final method of determining particle properties is the actual observation of individual, fluorescently labeled particles over time. This form of analysis has many drawbacks and is not widely used, but it has the advantage that it allows the tracking of single particles in a population. By placing particles in a larger array with well-defined electric field gradient — typically a larger polynomial array with an interelectrode gap of 25–50 μm across the center of the chamber — and applying an electric field, the motion of the particles can be captured, for example by using a microscope, camera, and video tape recorder. By analyzing the videotape frame by frame, individual particles can be tracked across the interelectrode gap as they are attracted to, or repelled from, the electrodes. Particles on this scale are subjected to a significant amount of Brownian motion — which can cause displacements of several diameters between one video frame and the next — by comparing the measured data with plots of the local electric field gradient. An example of this is shown in Figure 5.6, where the displacements of a large number of 282-nm latex beads are shown together with the magnitude of the

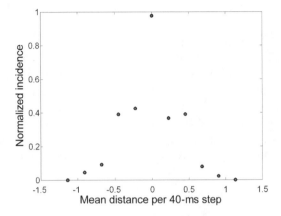

Figure 5.7 The statistical analysis of particle movement in Figure 5.6, showing the mean movement of a particle toward or away from an electrode edge, in the time between camera frames (40 ms apart). Movement toward the electrode is given a positive value, away from the electrode is negative.

dielectrophoretic force predicted by simulation. If we analyze the motion of particles on a frame-by-frame basis (as shown in Figure 5.7), we see that the statistical variation of particle movement per camera frame (40 ms) is on average zero; the direction vector indicates movement in the direction toward or away from the electrode (and 90° to the electrode edge), with direction toward the electrode indicating the positive direction. Note that the data were taken at the frequency at which the thermally induced fluid flow (as discussed in Chapter 3) is zero; if this is not the case, the additional effects of fluid flow must be taken into account. At present, the majority of study using this method is directed at understanding the nature of fluid flow itself and the way in which it interacts with the dielectrophoretic force.

5.5 Examining virus structure by dielectrophoresis

While the multishell model is useful for the interpretation of experimental data, it does have the drawback that there are too many parameters in the model for there to be a single, unique solution for any given data set; there are usually a number of possible combinations of values for permittivity and conductivity that could produce the same net crossover spectrum. This problem is compounded as more layers are added to a model. In order to extract some useful information about a particle, it is necessary to make some assumptions about the properties of some parts of the particle as a basis for further model refinement.

For naked virus particles consisting of naked capsids, the modeling of the particle structure is straightforward; where holes exist between coat proteins, it is possible for interior and exterior fluids to exchange, so that we can model the particle as a protein shell surrounding an internal volume with properties similar to the suspending medium. An example of this is described toward the end of this chapter.

If the particle is more complex than this, such as an enveloped layer containing capsid, tegument, and envelope, we need to make more assumptions in order to satisfactorily reduce the number of possible parameter combinations. The first layer about which such assumptions can be made is the lipid envelope, if there is one. As in the case of latex beads, the response of the particle is dominated by surface effects. For any outer layer, the net conductivity is composed of two parts; the conductivity through the layer (the *transmembrane* conductivity) and that around the layer (the *surface* conductivity). These are related by the equation

$$\sigma_{mem} = \sigma_b + \frac{2K_s^i}{r} + \frac{2K_s^d}{r} \qquad (5.8)$$

where σ_b is the conductivity through the membrane, a is the particle radius, and K_s^i and K_d^i are the surface conductances due to the Stern and diffuse layers, respectively. Since lipid membranes are insulating, they usually have a low value of σ_b, such that the conductivity of the outer layer is dominated by the surface components. For particles on the scale of viruses, σ_b can be treated as zero unless the membrane is damaged or has been treated with an agent that raises the membrane conductivity (such as valinomycin — see Section 5.6). Even though we now have two new variables to replace the one we had previously, the effect of both of them is distinct; where the particle exhibits a rise in crossover frequency as a function of medium conductivity, it can only be attributed to the diffuse layer conductivity, as discussed in the previous chapter. Similarly, the value of K_s^i is related directly to the Stern layer dispersion in high-conductivity media, which allows us to double-check our result. Finally, we are assisted in assigning values to the permittivity of viral envelopes by the fact that they are made from the same material as the cell membranes of the cells from which they came, and the dielectric properties of cells have been studied in some detail, with values in the range between $8\varepsilon_o$ and $17\varepsilon_o$ being quoted.[16] Note that the permittivity of the membrane is often expressed as a capacitance per unit area for the area of the surface, which is equivalent to the permittivity divided by the membrane thickness (usually 7 nm). Moving within the envelope we may find a gel-like tegument. Since this is enclosed by the envelope when the virus was constructed inside the cell, we might expect the internal conductivity to be of a similar order to the values found in cell interiors. Cell values are typically 300–400 mS m^{-1},[16] though this can be lower in the virus due to ion leakage across the membrane after the virions are harvested. Typical values for cell internal relative permittivity are of the order 50–70, though this is arguably less relevant when comparing to internal viral structures. Where no tegument is present, the viral interior contains a capsid structure.

Within the tegument (if present), we find the capsid. If the capsid is naked, then the structure may still play an important role in the dielectro-phoretic response. Capsids such as that in herpes simplex are icosahedral structures surrounding a core containing the viral DNA. The protein shell is often porous, to allow access to the DNA when the capsid is constructed. These pores then allow an exchange of materials between the interior and exterior, so that the conductivity of the interior is closely related to the suspending medium. The response of a naked capsid will be dominated by surface effects, but capsids enclosed within a tegument may become "invisible" because its internal and external properties are the same, and the capsid (being made of protein, just as the tegument is) is not sufficiently different from the material surrounding it to have a significant impact on the properties of the tegument. Indeed, in many cases where the size of the capsid is small in proportion to the size of the tegument, the properties of the capsid may not affect the net response of the virus in any appreciable way. In either case, the virus can be treated as having a core (tegument) and membrane. This simplifies the problem to the extent that values can be determined with some confidence.

Ultimately, such simplification may not be necessary if the viral structure is sufficiently homogeneous that it can be treated as one single solid mass, as with our particles in Chapter 4. One example of such a virus is tobacco mosaic virus, which resembles a long, thin protein tube. However, this virus presents other problems, namely, how to adapt our model for nonspherical objects. We will deal with this problem later in the chapter.

5.6   The interpretation of crossover data

5.6.1   Clarifying assumptions

By way of example of the interpretation of the data collected by dielectro-phoresis, we can examine the properties of herpes simplex virus type 1 (HSV-1), both in its native state and after having been subjected to various modifying agents. The structure of HSV-1 is shown in Figure 5.8; as can be seen, it consists of most of the components that have been discussed, includ-ing a capsid, tegument, envelope, and glycoproteins.

In order to derive the dielectric properties of the virus from the data, a crossover spectrum was generated based on a hypothetical Clausius–Mos-sotti response; the best-fit match between observed and predicted response was determined by generating a first estimate on assumed values, which was then refined iteratively as the various spectra were examined. The model was based on a simplified version of the virus as an insulating envelope surrounding a conducting tegument; if we vary the parameters in a model of the complete virus, it is evident that the properties of the capsid do not affect the overall predicted dielectrophoretic response of the virion.

In order to obtain an understanding of how the various components of the virus contribute to the overall dielectric response, it is possible to use

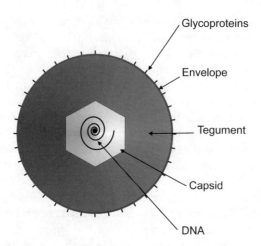

Figure 5.8 Schematic diagram of the HSV-1 virion showing the DNA, capsid, tegument, and membrane.

chemical agents that only modify one aspect of the virus' electrical makeup. Three such agents are saponin (which causes small holes to open in the membrane, allowing some of the interior to leak out); trypsin (which digests the glycoproteins on the surface but leaves the rest of the virus intact); and valinomycin (a potassium *ionophore* — a chemical that is incorporated into the membrane and allows transport of potassium ions across the membrane, between interior and exterior). The model was constructed using the following premises: the dielectric properties of the trypsinized virions should differ from untreated virions only in the surface properties, and that saponin would make the virus "leaky," affecting only the internal conductivity.

These assumptions allowed the results to be compared and a unique solution that satisfied all the above conditions to be found. These solutions are summarized in Table 5.2. Throughout this analysis, it was assumed that the virus particle has a total radius 125 nm and the lipid envelope is 7 nm thick.

Table 5.2 Dielectric Parameters for Viruses Estimated by Fitting the Data Shown in Figures 5.9 to 5.14 Using the Single Shell Model

	K_s (nS)	σ_{int} (mS m^{-1})	ε_{int}	ε_{mem}	ζ (mV)
Fresh	0.6 ± 0.1	100 ± 5	75 ± 25	7.5 ± 1.5	40 ± 5
+1 day	0.4 ± 0.02	85 ± 1	75 ± 15	7 ± 0.5	40 ± 2
+10 days	0.3 ± 0.1	24 ± 2	75 ± 3	35 ± 5	Less than 30
Trypsin	0.3 ± 0.1	83 ± 3	75 ± 20	7.5 ± 0.5	40 ± 5
Saponin	0.55 ± 0.05	40–60	75 ± 10	10 ± 5	40 ± 20
Valinomycin	$0.6 + \sigma_{mem} = 18$ mS m^{-1}	$\sigma_{med} + 40$	78 ± 2	26 ± 2	40 ± 4

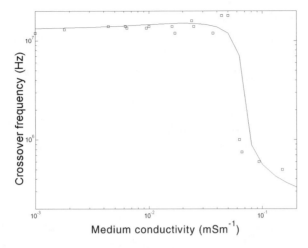

Figure 5.9 Dielectric crossover spectrum of HSV-1 taken 5 hours after virus preparation. The circles denote the frequencies at which the majority of the particles were observed to change from positive to negative dielectrophoresis. The solid line indicates the best-fit result to the data using the model described in the text.

5.6.2 Interpretation of results

Figure 5.9 shows the response of fresh, untreated virus. The crossover spectrum exhibits a gentle rise as the medium conductivity is increased, from 12 MHz at low conductivities to a peak of 18 MHz at 50 mS m^{-1} medium conductivity. Thereafter the response falls sharply to below 1 MHz. The best fit between predicted and recorded dielectrophoretic response indicates that a fresh virion has internal conductivity 0.1 S m^{-1}, internal permittivity $75\varepsilon_o$, membrane permittivity $7.5\varepsilon_o$, surface conductance (K_s^i) 0.6 nS, and ξ 40 mV. As described above, the magnitude of the surface conductance when dealing with particles of this size effectively masks the value of membrane conductivity, which is typically less than 0.1 mS m^{-1} but may be larger in the event of the membrane losing its integrity. The calculated parameters are consistent with values previously presented for cells; the value of K_s^i is comparable with a value of 0.54 nS determined for erythrocytes in the work of Gascoyne et al.[17] Similarly the membrane permittivity compares well with the value $6.8\varepsilon_o$ for erythrocytes determined by Gascoyne et al.[17] These are reasonable bearing in mind the cellular origin of the virus particles and the production mechanisms within infected cells. Of interest is the internal conductivity, which is somewhat lower than that for a typical cell. This may be due to ion loss across the membrane during the harvesting process, or due to the exclusion of ions from the forming virion during maturation within the host cell.

Trypsin acts to remove the surface glycoproteins while leaving the remainder of the virus intact. The response in Figure 5.10 shows that the crossover frequency below 10 mS m^{-1} solution has dropped to a constant 7–8 MHz, rising through a small peak before falling to below 200 kHz. The

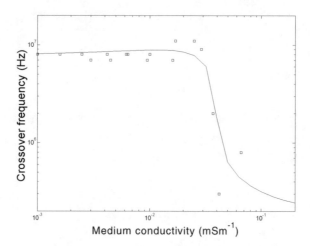

Figure 5.10 Dielectric crossover spectrum of HSV-1 taken after treating the virus with trypsin. Trypsin removes the outer glycoproteins from the virus, leading to a reduction in surface charge.

absence of the rise in response is indicative of a small double-layer surface conductance, and, indeed, the best-fit model indicates that K_s^i has fallen to 0.3 nS. This is in line with the predicted effect of trypsinization, which affects the surface glycoproteins but not the viral membrane or interior.

The virions treated with saponin also show a fall in the DEP spectrum, though the fall is less uniform. Furthermore, the particles exhibit a rise in crossover frequency with medium conductivity that is significant and can be ascribed to either an increase in surface charge or, more likely, to an increase in internal conductivity as a function of suspending medium conductivity. This can be seen in Figure 5.11. An explanation for this behavior can be derived thus: saponin permeabilizes the membrane, causing the interior and exterior ionic environments to mix. This has the consequence of causing the internal conductivity to ultimately equilibrate to that of the suspending medium. However, as these experiments were performed shortly after saponinization, it appears that the conductivity of the interiors of the virus particles had not achieved equilibrium, particularly at lower ionic concentrations. In their work with mouse erythrocytes, Gascoyne et al.[17] showed that treatment with saponin causes cells to lose interior ions slowly, resulting in a loss of internal conductivity. It is difficult to find a single best fit line for these data; fitting the range of conductivities indicates that the internal conductivity lies somewhere between 40 and 60 mS m^{-1}, which define the upper and lower limits of the scattered points at low frequency. Note that at higher conductivity, the graph becomes much more regular and indicates a value of K_s^i of approximately 0.55 nS, similar to that of fresh virus, with all other parameters remaining equal. We would also anticipate an increase in transmembrane conductivity, but in particles of this size, such an effect is masked by the much larger contributions due to double layer conduction.

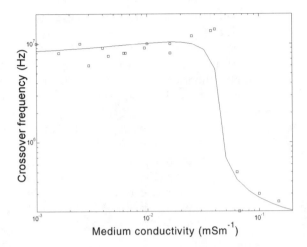

Figure 5.11 Dielectric crossover spectrum of HSV-1 taken after treatment of fresh virus with saponin. Saponin permeabilizes the viral membrane, causing a reduction in internal conductivity. Since this occurs at different rates in each virus, the results are more scattered than is seen in other spectra. The circles and line denote experimental and simulated results, respectively.

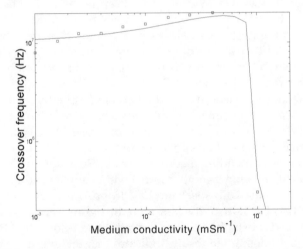

Figure 5.12 Dielectric crossover spectrum of HSV-1 taken after treatment of the viruses with valinomycin. Valinomycin is a K+ ionophore, which acts to equalize the internal and external K+ concentrations, leading to a strong interdependence between internal and external conductivities. The circles and line denote experimental and simulated results, respectively.

As seen in Figure 5.12, the spectrum of particles treated with valinomycin is different from the other results in that it exhibits a sharp rise in crossover frequency from 8 MHz in low conductivity media to in excess of 20 MHz at higher conductivities. Valinomycin is a K+ ionophore, that is, it acts to transport potassium ions across the membrane. This leads to equilibrium of K+

ions on either side of the membrane, with other ions contained within the virion being retained. The gradient of the slope of the dielectrophoretic spectrum is therefore due to the interior conductivity following the exterior conductivity, with an extra component corresponding to the non-K^+ ions. The model indicates that if the internal baseline conductivity is 40 mS m^{-1}, then the best-fit model matches the response if the membrane relative permittivity is increased to 26. This is a high value for a lipid membrane but may reflect the action of the ionophores suspended within the membrane. All other parameters are within the ranges indicated for fresh viruses. Furthermore, there is a high transmembrane conductivity component σ_{mem} of 18 mS m^{-1}, indicative of the large effect of ion transport across the membrane by the ionophore. Such an effect could be mistaken for an unusually high value of K_s^i though the low data point at higher conductivity would count against this theory.

5.6.3   The effects of storage

In order to investigate the ageing of virus particles, viruses were stored for 1 day at 4°C, divided into 13 aliquots of experimental conductivity prior to storage. The spectra of these samples are shown in Figure 5.13. As can be seen, the overall crossover frequency has been reduced. The best-fit curve indicates that all parameters except the internal conductivity are essentially the same as for the fresh virus. However, the samples demonstrated a rise in conductivity of 0.4× suspending medium conductivity in addition to a basic conductivity of 50 mS m^{-1}. This change in behavior can be explained in terms of the work of Gascoyne et al.[17] in which cells suspended in nonionic, iso-osmotic media experienced a gradual exchange of internal ions with the surrounding media over time. The rate of exchange will have been slowed in these samples due to the low temperature of storage. Virus particles stored in experimental, ionic solutions not only lose ions at the same rate, but also gain ions from the suspending medium, resulting in an internal conductivity with components relating to both internal and external conductivities.

After 10 days, the DEP spectrum has dropped further, with the crossovers in media of conductivity less than 10 mS m^{-1} being at a constant 8 MHz with no rise exhibited before the fall-off to below 200 kHz, as seen in Figure 5.14. Modeling the results indicates that the internal conductivity has reached a value of approximately 24 mS m^{-1} and the membrane permittivity has risen to $42\varepsilon_0$. The vale of K_s^i has dropped to 0.3 nS, indicating further membrane damage and glycoprotein loss; ξ is no more than 30 mV but the model is insensitive to values below this. The drop in internal conductivity is again consistent with the explanation that internal ions have largely been lost either through diffusion across the membrane or through the membrane becoming ruptured. The internal permittivity remains unchanged, which may be expected due to the robust nature of the tegument's protein gel. However the significant rise in the membrane permittivity is surprising; it may be due to a number of factors including disintegration, invagination or

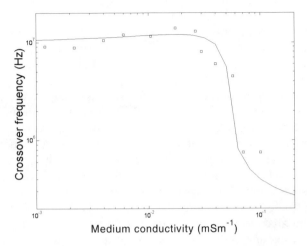

Figure 5.13 Dielectric crossover spectra of HSV-1 taken 24 hours after virus preparation, with the virus having been stored at +4°C. The circles and line denote experimental and simulated results, respectively. Virus particles were stored in separate mannitol solutions, each containing KCl of conductivity equal to that in which the experiments were conducted.

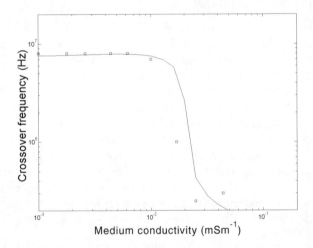

Figure 5.14 Dielectric crossover spectrum of HSV-1 taken 10 days after virus preparation, with the virus having been stored at +4°C for that duration. The circles and line denote experimental and simulated results, respectively.

thickening of the membrane, or the membrane proteins being more freely able to move.

In order to assess the effects of the treatments of the virus by other means, gel electrophoresis was used to examine the effect of the treatment of virus particles described in Section 5.6.2. Infectivity tests (titrations) were used to assess the infectivity of the control virus; samples of the control virus after 1 day at 4°C and samples treated with saponin and valinomycin exhibited equal titration results. This indicated that the treatments do not have any significant effect on the infectivity of HSV-1, despite the demonstration that the anticipated biophysical changes in the virus due to these agents have in fact taken place. However, treatment with trypsin caused a reduction in infectivity of five orders of magnitude.

This result is significant in that it demonstrates that chemical agents that affect the membrane integrity and allow the internal viral ions to escape from the tegument do not affect the particle infectivity. However, treatment with trypsin, which was demonstrated by gel electrophoresis to have removed only the surface glycoproteins, resulted in a significant drop in infectivity. By comparing the electrical properties with the change in virus infectivity, we can make assessments about the virus, its function, and the means by which it might be treated; for example, the above study demonstrates that while removing the glycoproteins has a significant effect on infectivity, treatment of the membrane and reduction of the tegument conductivity do not have any effect (unlike some other viruses). Such a result has significance in both the understanding of the function of the virus components and in the basis for any possible future development of a treatment.

5.7   Studying nonspherical particles

While many viruses are sufficiently spherical in shape to be approximated to spheres using the smeared-out shell model presented earlier, many others are elongated or flattened ellipsoids, or long cylinders. It is possible in these cases to adapt our model in order to compensate for the change in shape by deriving a general expression for the Clausius–Mossotti factor for elliptical particles, of which the spherical model is a special case.

When an elliptical particle polarizes, the magnitude of the dipole moment is different along each axis; for example, a prolate (football or rugby ball shaped) ellipsoid will have a dispersion along its long axis of different relaxation frequency to the dispersion across its short (but equal) axes. The dispersion frequency of the dipole formed along the long axis will be of lower frequency than that formed across the shorter axes, but the magnitude of the dipole formed will be greater due to the greater separation between the charges.

Consider an elliptical multishelled particle such as that shown in Figure 5.15. It consists of two axes in projection, x and y, plus a third axis projecting from the page, z. The dimensions of the object along these axes are a, b, and c, respectively. As indicated in Chapter 2, the particle will undergo three dispersions at different frequencies according to the thickness of the ellipsoid along each axis.[18,19] In addition to the dielectrophoretic force experience by the particle, it will also experience a torque acting to align the longest nondispersed axis with the field (electro-orientation).

When a nonspherical object is suspended in an electric field (for example, but not solely, when experiencing dielectrophoresis) any induced dipole will

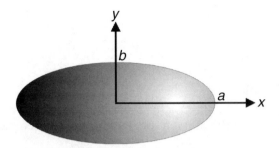

Figure 5.15 A schematic diagram of an elliptical particle, showing axes x and y, along which the particle extends by distances a and b. The particle extends along the z axis (out of the page) by a length c. If $c = b$, the particle is prolate; if $c = a$, the particle is oblate.

have a different time constant according to its alignment to the field. Since each axis has a different dispersion, the particle orientation will vary according to the applied frequency. For example, at lower frequencies, a rod-shaped particle experiencing positive dielectrophoresis will align with its longest axis along the direction of electric field; the distribution of charges along this axis has the greatest moment and therefore exerts the greatest torque on the particle to force it into alignment with the applied field. As the frequency is increased, the dipole along this axis reaches dispersion, but the dipole formed *across* the rod does not and the particle will rotate 90° and align perpendicularly to the field. However, the shorter distance means the dipole moment is smaller. This will result in the force experienced by the particle being smaller in this mode of behavior. Particles with three different axial radii will experience three different dispersion frequencies and will thus align along each axis as the frequency is raised, producing an "orientational spectrum," as described by Miller and Jones.[19]

This alignment force may also be observed in the electrorotation of elliptical objects, for example latex beads, in circumstances where the field strength is greater in one direction than in the others, which can lead to an object exhibiting a wobble when rotating. The mathematical treatment of electro-orientation is complex, and we do not need to explore it in detail here; readers are referred to the work of Jones for a mathematical exploration of the process.[20]

An important consideration is the calculation of the dielectrophoretic force experienced by the particle, both in terms of the determination of crossover frequencies for ellipsoids (and particles whose shape can be approximated as such), and for use in analyses such as collection rate studies. When aligned with the field, a prolate ellipsoid experiences a force given by the equation:

$$\mathbf{F}_{\text{DEP}} = \frac{2\pi abc}{3} \varepsilon_m \operatorname{Re}[X(\omega)] \nabla E^2 \qquad (5.9)$$

where

$$X(\omega) = \frac{\varepsilon_p^* - \varepsilon_m^*}{\left(\varepsilon_p^* - \varepsilon_m^*\right)A_\alpha + \varepsilon_m^*} \tag{5.10}$$

where α represents either the x, y, or z axis and A is the *depolarization factor*. This factor represents the different degrees of polarization along each axis, such that:

$$A_x = \frac{abc}{2} \int_0^\infty \frac{ds}{(s+a)^2 \sqrt{(s+a^2)(s+b^2)(s+c^2)}}$$

$$A_y = \frac{abc}{2} \int_0^\infty \frac{ds}{(s+b)^2 \sqrt{(s+a^2)(s+b^2)(s+c^2)}} \tag{5.11}$$

$$A_z = \frac{abc}{2} \int_0^\infty \frac{ds}{(s+c)^2 \sqrt{(s+a^2)(s+b^2)(s+c^2)}}$$

where s is the variable of integration. The polarization factors are interrelated such that $A_x + A_y + A_z = 1$. The most useful version of these expressions is that simplified for the case of prolate ellipsoids ($a > b$, $b = c$). This expression is useful because many viral particles can be approximated to a prolate ellipsoid, including needle-shaped viruses such as tobacco mosaic viruses.[7] In that case, Equation 5.9 may be rewritten as

$$\mathbf{F}_{\text{DEP}} = \frac{2\pi abc}{3} \varepsilon_m \operatorname{Re}\left[\frac{\varepsilon_p^* - \varepsilon_m^*}{1 + \left(\dfrac{\varepsilon_p^* - \varepsilon_m^*}{\varepsilon_m^*}\right)A} \right] \nabla E^2 \tag{5.12}$$

where A is given by the expansion[21]

$$A = \frac{1}{3\gamma^{-2}}\left[1 + \frac{3}{5}\left(1 - \gamma^{-2}\right) + \frac{3}{7}\left(1 - \gamma^{-2}\right)^2 + \ldots \right] \tag{5.13}$$

and where $\gamma = a/b$. For a spherical particle, $\gamma = 1$ and $A = 1/3$, and Equation 5.12 can be rearranged to the expression for the force on a sphere as derived in Chapter 2.

As with spherical particles, multishell prolate ellipsoids may also be modeled provided their dimensions are known. The procedure is exactly as demonstrated earlier for spheroids, but with the expression for the Clausius–Mossotti factor being replaced by Equation 5.10, thus,

$$\varepsilon^{*}_{(i)eff} = \varepsilon^{*}_{i} \frac{\varepsilon^{*}_{i} + \left(\varepsilon^{*}_{(i+1)eff} - \varepsilon^{*}_{i}\right)\left[A_{i\alpha} + v_{i}\left(1 - A_{(i-1)\alpha}\right)\right]}{\varepsilon^{*}_{i} + \left(\varepsilon^{*}_{(i+1)eff} - \varepsilon^{*}_{i}\right)\left(A_{i\alpha} + v_{i}A_{(i-1)\alpha}\right)}$$

(5.14)

for $i = 1$ to $N - 2$ as before, and where

$$v_{i} = \frac{a_{i}b_{i}c_{i}}{a_{i-1}a_{i-1}a_{i-1}}$$

(5.15)

Multishell ellipsoid particles include viruses such as *vaccina*, or helical enveloped viruses such as rhabdoviruses. For a more complete exploration of the mathematics underlying the dielectrophoresis of elliptical particles, readers are again pointed toward the excellent book by Jones.[20]

5.8   Separating viruses

Just as latex beads of different sizes, properties, or surface functionalities can be separated into subpopulations on an electrode array, so we can separate virus particles with different properties. An example of this is shown in Figure 5.16, which illustrates the separation of herpes simplex virus particles, which have collected in the arms of the electrodes and here appear pale, from herpes simplex capsids that appear here as a bright ball at the center of the array. Similar separations have been performed for herpes simplex virions and tobacco mosaic virions.[22] The procedure for such separations usually follows a categorization of the dielectric response of the two particle species in order to find the optimum frequency and suspending medium conductivity.

An awareness of the change in properties of the virus over time is important. For example, the day after the photograph in Figure 5.13 was taken, the crossover frequency of the virions had dropped (as described in Section 5.6). Separation of particles could still be achieved, but the optimum separation occurred at a lower frequency, and the (highly stable) capsids were trapped by positive dielectrophoresis while the virions were repelled.

Virus separation and identification is perhaps the most important application of dielectrophoresis to nanomedicine — allowing, for example, the point-of-care analysis of blood samples to determine the cause of an infection without the need for lengthy analyses at remote laboratories. We will examine possible separation techniques for performing such analyses on larger samples such as blood samples in Chapters 8 and 9.

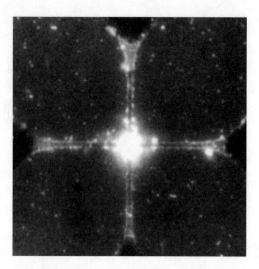

Figure 5.16 A separation of herpes simplex virions (in the electrode arms) from herpes simplex capsids (in the ball at the center of the array).

5.9   Unexpected charge effects

This last section of the chapter deals with the fact that, ultimately, virus particles are highly intricate particles whose interactions with the double layer, applied electric field and suspending medium can produce unexpected results. The capsids of herpes simplex virus can be harvested prior to the addition of the tegument and envelope, enabling their properties to be measured independently. Three forms of the capsid can be isolated and are referred to as A (comprising only the capsid shell), B (in addition to the capsid shell, it contains an internal protein core, the scaffold, which is needed during capsid assembly), and C (containing the viral DNA in place of the internal scaffold).

Capsid particles were examined both to study the contribution of capsids to the dielectrophoretic response of the virion and to establish its own crossover spectrum for the separation between capsids and virions described in Section 5.8. The virus was studied in two media: in ultrapure water to which KCl was added to vary the conductivity and in solutions to which the sugar mannitol was added in iso-osmotic quantities (in which the separation would take place).

Figure 5.17 shows the dielectrophoretic response of capsids in KCl solutions without (Figure 5.17a) and with (Figure 5.17b) mannitol present. In both solutions, the low-conductivity (less than about 0.5 mS m^{-1}) response is similar, with crossover frequencies observed at approximately 3–4 MHz. However, above this conductivity the graphs begin to diverge. The capsids in the solution with added mannitol behave as most other particles do (including viruses and beads, with and without mannitol present) in exhibiting a rise in crossover frequency at a rate that can be described by our existing

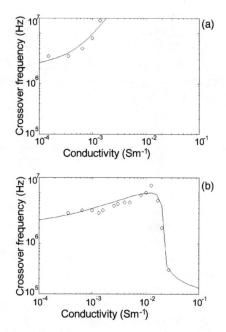

Figure 5.17 The crossover frequency spectrum of herpes simplex capsids as a function of medium conductivity. Capsids were suspended in solutions of KCl (a) or KCL and mannitol (b). Points indicate experimental data; lines indicate best fits.

model, before peaking at 13 MHz and declining to its Stern layer dispersion at approximately 250 kHz for conductivities above 16 mS m^{-1}. Modeling this, we must consider the physical form of the capsid. An icosohedral protein structure, the capsid is largely hollow save for a 30-nm-radius protein structure responsible for packing DNA into the capsid (the so-called scaffolding proteins). The proteins themselves are designed with a large number of pores connecting the inside and outside of the capsid, allowing the entry of the DNA, which is produced separately during virus replication.[23] We can therefore model the capsid as a shelled sphere, on average 62.5 nm in radius and with a shell thickness of 15 nm. Since there is access between exterior and interior, we can consider the volume between the scaffold and the shell as having properties equal to the medium, thick (inner radius 47.5 nm), containing channels that connect the interior and exterior of the capsid. This would mean that the dielectric properties of the capsid interior are equivalent to those of the suspending medium. Fitting data with the model described previously gives the result that the scaffold has a relative permittivity 30, conductivity 30 mS m^{-1}, $K_s^i = 0.082$ nS, and $\zeta = 80$ mV, and that the shell has relative permittivity 60, $K_s^i = 0.1$ nS, and $\zeta = 65$ mV.

For KCl without mannitol, however, the crossover behavior does not obey that predicted by our model. From its low-conductivity value of 4 MHz, the crossover frequency rapidly increases in excess of 20 MHz within a one decade change in conductivity, at which point the crossover was greater than the measuring equipment could determine; the capsids only demonstrated positive dielectrophoresis for frequencies up to 20 MHz and continued to do so for medium conductivities up to 160 mS m^{-1}.

No reason has been found for this behavior. Modeling the response empirically, there appears to be a relationship between the diffuse layer conductance of either the shell or the core that has a relationship to the medium conductivity of a higher power than the $(\sigma_m)^{1/2}$ we might expect from the expression for diffuse layer conductance determined in Chapters 3 and 4. If the diffuse layer conductance scales as a function of $(\sigma_m)^{3/2}$, then we can at least obtain a fit to the data; however, this is entirely empirical, and not dimensionally sound.

These data raise two questions — why do the capsids in solutions with mannitol behave so differently to capsids in solutions without mannitol, and why do capsids in solutions without mannitol behave in this inexplicable manner? Mannitol is a simple, noncharged, nonpolar sugar molecule, whose presence has not made herpes viruses behave differently from latex beads, which have been measured without mannitol. Although its presence may cause the medium permittivity to change slightly,[24] it is unlikely to cause any change as significant as the one observed here. Furthermore, the change observed in the absence of mannitol defies any attempt thus far to explain it. As with many branches of science, the more we learn about the dielectrophoretic behavior of nanoparticles, the more we find things we do not understand.

References

1. Pohl, H.A., *Dielectrophoresis*, Cambridge University Press, Cambridge, 1978.
2. Irimajiri, A., Hanai, T., and Inouye, V.A., Dielectric theory of "multi-stratified shell" model with its application to lymphoma cell, *J. Theor. Biol.*, 78, 251, 1979.
3. Huang, Y., Holzel, R., Pethig, R., and Wang, X.B., Differences in the AC electrodynamics of viable and nonviable yeast-cells determined through combined dielectrophoresis and electrorotation studies, *Phys. Med. Biol.*, 37, 1499, 1992.
4. Gimsa, J., New light-scattering and field-trapping methods access the internal electric structure of submicron particles, like influenza viruses, *Ann. New York Acad. Sci.*, 873, 287, 1999.
5. Kaler, K.V.I.S., Fritz, O.G., and Adamson, R.J., Dielectrophoretic velocity measurements using quasi-elastic light scattering, *J. Electrostatics*, 21, 193, 1988.
6. Hughes, M.P., Morgan, H., Rixon, F.J., Burt, J.P.H., and Pethig, R., Manipulation of herpes simplex virus type 1 by dielectrophoresis, *Biochim. Biophys. Acta*, 1425, 119, 1998.
7. Morgan, H. and Green, N.G., Dielectrophoretic manipulation of rod-shaped viral particles, *J. Electrostatics*, 42, 279, 1997.
8. Green, N.G., Morgan, H., and Milner, J. J., Manipulation and trapping of sub-micron bioparticles using dielectrophoresis, *J. Biochem. Biophys. Meth.*, 35, 89, 1997.

9. Müller, T., Fiedler, S., Schnelle, T., Ludwig, K., Jung, H., and Fuhr, G., High frequency electric fields for trapping of viruses, *Biotechnol. Techniques*, 4, 221, 1996.

10. Schnelle, T., Müller, T., Fiedler, S., Shirley, S.G., Ludwig, K., Hermann, A., Fuhr, G., Wagner, B., and Zimmermann, U., Trapping of viruses in high-frequency electric field cages, *Naturwissenschaften*, 83, 172, 1996.

11. Hughes, M.P., Morgan, H., and Rixon, F.J., Dielectrophoretic manipulation and characterisation of herpes simplex virus-1 capsids, *Eur. Biophys. J.*, 30, 268, 2001.

12. Talary, M.S. and Pethig, R., Optical technique for measuring the positive and negative dielectrophoretic behavior of cells and colloidal suspensions, *IEE Proc. — Sci. Meas. Technol.*, 141, 395, 1994.

13. Jones, T.B. and Bliss, G.W., Bubble dielectrophoresis, *J. Appl. Phys.*, 48, 1412, 1977.

14. Kaler, K.V.I.S. and Jones, T.B., Dielectrophoretic spectra of single cells determined by feedback-controlled levitation, *Biophys. J.*, 57, 173, 1990.

15. Roizman, B., *The Herpesviruses*, Kluwer Academic, New York, 1982.

16. Archer, S., Morgan, H., and Rixon, F.J., Electrorotation studies of baby hamster kidney fibroblasts infected with herpes simplex virus type 1, *Biophys. J.*, 76, 2833, 1999.

17. Gascoyne, P.R.C., Pethig, R., Burt, J.P.H., and Becker, F.F.,Membrane-changes accompanying the induced-differentiation of friend murine erythroleuke-mia-cells studied by dielectrophoresis, *Biochim. Biophys. Acta*, 1149, 119, 1993.

18. Kakutani, T., Shibatani, S., and Sugai M., Electrorotation of non-spherical cells: theory for ellipsoidal cells with an arbitrary number of shells, *Biochem. Bioenerg.*, 31, 131, 1993.

19. Miller, D. and Jones, T.B., Electro-orientation of ellipsoidal erythrocytes: theory and experiment, *Biophys. J.*, 64, 1588, 1993.

20. Jones, T.B., *Electromechanics of Particles*, Cambridge University Press, Cambridge, 1995.

21. Garton, C.C. and Krasucki, Z., Bubbles in insulation liquids: stability in an electric field, *Proc. Roy. Soc. London*, A280, 211, 1964.

22. Morgan, H., Hughes, M.P., and Green, N.G., Separation of submicron bio-particles by dielectrophoresis, *Biophys. J.*, 77, 516, 1999.

23. Zhou, Z.H., MacNab, S.J., Jakana, J., Scott, L.R., Chiu, W., and Rixon, F.J., Identification of the sites of interaction between the scaffold and outer shell in herpes simplex virus-1 capsids by difference electron imaging, *Proc. Natl. Acad. Sci.*, 95, 2778, 1998.

24. Arnold, W.M., Gessner, A.G., and Zimmermann, U., Dielectric measurements on electro-manipulation media, *Biochim. Biophys. Acta*, 157, 32, 1993.

Supplementary reading

1. Dimmock, N.J. and Primrose, S.B., *Introduction to Modern Virology*, Blackwell Science, Oxford, 1974.

2. Tortora, G.J., Funke, B.R., and Case, C.L., *Microbiology: An Introduction*, Benjamin Cummings, Redwood City, CA, 1992.

chapter six

Dielectrophoresis, molecules, and materials

6.1 Manipulation at the molecular scale

Nanotechnology is one of the technological buzzwords of our age. The idea of being able to manipulate molecules in precise structures in order to define new materials and construct elaborate devices has caught the public imagination. However, this is in many cases a rebranding of old ideas — the idea of performing such operations has been around for centuries, but has gone under different names such as chemistry or, later on, materials science. If we consider the development of such devices without those defined by human ingenuity, we find that the arrangement of molecules, each shaped specifically to perform an operation on the nanometer scale, has been continuing for billions of years. Biological macromolecules have evolved over millennia to perform a huge array of tasks with speed and precision still beyond us today, from the tiny molecular rotary motors that power bacterial flagella to the mighty DNA, which contains all the knowledge of how to build and run a human being — with all our characteristics, right down to our instincts and thought processes — in a space a few hundred nanometers across.

So far we have considered the applications of electrostatic fields for the manipulation of nanometer-scale particles but on a fairly large nanometer scale; latex beads 200 nm across, or complex virus particles. While these studies are both important in themselves and necessary for understanding how the processes of charge movement around and through complex colloidal particles are realized, they do not define the limits of smallness where particle manipulation is concerned. In this chapter we will examine the manipulation of molecules themselves — in the form of macromolecules such as DNA, proteins, and carbon nanotubes.

6.2 Manipulating proteins

Proteins are biological macromolecules that perform a bewildering array of biological functions — from forming structures such as hair, to powering muscles, to providing the catalysts for the chemical reactions in cells. They are formed from sequences of amino acids in long chains, which, under the correct conditions, fold into complex three-dimensional shapes; the shape of the protein defines its function.

There are a number of reasons why we may wish to manipulate proteins by dielectrophoresis. Proteins have very precisely defined (and often relatively large, on a molecular scale) structures, dimensions, and masses, and the charge on the molecule can be determined accurately, enabling predictions of electrical properties; this makes them useful models for more general forms of molecular manipulation. Second, many proteins are known to interact in a specific way, such as the way in which two complementary antibodies will bind, which has potential use in the assembly of complex nanostructures. Third, specific proteins are often markers for disease, and the ability to trap and analyze them from a sample would be a useful tool in diagnostic medicine. Finally, the detection and separation of proteins is of great benefit to the science of proteomics, the detection of proteins with potential use in drug discovery.

The manipulation of proteins was first demonstrated by Washizu and co-workers in 1994,[1] in their paper "Molecular dielectrophoresis of biopolymers." In this paper, it was demonstrated for the first time that positive dielectrophoretic force could be used to trap proteins and nanoscale DNA fragments from solution, overcoming the action of Brownian motion that had been previously believed to be sufficiently great as to overcome any applied dielectrophoretic trapping force on particles of this scale. Furthermore, it was demonstrated that by taking account of the different magnitudes of force imparted to molecules of different size (the smallest being a mere 25 kDa), it is possible to separate proteins using field-flow fractionation methods, discussed in more detail in Chapter 8. Since then, other demonstrations have been made of protein dielectrophoresis, with target molecules including antibodies[2] and the flagella from bacteria.[3]

Although Washizu et al. demonstrated the use of positive dielectrophoresis for macromolecules, negative dielectrophoresis of proteins was not demonstrated until much later.[4] As we have already found, where both positive and negative dielectrophoresis are available, it is possible to perform crossover analysis for particle investigation, to separate particles, and to manipulate them singly — which in the case of single molecules has benefits for the type of nanoconstruction described by Drexler[5] and may give rise to useful manipulation technologies for nanoassembly.

6.3   Dielectrophores for protein analysis

As an example of the dielectrophoretic manipulation of proteins in solution, we shall consider here the analysis of avidin molecules in solution by dielectrophoretic methods. We can augment the dielectrophoretic crossover response by altering the suspending medium pH; since we are dealing with single molecules we can take advantage of a more chemistry-oriented approach in order to gain more information. Avidin is a protein molecule

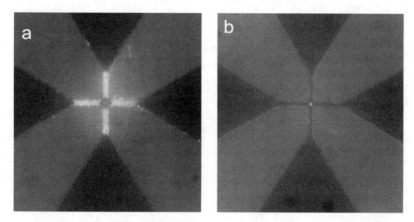

Figure 6.1 Fluorescence photographs showing (a) positive and (b) negative dielectrophoresis of avidin molecules.

found in the whites of chicken eggs. It has a molecular weight of 64 kDa and comprises four identical subunits that interlock into an approximately cuboid shape 5 nm along each edge. It has an isoelectric point — the medium pH at which it is electrically neutral — of 10. It is most notable for forming one of the strongest molecular bonds in nature when attached to the molecule biotin, or Vitamin H. This strong bonding potential has demonstrated great advantages when both molecules are conjugated onto micrometer-scale objects that can be brought together and assembled.

6.3.1 Qualitative description

Figure 6.1 shows the results of positive and negative dielectrophoretic collection of avidin molecules in a quadrupolar electrode array. Before the electric field is applied, the avidin solution appears as a uniform glow emanating from the solution, with the lack of visible clumping indicating that the molecules were evenly dispersed through the solution. When a 20 V_{pk-pk} signal is applied, avidin is observed to collect. As with other colloidal particles, there is a single crossover frequency exhibited, with positive dielectrophoresis exhibited below and negative dielectrophoresis exhibited above. Intriguingly, the collection process for such small particles means that when collection occurs, the protein appears as if from nowhere (since it is not visible in solution) and "grows" from the point of first collection; below the crossover frequency, avidin accumulates along the edges of the interelectrode gaps, with the collection broadening and filling the gap after a period of about a minute as molecular pearl-chains form along the interelectrode gap. Similarly, when the applied field frequency was greater than the crossover frequency, an approximately spherical aggregate formed and grew at the center of the electrode array. When the field was removed, these aggregates remained intact and drifted into solution. As described elsewhere in this book, fluid flow can be found to be significant at lower frequencies, although it is only

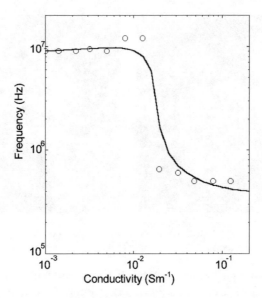

Figure 6.2 Experimental crossover frequencies for avidin molecules in KCl solutions (pH 6.3) with best-fit line according to the procedure outlined in Chapter 4, amended as described in the text below.

a problem under certain conditions where the crossover frequency is within the same frequency window where fluid flow is present, such as those of high frequency and conductivity and medium pH (as described later).

6.3.2   Crossover as a function of conductivity

Figure 6.2 shows the crossover spectrum of fluorescently labeled avidin in solutions of varying KCl concentration (with measured pH of approximately 6.3). As can be seen, the dielectrophoretic behavior follows similar trends to those exhibited by both latex beads and virus particles, with a low-conductivity crossover of approximately 9 MHz, rising to 12 MHz in media of conductivity 10–16 mS m^{-1} before dropping to a lower constant frequency of 650 kHz.

In previous chapters, we have applied models based on Maxwell–Wagner interfacial polarization theory for either homogeneous or heterogeneous (concentric) spherical or ellipsoidal solids, modified in order to compensate for additional charge movement and dispersion in the Stern and diffuse electrical double layers. However, we may wish to question how appropriate it is to apply these theories directly to the modeling of proteins such as avidin.

The first difference between our spherical model and the structure of avidin is the fact that the molecule is not spherical, nor can it be easily approximated to any other ellipsoid. Second, unlike previous particles that are either electrically neutral or have charges free to move and take part in the polarization process or are distributed equally across the particle when not exposed to an external electric field, protein molecules contain fixed areas of positive and negative charge that contribute to the molecule possessing and additional *permanent* dipole, as well as any field-induced dipole. For further details of the electrical properties of biomolecules in solution, see Reference 6.

However, since the model we have is the only one that can be used to predict the dielectric response analytically, we must approximate the behavior of avidin to that of a spheroidal particle and draw conclusions from any deviations from that model required in order to make the predicted results match the observed data. In this, we have an advantage as our data conform to the pattern observed (and described previously) for spherical particles; the flat region at lower conductivity and the steep drop in crossover frequency are both characteristic of the Maxwell–Wagner model. However, the rise in crossover frequency with medium conductivity is small, consisting of a sharp rise very near to the conductivity where crossover frequency drops; similar behavior can be seen in the crossover spectra presented in previous chapters. Below this conductivity, the response is quite flat, indicating very little conductivity dependence (and hence, diffuse layer conduction). If the brief rise in crossover is disregarded, a good fit to the data can be obtained by treating the particle as a homogeneous sphere with negligible internal conductivity, relative permittivity of the molecule of 71, and a surface conductance of value 25 pS. The zeta potential was found to have negligible value. This best-fit response is shown as the solid line on Figure 6.2.

A number of models have been suggested for protein polarization (see Chapter 3 of Pethig[6] for a review). However, those molecules do not exhibit a conductivity dependence of the kind observed here. What may be occurring is that, on the imposition of the electric field, the molecules are attracted to one another by dipole–dipole interaction, forming clusters through which the solution can permeate.

We can also provide approximate values for the surface properties of the molecule. For example, it is possible to examine the effect of the charge on the surface of the molecule that is free to take part in conduction. Using the guide of approximately 10^{17} charges per square meter from Pethig[6] and the values derived for latex beads in KCl in Section 4.3.3, we find a value of ion mobility of 2.5×10^{-9} m^2 s^2 V^{-1}, approximately one order of magnitude lower than that for latex beads. However, at particle dimensions such as these — where the electrical double layer contains a handful of charges — the basic model of Maxwell–Wagner polarization becomes stretched and will need to be adapted. Further work will be needed in this area of dielectrophoretic manipulation before it can be fully understood.

6.3.3 *Crossover as a function of conductivity and pH*

In order to further test the dielectrophoretic response of avidin, we can also investigate the behavior of the molecule as a function not only of medium conductivity, but also of medium pH. The charge on molecules in solution is dependent on the pH of the medium in which it is suspended, with the net charge given by the Henderson–Hasselbalch equation:

$$pH \quad pK_a \quad \log\left[\frac{[A^+][B^-]}{[AB]}\right] \tag{6.1}$$

where *A* and *S* are the molar concentrations of acid and base in the solution, respectively, and pK_a is an acidity constant that describes the pH where 50% of the acid sites on the molecule are ionized, given by

$$pK_a = -\log\left[\frac{[A^+][B^-]}{[AB]}\right] \tag{6.2}$$

where A^+ is the molar concentration of proton donors on the molecule, B⁻ is the molar concentration of proton acceptors, and AB is the molar concentration of the molecule. A similar equation describes the basicity constant pK_b; between these values exists the isoelectric point pI at which the molecule is effectively charge neutral, with equal numbers of positive and negative sites. As the pH of the solution changes, the number of ionized acid and basic sites alter, leading to a change in the effective conductivity of the molecule. This change is due to proton donors ceasing to ionize where the pH is low (and therefore there is an excess of protons in the solution) and proton acceptors remaining unionized due to a shortage of available protons at high pH.

Since the molarity of the ions in a solution affects both the conductivity and the pH, a limited number of pH/conductivity combinations can be investigated; adding ionic species to raise or lower the pH from 7 (neutrality) inevitably raises the conductivity. However, by examining the variation in crossover with medium pH, we can infer the data from the surface generated by combinations of pH and conductivity. Higher values of conductivity allow a broader range of pH values to be investigated. For example, Figure 6.3 shows the variation in crossover frequency as a function of varying pH but constant medium conductivity (10 mS m⁻¹). The range of pH was created by using a combination of monobasic (acidic), dibasic (alkaline), and tribasic (highly alkaline) KPO_4 and (pH neutral) KCl. In this case the counterion is different between the chloride and phosphate molecules; however, the inclusion of the KCl data point allows us to examine the effect of varying this parameter, if any.

The crossover frequency is approximately 10 MHz in media less than pH 8, consistent with the constant-pH graph shown earlier. Above pH 8, the crossover starts to decline until no behavior can be observed at pH 9.5. When the medium pH is above 9.5, the crossover frequency is too low to see, being

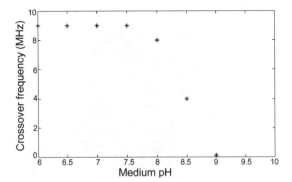

Figure 6.3 A graph showing the change in crossover frequency of avidin for a constant medium conductivity of 10 mS m^{-1}, but varying medium pH. The crossover declines as the pH approaches the pI of avidin.

obscured by electric-field-induced fluid flow. This behavior is in line with the change in the charge on the molecule as the pH moves toward the pI (which in avidin is 10.0), as described by the Henderson–Hasselbalch relationship (where the numbers of acids and bases change as a function of the change in pH). We can adapt the model used for the pH 6.3 case in order to account for the change in surface conductance, which probably takes place in the form of charges moving from site to site along the molecule. The values of conductance required to model the reduction in charge are 22, 17, 13, and 10 pS for surface conductance in media with pH 7.5, 8, 8.5, and 9. This loosely follows the trend predicted by the Henderson–Hasselbalch equation, though not precisely since our approximation does not take into account the fact that some of the charged sites are involved in molecular bonds that allow avidin to retain its shape and that do not therefore participate in the accepting and donating of protons from solution.

We can get a more accurate picture of the effect of pH and conductivity variation of crossover frequency by plotting the data as a surface, such as the one shown in Figure 6.4a. Using only the models described above, we can form a predicted crossover map as shown in Figure 6.4b. As can be seen, the model correlates very well with the observed results.

Although protein molecules such as avidin feature permanent dipoles due to the localization of acid and base sites in different parts of the molecule, the crossover spectrum is essentially the same as would be seen for latex beads and spherical viruses. We would anticipate that avidin might exhibit a dielectric dispersion with a frequency similar to those of other proteins with similar dimensions and molecular weights. For example, hemoglobin has a dispersion frequency of about 10 MHz,[6] and we might expect to see only positive dielectrophoresis below that frequency. There are two possible explanations for this; either the dielectric dispersion frequency of the permanent dipole is sufficiently low for it not to affect the dielectrophoretic behavior or the contribution of the permanent dipole may be small compared to the

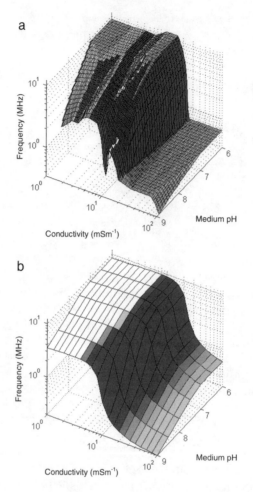

Figure 6.4 Three-dimensional surfaces similar to those in Figure 4.5, showing how the crossover frequency of avidin varies as a function of both medium conductivity and pH. The figures show (a) the experimental response and (b) the theoretical response according to the model described in the text.

magnitude of the induced dielectrophoretic force. Whichever is the case, the permanent dipole does not appear to play any significant role in determining the dielectrophoretic response of avidin molecules in solution.

6.4   DNA

Perhaps the most important molecule in biological science is deoxyribonucleic acid — DNA — which controls the heredity of living organisms. As the unique marker that identifies all organisms and as the basis of many human diseases, the identification and study of DNA molecules is of paramount importance for biotechnology.

At present, identification and analysis of DNA is performed by biochemical methods where corresponding DNA segments are attached to markers such as magnetic beads or fluorescent molecules. However, the problem with such methods is that they require the chance meeting of molecule and detector within the sample solution — which potentially requires a long time for these to come into contact. The most common method to overcome this problem is to increase the amount of DNA being analyzed, using so-called amplification methods such as PCR (polymerase chain reaction) or FISH (fluorescence *in situ* hybridization), but these methods are demanding both in terms of the resources and time they require. There is therefore a requirement for enhancing the effectiveness of DNA sensing for identification purposes. One possible method for overcoming this is the dielectrophoretic concentration of DNA molecules at the sensor surface, effectively forcing the meeting that would otherwise be left to chance.

Although a single molecule, DNA could be considered to be the largest physical object this book will discuss. A single DNA molecule consists of two interlocked phosphate chains (or *backbones*) to which are connected molecular pairs that form millions of regular links between the backbones, which appear as the rungs of a ladder. Each rung is formed from two molecules that have specific affinity for binding to one another; these molecules are adenosine, cytosine, guanine, and thymine (represented by A, C, G, and T), but A will only form a connection with T, and C will only form a connection with G. In this manner, with only one phosphate backbone populated with a sequence of A, C, G and T along it, it is possible to reconstruct the other half of the molecule, forming the basis of the means by which genetic information can be passed from one generation to the next.

The DNA molecule is organized such that rather than the backbones being straight and parallel, they are twisted into an interlocking spiral pattern — the famous double helix. This helical pattern is then further wrapped on itself, or supercoiled, under certain conditions to form a structure called a chromosome. This is shown schematically in Figure 6.5. The molecule has a uniform width of 2 nm between the backbones and 0.34 nm between each rung; the helix is such that each backbone completes a circuit every 3.4 nm.

DNA can be found in a number of states. The most well known is in the single supercoiled thread known as a *chromosome*; this only appears during the process of cell division. A second state, in which DNA spends most of its time in the cell, is in a loose thread referred to as *chromatin*. In some species, particularly those with small amounts of DNA, the molecule can be looped back on itself to form a continuous thread called a *plasmid*. Finally, in biotechnology and in DNA identification it is common for the large DNA molecule (which if completely uncoiled, would be over one meter long in humans) to be cut into *fragments*. These fragments can then be analyzed, cloned, or otherwise used more easily. The cutting process is usually performed by so-called "restriction enzymes," molecules that cleave DNA at points where the base-pair sequence follows a specific pattern for perhaps

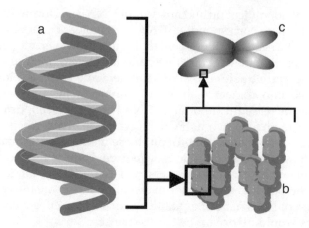

Figure 6.5 A schematic of the structure of the DNA molecule. (a) The molecule consists of two helical backbones (dark and medium gray) with joined base pairs (light gray) between them. The molecule goes through one turn every 34 nm. The structure is coiled on itself to produce a supercoil (b) which is further coiled into the structure of a chromosome (c).

4–6 base pairs. Many hundreds of restriction enzymes have been identified, giving a high degree of control over the fragmentation process. These specific fragments can be used to identify the organism of origin or inserted into new genes in order to affect the molecular behavior of the target organism.

From an electrostatic perspective, perhaps the most important aspect to DNA is that it carries a very large negative charge, resulting in large surface conduction properties and a large degree of water adsorption. This contributes to a high degree of effective particle conductivity, with DNA molecules in electric fields polarizing readily, and the degree of polarization being related to the length of the molecule.[7] At sufficiently high field strengths this effect saturates, such that the molecule appears to have a permanent dipole.

Large DNA fragments and plasmids can actually be observed with conventional light microscopy when supercoiled, but when in less clustered forms it is effectively invisible. To overcome this, as with most of the subjects in this book, fluorescent staining of the molecule must be employed, the most common of which is 4′,6′-diamidino-2-phenyl-indole (DAPI),[7] which specifically binds to DNA.

6.5 Dielectrophoretic manipulation of DNA

Since it is a large and heavily charged molecule, it is perhaps unsurprising that single DNA molecules can be induced to move by dielectrophoretic forces. Indeed, this was among the first molecules to be studied using dielectrophoretic methods. In 1990, Washizu and Kurosawa[7] first described the manipulation of DNA molecules using electrodes with interelectrode spacing of 60 μm and applied peak voltage 60 V; the high voltage compensated for the relatively coarse dimensions of the array and allowed an applied peak field strength of 1 MV m^{-1}.

When no electric field is applied, the natural condition for DNA molecules is to be in a coiled state. However, when an electric field is applied, DNA is observed to uncoil and align along the electric field lines in strands of many molecules, attracted by mutual dielectrophoresis. These strands are observed to coalesce until stripes are formed between electrodes. With sufficient magnification available, it is possible to observe this process occurring with single DNA molecules. The time taken to uncoil a 48,500 base-pair DNA molecule, approximately 17 µm long, is about one third of one second in an electric field of 10^6 V m^{-1}.[7]

One consideration in the dielectrophoretic manipulation of DNA is that since the molecule is so polarizable, it remains more polarizable than the suspending medium under all conditions thus far employed, such that even in high conductivity media and very high frequencies, only positive dielectrophoresis of loosely coiled DNA molecules has thus far been observed. However, while no dispersion that changes the observed behavior from positive to negative dielectrophoresis has been observed, the change in the direction of alignment of the molecules (from being aligned parallel to the field lines to being aligned orthogonal to them) has been observed. Such alignment changes are indicative of a dielectric dispersion of the dipole formed along the molecule (for further information refer to the dielectrophoresis of ellipsoids in Chapter 5). At these higher frequencies, the dipole forms across the axis of the molecule and can realign as the charges move from one side of the molecule to the other — and when one considers that the distance across the molecule is a mere 2 nm, it is perhaps not surprising that frequencies have not yet been found where repulsion of the molecule is achievable. Where the alignment is along the field lines, the molecule is more-or-less straight, but when alignment is perpendicular to the field lines, the molecule does partially recoil in this plane. This is because while the action of parallel alignment leaves the molecule only one direction to occupy, alignment orthogonal to this line allows the molecule to occupy any position in a plane to which the electric field is normal. At low frequencies, it has been observed that the alignment of DNA strands is neither parallel nor perpendicular to the electric field lines but acquires a mesh-like quality. Moving to this frequency window from higher electric fields causes a change from parallel to perpendicular alignment. For the molecules described above, parallel alignment was observed between 100 kHz and 2 MHz, with perpendicular alignment observed at other frequencies.

While negative dielectrophoresis of loosely coiled DNA has not been reported, it has been demonstrated for supercoiled plasmid DNA. Bakewell et al.[8] demonstrated in 1998 that small supercoiled plasmids (which retained their coiled shape and were thus unable to align with a 2-nm-long dipole, as seen in loose DNA) could be trapped by negative dielectrophoresis in a typical quadrupole electrode array. The plasmids, consisting of 12,000 base pairs suspended in an 80 mS m^{-1} electrolyte and labeled with ethidium bromide, were observed to exhibit negative dielectrophoresis between 1 and 20 MHz in electric field strengths similar to those used by Washizu and co-workers.

As stated previously, the ultimate limit of dielectrophoresis in the nanometer scale is for the manipulation of single molecules. Thus far, the only single molecule to have been successfully manipulated in this way is DNA. Tsukahara and colleagues[9] used an electrode geometry similar to that developed by Gimsa (and described in Chapter 5) for the detection of electro-rotation by light scattering — that of a cylindrical capillary drilled through a solid block with four wire-shaped electrodes along the side of the capillary. Tsukahara et al. used a capillary 82.5 μm and 2 cm long, and the solution (of similar composition to those described above) had conductivity 4 mS m^{-1}. The DNA molecules could be observed as single approximate spheroids about 0.8 μm across and were observed to undergo both positive and negative dielectrophoresis. The extracted data seem to imply a double dispersion (as anticipated from work on latex beads), with the main (Maxwell–Wagner) dispersion occurring at approximately 100 kHz.

6.6   Applications of DNA manipulation

6.6.1   Electrical measurement of single DNA molecules

Perhaps the simplest use of dielectrophoresis for the investigation of the electrical properties of single molecules is to use it to trap a molecule onto electrical contacts and measure the properties conventionally, as demonstrated by Porath et al.[10] in 2000. Porath, working with the same research team that had trapped a single 3-nm-diameter palladium sphere in 1997 (described in Chapter 4), used suspended electrodes to attract and stretch a single 10.4-nm-long DNA fragment into the interelectrode gap, removing the electric fields as soon as a single molecule was detected to have bridged that gap. The molecule thus held, it was possible to investigate its conductive properties using conventional voltage and current measuring equipment.

As described in Chapter 4, the electrode configuration used was a pair of point electrodes, in this case 8 nm apart, suspended over a gap etched through the silicon dioxide substrate and fabricated using methods described in Chapter 9. The DNA used was 30 base pairs (10.6 nm) long. To verify that only single molecules were trapped, the experimenters also examined solutions containing no DNA and solutions with the same DNA but using electrodes with a gap of 12 nm between the opposing tips, and in both cases no trapping was observed. Since DNA at this length is mechanically fairly rigid, it was assumed that the molecule formed a straight line between the opposing tips. The molecules were suspended in a solution containing 5 nM EDTA, 10 mM sodium citrate, and 300 mM sodium chloride at a concentration of about 10^{21} per liter. A 1-μl droplet of the solution was applied to the electrodes, which were energized with a 5 V DC potential. As before, when the single molecule falls between the electrode tips, the resultant current flow through the molecule cuts the electric potential, preventing any further collection of molecules.

In order to study the molecule in the absence of charge carriers in the electrical double layer, the solution surrounding the trapped molecule was removed by drying in nitrogen, so that only the suspended molecule remained. The measurements were taken in vacuum and at temperatures below −20°C to ensure that measurements made were exclusively the effect of conduction along the molecule itself. By doing this, the group demonstrated that charge moves along the bare DNA molecule in the same manner of charge moving along quantum dots (where each base pair corresponded to one dot), with charges hopping between base pairs.

6.6.2 Stretch-and-positioning of DNA

As described above, Washizu and co-workers[12] demonstrated that when held in an electric field, DNA molecules unravel and stretch along the axis of the largest nondispersed induced dipole. In subsequent work[7,11–14] they explored the concept further, demonstrating a number of applications for the technique. The most obvious application of stretching DNA molecules is to measure their size. Where the molecules have been stained with a fluorescent dye, the intensity of the fluorescence at any given distance from the edge of the electrodes is directly proportional to the amount of DNA at that distance, so by measuring the variation in fluorescence intensity it is possible to determine how many molecules of a given length are in the sample.

Another application is in the immobilization of DNA molecules at an electrode edge prior to enzymatic treatment. It has been observed[12] that where the electrodes are fabricated from vacuum-deposited aluminum, DNA accumulating at the electrode edge by the stretch-and-position method adhere to the electrode edge permanently. This overcomes a major problem with the application of dielectrophoretic manipulation for biochemical applications, *vis.*, that biochemical agents such as enzymes require specific conditions such as pH in order to function correctly, thereby requiring a high medium conductivity that may not be amenable to the manipulation of molecules. While the replacement of the medium would ordinarily release the DNA molecule back into a coiled state, Washizu and co-workers[12] showed that the addition of divalent positive ions could overcome this; the surface of the glass that covers the interelectrode space (across which the DNA is stretched) is strongly negatively charged, but the divalent ions interact with both the negatively charged DNA backbone and the glass surface, allowing the DNA to be fixed onto the glass and thereby retain its stretched structure after the electric field is removed. Alternatively, there are circumstances where both ends of the DNA must be tethered (e.g., to an aluminum electrode), but the remainder of the molecule must remain free. One such application is where the DNA is used with enzymes such as RNA polymerase, the chemical agent responsible for transcribing DNA information into RNA for transport out of the cell nucleus. This transcription process requires that the enzyme be free to rotate around the DNA helix, and fixing the molecule along its length would inhibit this. In order to ensure the molecule remains secure, additional metal "islands" can be fabricated between the energizing electrodes in order to fasten the molecule in position.

Another application demonstrated for the technique concerns the identification of specific genes on a strand of DNA. By using a single-back-boned fragment of DNA (where the two helices have been "unzipped"), containing only the gene in question and conjugated to a fluorescent dye, it is possible to detect the presence of that gene on a stretched strand of DNA. The gene strand will bind only to the stretched molecule at the point where all the base pairs are in the same sequence (that is, where the same gene is expressed). The binding of the two molecules can be detected, and its position determined, with application in the rapid detection of genes in medicine — such as the detection of faulty genes prior to the use of gene therapy or for the rapid diagnosis of genetic disease. Recent work has demonstrated this[13] using a specific enzyme (*Eco*) that bonds only to sections of DNA with the base-pair sequence GAATTC. By introducing a flow of fluorescently labeled enzymes across immobilized stretched DNA, the sections relating to that specific sequence can be identified.

6.6.3   Molecular laser surgery

In addition to the manipulation of DNA in the 1990 paper,[7] Washizu and Kurosawa also described how a focused spot of UV light could be used to cut the DNA strands tethered to the electrode edge. By scanning the beam along the space adjacent to the electrode edge, they were able to trim the molecule to equal 9 μm lengths as shown in Figure 6.6. It was suggested that laser surgery of DNA may have applications for DNA sequencing, the name given to the determination of the genetic content of a DNA strand for a number of common uses, including (most famously) genetic fingerprinting for forensic science or for the determination of a genetic link between two relatives. In order to achieve this, DNA fragments are separated by electrophoresis (i.e., according to their net charge, and hence size) to form the distinct bands that are widely recognized in DNA fingerprinting. However, the problem with electrophoresis is that it only works effectively for very small fragments of DNA (a few hundred base pairs or less). When the DNA is fragmented, the order of the fragments is lost and repetitions cannot be detected, leading to possible errors in the sequence.

The significant problem with this method of "molecular surgery" is that in molecular terms, it is quite imprecise — a 300-nm focused spot covers a length of DNA equal to 1,000 base pairs, so precision to a single base pair is not achievable. This is limited by the smallest spot that can be focused by a laser, which in turn is dictated by the wavelength of the laser light used. In order to overcome that limitation, Washizu and colleagues[14] have recently used laser tweezers to manipulate enzyme-coated latex beads onto the stretched DNA molecule. Where the bead and molecule touch, the DNA is broken, dramatically improving the precision of the cutting process.

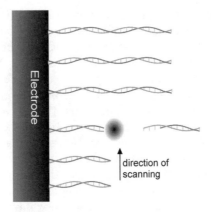

direction of
scanning

Figure 6.6 A schematic of the laser surgery technique pioneered by Washizu and co-workers. DNA molecules are attracted to the electrode surface and stretched by dielectrophoresis; a laser beam is scanned along the molecules allowing them to be cut to size. The accuracy of the technique can be enhanced using enzyme-coated beads.

Another group[15] in Japan have also developed this DNA-cutting method to allow the isolation and cutting of single DNA molecules. The DNA molecules are immobilized on latex beads within an agarose gel. When a DC field is applied, the molecules are attracted to the electrode edge, but since they are attached to the bead surface, they are unable to move, as the viscosity of the gel prevents the bead from moving. The molecules uncoil and can be cut in position using a pulsed N_2 laser. By using beads that are sufficiently small (3 μm), it is possible to identify beads that are attached to only one molecule, thereby allowing a degree of selectivity not available in the electrode systems of Washizu's group using DNA collection at continuous electrode edges. Following the cutting procedure, the ends of the molecule not bound to the bead could be collected by dielectrophoretic force.

6.7 Nanotubes, nanowires, and carbon-60

Another category of highly important molecular structures are those made of carbon, which have recently become the subject of intense study due to their remarkable physical properties, including great tensile strength, lightness, resilience, heat transmission, and electrical conductivity. They have been postulated to be the new standard for electronic materials, a way of storing hydrogen for fuel cells, a source of superstrong materials, and more.

Key contenders for the crown of replacement technology in what is now being referred to as the era of postsilicon electronics are carbon nanotubes, carbon fullerenes, and nanowires. The last of these three differ from the first two in that they are chemically formed fibers of material, often micrometers long but only a few nanometers wide, and made of semiconducting material. The other two — nanotubes and fullerenes — both belong to a class of material first described by Kroto and co-workers.[16] Fullerenes consist of carbon atoms arranged in regular forms not unlike soccer balls, consisting of 60 atoms of carbon arranged in either hexagons or pentagons. This molecule is often referred to as carbon-60, but has also been given the name buckminsterfullerene after the architect Buckminster Fuller; this is often abbreviated either to fullerene or Buckyball.

If this structure is then separated into two halves and a rolled-up sheet of graphite carbon (with a flat structure consisting of carbon atoms in hexagons, like chicken wire) is introduced in between the halves, we have a nanotube[17] — like nanowires, a structure that can be many micrometers long but is only a few nanometers wide. Carbon nanotubes come in many forms, which can be classified in different ways. One division is between those that are structures as described above, with a single carbon tube capped with a hemisphere at either end; this is called a single-walled nanotube (SWNT).[18] Alternatively (and more commonly), several such structures can be formed, one around another in the form of concentric shells, like a nanotube Russian doll. These are called multiwalled nanotubes (MWNTs). Another means of dividing populations of nanotubes is according to the structure of the hexagons that form the main body of the nanotube. Where these hexagons are symmetrical along the axis of the tube, the nanotube acts like an electrical conductor. Where that arrangement is skewed slightly, the nanotube is electrically semiconducting. Skewed further, the nanotube is an electrical insulator. When nanotubes are fabricated, all forms are produced simultaneously and, in the case of MWNTs, a nanotube can contain different types of electrical species in each individual wall. At present there is no efficient way of separating these into different populations.

However, the properties of these various electronic nanocomponents have potentially huge benefits for the electronics industry. First, they overcome the problem of ill-formed lines of conductors formed by chemical doping in silicon chips, since they are already mechanically sound and conduct along their length. Similarly, because their dimensions are exceptionally well defined, it should be possible to achieve a much higher packing density with nanotubes than with silicon. Other advantages are related more to the molecular structure, particularly for nanotubes, which can dissipate heat far more effectively than other materials used in electronics. Finally, the use of such thin wires for electronics applications means that electrons traveling along them cannot deviate from a path from one end of the nanotube or nanowires to the other — they are effectively "one-dimensional" conductors. This means that current flows through them *ballistically* — that is, it travels in only one direction without diversion (for a review, see McEuen[19]). Nanotubes also have excellent mechanical properties, being exceptionally rigid and also having excellent properties as lubricants. An electron micrograph of a carbon nanotube is shown in Figure 6.7.

The dielectrophoretic manipulation of such particles was first described by Bezryadin et al.[20,21] in 1997, in which the trapping of carbon-60 and nanotubes was reported, as well as the trapping of single palladium colloids already covered in Chapter 4. In order to achieve this, electrodes with interelectrode

Figure 6.7 A scanning electron micrograph showing a single multiwalled carbon nanotube. Scale bar: 20 nm. Courtesy Prof. S.R.P. Silva, University of Surrey, UK.

gaps a few nanometers across were energized with a DC electric field to attract particles from solution. By dissolving nanotube soot in cyclohexane and dispersing it through the medium ultrasonically, nanotubes were applied to the electrodes and a 4.5 V DC signal was applied across a 15-nm gap. Following trapping, the cyclohexane was removed by drying, leaving the nanotube held in place. Trapping was reported to take a few tens of seconds, during which the current flowing across the nanotube was reported to change greatly, with a few exceptions when the resistance between the electrodes diminished by an order of magnitude.

The following year, Yamamoto and co-workers[22] demonstrated that carbon nanotubes could also be attracted by AC fields (which they referred to as AC electrophoresis) and examined the frequency-dependent collection of MWNTs of length 1–5 μm and diameter 5–20 nm. The planar, parallel aluminum electrodes had an interelectrode gap of 400 μm, and the applied voltage was approximately 88 V_{rms}. The frequency of the applied potential was varied from 10 Hz to 10 MHz. The degree of orientation was found to correspond to the frequency, with nanotubes becoming more aligned as frequency was increased. Furthermore, it was demonstrated that as the frequency increased, the number of nanotubes collecting at the electrode edge as a function of the total amount of materials (nanotubes and contaminant particles) collecting at the electrodes rose as a function of frequency, indicating that the collection of contaminant decreased. The contaminants remained in solution by negative dielectrophoresis. Similar work has been performed with nanowires, both metallic and semiconducting in origin. For example, studies by Smith et al.[23] characterized gold nanowires of diameter 350 nm and 70 nm and length 5 μm using dielectrophoresis on interdigitated electrodes 5 μm apart. They found that nanowire alignment required voltages of at least 25 V_{rms} at 1 kHz; below that voltage, the induced dipole is of insufficient magnitude to induce movement. Furthermore, investigations as a function of frequency show that polarization time is reduced as a function of applied field frequency from 20 Hz to 20 kHz. Work by van der Zande et al.[24] examined the orientation of gold nanowires with diameters of 15 nm and a range of lengths (between 40 nm and 735 nm) in uniform electric fields and with frequencies of 10 kHz. This work demonstrated the electro-orientation effect and its contribution to the optical properties of the colloidal solution, with the solution exhibiting different absorbance according to the alignment of the nanotubes with respect to the applied field.

Although the study of the processes of polarization and the full understanding of the manipulation of these particles is still in its infancy, the positioning and orientation of nanotubes and nanowires is presently undergoing a massive expansion for the directed self-assembly of nanoelectronic devices. Applications such as the dielectrophoretic assembly of diodes, transistors, and logic gates from nanotube and nanowires components are of great significance, and for this reason they are discussed separately in Chapter 7.

References

1. Washizu, M., Suzuki, S., Kurosawa, O., Nishizaka, T., and Shinohara, T., Molecular dielectrophoresis of biopolymers, *IEEE Trans. Ind. Appl.*, 30, 835, 1994.
2. Huang, Y., Ewalt, K.L., Tirado, M., Haigis, R., Forster, A., Ackley, D., Heller, M.J., O'Connell, J.P., and Krihak, M., Electronic manipulation of bioparticles and macromolecules on microfabricated electrodes, *Anal. Chem.*, 73, 1549, 2001.
3. Washizu, M., Shikida, M., Aizawa, S.I., and Hotani, H., Orientation and transformation of flagella in electrostatic field, *IEEE Trans. Ind. Appl.*, 28, 1194, 1992.
4. Hughes, M.P. and Morgan, H., Positive and negative dielectrophoretic manipulation of avidin, *Proc. 1st Eur. Workshop on Elelectrokinetics and Electrohydrodynamics*, Glasgow, 2001.
5. Drexler, K.E., *Nanosystems: Molecular Machinery, Manufacturing and Computation*, Wiley, New York, 1992.
6. Pethig, R., *Dielectric and Electronic Properties of Biological Materials*, Wiley, Chichester, 1979.
7. Washizu, M. and Kurosawa, O., Electrostatic manipulation of DNA in microfabricated structures, *IEEE Trans. Ind. Appl.*, 26, 1165, 1990.
8. Bakewell, D.J.G., Hughes, M.P., Milner, J.J., and Morgan, H., Dielectrophoretic manipulation of avidin and DNA, in *Proc. 20th Annual International Conference of the IEEE Engineering in Medicine and Biology Society*, 1998.

9. Tsukahara, S., Yamanaka, K., and Watari, H., Dielectrophretic behavior of single DNA in planar and capillary quadrupole microelectrodes, *Chem. Lett.*, 3, 250, 2001.

10. Porath, D., Bezryadin, A., de Vries, S., and Dekker, C., Direct measurement of electrical transport through DNA molecules, *Nature*, 403, 635, 2000.

11. Washizu, M. and Kurosawa, O., Electrostatic manipulation of DNA in microfabricated structures, *IEEE Trans. Ind. Appl.*, 26, 1165, 1990.

12. Washizu, M., Kurosawa, O., Arai, I., Suzuki, S., and Shimamato, N., Applications of electrostatic stretch and positioning of DNA, *IEEE Trans. Ind. Appl.*, 31, 447, 1995.

13. Kabata, H., Okada, W., and Washizu, M., Single-molecule dynamics of the *Eco* RI enzyme using stretched DNA: its application to *in situ* sliding assay and optical DNA mapping, *Jpn. J. Appl. Phys.*, 39, 7164, 2000.

14. Yamamoto, T., Kurosawa, O., Kabata, H., Shimamato, N., and Washizu, M., Molecular surgery of DNA based on electrostatic manipulation, *IEEE Trans. Ind. Appl.*, 36, 1010, 2000.

15. Nishioka, M., Tanizoe, T., Katsura, S., and Mizuno, A., Micromanipulation of cells and DNA molecules, *J. Electrostatics*, 35, 73, 1995.

16. Kroto, H.W., Heath, J.R., O'Brien, S.C., Curl, R.F., and Smalley, R.E., C_{60}: Buckminsterfullerene, *Nature*, 318, 162, 1985.

17. Iijima, S., Helical microtubules of graphitic carbon, *Nature*, 354, 56, 1991.

18. Iijima, S. and Ichihashi, T., Single-shell carbon nanotubes of 1-nm diameter, *Nature*, 363, 603, 1993.

19. McEuen, P.L., Single-wall carbon nanotubes, *Phys. World*, 13(4), 31, 2000.

20. Bezryadin, A., Dekker, C., and Schmid, G., Electrostatic trapping of single conducting nanoparticles between nanoelectrodes, *Appl. Phys. Lett.*, 71, 1273, 1997.

21. Bezryadin, A. and Dekker, C., Nanofabrication of electrodes with sub-5 nm spacing for transport experiments on single molecules and metal clusters, *J. Vac. Sci. Technol. B.*, 15, 793, 1997.

22. Yamamoto, K., Akita, S., and Nakayama, Y., Orientation and purification of carbon nanotubes using AC electrophoresis, *J. Phys. D: Appl. Phys.*, 31, L34, 1998.

23. Smith, P.A., Nordquist, C.D., Jackson, T.N., Mayer, T.S., Martin, B.R., Mbindyo, J., and Mallouk, T.E., Electric-field assisted assembly and alignment of metallic nanowires, *Appl. Phys. Lett.*, 77, 1399, 2000.

24. van der Zande, B.M. I., Koper, G.J.M., and Lekkerker, H.N.W., Alignment of rod-shaped gold particles by electric fields, *J. Phys. Chem. B*, 103, 5754, 1999.

chapter seven

Nanoengineering

7.1 Toward molecular nanotechnology

Thus far, we have considered the application of AC electrokinetic techniques for the manipulation of individual particles and ensembles in order to assess their electrical properties and thence make assessments about their physical or biophysical state. We have considered how we can attract larger quantities of particles from heterogeneous mixtures in order to allow particle separation. We have also considered how to trap single particles, such as DNA molecules, in order to determine their electrical properties.

We may consider these applications to be of value to fields of microbiology and biochemistry, or chemistry and physics; however, the most recent trend in the development of practical uses of forces such as dielectrophoresis is in its use as an engineering tool. As stated in Chapter 1, there has in recent years been a tremendous expansion in research in the development of "nanotechnology," an expansion that spans the boundaries of physics, chemistry, and biology and yet ultimately serves the applications more closely related to engineering and computing science. Most usually, the focus of this research is in the development of electronic devices for near-single-electron computing on the nanometer scale (nanoelectronics), or in the construction of nanometer-scale mechanical devices for molecule-scale chemistry (nanomechanics) and the futuristic (and, many would say, impossible) speculation of the development of independent micrometer robots for a vast range of applications, from swimming about the blood stream and performing surgery from within to mass producing any object from the molecules up. Whatever the ultimate application, there is a degree of convergence between these two approaches that lends itself to a single encompassing term — nanoelectromechanics. This can involve the simple positioning of nanoparticles by electric fields for engineering applications, through to the study of nanoelectromechanical systems (NEMS). Another term we can use to describe the engineering of materials on the nanoscale is nanoengineering, the title of this chapter. In this chapter we will consider how electromechanical interactions can be used to assemble electronic components, and organize materials to change their properties, and we will glimpse ahead to possible applications for nanometer-scale electric motors.

7.2   Directed self-assembly

The principal requirements in the development of new electronics technologies required for nanometer-scale engineering are the development of the nanoscale components themselves and the need to be able to manipulate them in such a way as to assemble functional devices that are also on the nanometer scale. Current microelectronic fabrication relies on photolithographic techniques (discussed in detail in Chapter 9 for the construction of microelectrode structures) to assemble devices from a silicon substrate. By exposing holes etched through protective layers, shapes can be defined into which impurities can be added (in order to make transistors), or layers of conductors and insulators can be placed to connect microelectronic components together.

However, there are limitations to this kind of technology. Photolithography, as its name suggests (it originates from the Greek "to write on stone with light"), uses light exposure (typically in the UV range) to define the areas where a process is to take place; the minimum feature size that can be defined is related to the wavelength of that light. Although strategies have been developed to overcome this limitation to a certain degree, it is generally assumed that in the longer term, electronic device manufacture requires alternative methods of device assembly.

For some years,[1] scientists have been examining the process of *molecular self-assembly*, the phenomenon of molecules organizing themselves automatically and forming complex structures; this is a common phenomenon in biology, where effectively the whole process of cell maintenance and reproduction can be described entirely in terms of complex molecules self-assembling. Work in this field has led to the development of many such structures, but usually the results are more remarkable for their uniformity than their complexity.

What is required is a technology that allows not only self-assembly, but self-assembly into hugely complex electronic devices such as microprocessors. It is logical to expect that such devices might be on the colloidal scale themselves, in order to ensure the resultant device will also be on that scale. As such, electrostatic manipulation can play an obvious role in the manipulation of such particles.

The use of dielectrophoresis for device assembly can be approached in a number of ways and used to perform a variety of functions. We will examine three approaches to the construction of nanoscale devices using electric fields; each demonstrates a different approach to nanoconstruction. These are the assembly of devices from multiple components, electrostatic self-assembly, and the development of nanoelectronics.

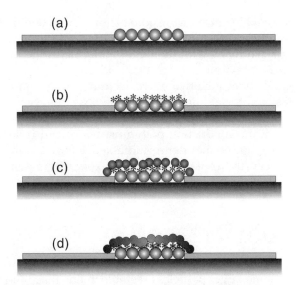

Figure 7.1 Dielectrophoretic assembly of a biosensor, as described by Velev and Kaler.[2] (a) Functionalized latex beads are attracted into the interelectrode gap and are fixed by adding surfactant to the solution. (b) The material to be sensed is introduced to the array. When the target molecules contact the functionalized surface they adhere. (c) After sensing is complete, functionalized colloids are added; these adhere to any target molecules stuck to the beads. (d) A silver enhancer is added, which attaches to the colloids; if sufficient target molecules are present, there will be enough colloids to form a conducting path between the electrodes.

7.3   Device assembly

A good example of the use of dielectrophoresis, not as a tool for direct measurement of particles so much as a means of assembling a device, was presented by Orlin Velev and Eric Kaler in 1999.[2] Their device consisted of two opposing, micropatterned wires with a 1-μm interelectrode gap, which acted as conventional electrodes when powered; these were used to attract antibody-labeled latex beads (such as those described in Chapter 4) to the interelectrode space.

The innovative step that allowed dielectrophoresis to move from manipulation tool to assembly tool is the addition of a surfactant to the solution after the particles have been collected; this reduces the surface charge of the spheres to the extent that the repulsive interparticle forces are insufficient to prevent van der Waals and hydrophobic forces (as described in Chapter 3) causing particle coagulation. This means that the interelectrode gap becomes filled by a permanent structure consisting of antibody-labeled spheres. This procedure is illustrated in Figure 7.1

The biosensor operates by introducing the test solution over the sensor area. The target molecules attach to the antibodies (selected to attach to that particular biomolecule), until the sensor is ready to be read. The testing process is performed by introducing gold colloids labeled with antitest molecule antibodies; these attach to the upper surface of the sensor area, effectively forming a sandwich with the target molecule as the filling. Addition of a silver enhancer solution fuses the gold colloids together, creating a conducting path between the two microelectrodes. If the biosensor has been in the presence of sufficient quantities of the target molecule, there will be a conducting path between the electrodes; otherwise there is not. Since the area across which the conducting path must be formed is so small, a very small quantity is required in order to successfully form a complete conducting bridge between the contact electrodes; experiments demonstrated the device could detect quantities of human IgG antibody at concentrations as low as 10^{-13} M.

7.4   Electrostatic self-assembly

One of the problems to be addressed by self-assembly is positioning; how does one make some target particles assemble in one place and not another? Some work has achieved successful results in micropatterning using self-assembled monolayers of particles,[1] particularly when molding processes are used to shape the resultant film. However, if we wish to pattern conductors into circuits, we must seek a different approach.

It is here that we can consider the applicability of electrostatic interactions to the problem. We have, over the last several chapters, seen that electrostatic forces can be used to position particles with great accuracy, but these demonstrations often rely on electrodes that are significantly larger than the particles being manipulated; ultimately, if we require complex geometries on the nanometer scale, we also require the absence of large, micrometer-scale objects between every assembled device!

An elegant solution to this problem is to avoid the use of electrodes altogether, an approach first proposed by Fudouzi and colleagues[3] in 2001. The concept is as follows: if an insulating substrate is irradiated with an electron beam, then the charge injected from the beam into the surface will remain, creating an electrostatic potential in accordance with Coulomb's law. When this insulating substrate is then placed in a solution containing conducting colloidal particles, the point potential (and its associated electric field) acts to attract the colloidal particles to the point; when the solution is removed, the conducting particles remain and form conducting tracks across the surface of the substrate. This process is illustrated schematically in Figure 7.2.

Using a substrate of polycrystalline calcium titanate ($CaTiO_3$), Fudouzi and co-workers assembled tracks of particles using aluminum colloids as small as 100 nm in diameter. Owing to particle concentration in solution, conducting tracks were not established, but use of the technique with larger (5 µm) particles indicates that conducting tracks are possible using this technique. This is significant since the diameter of a focused electron beam can be significantly smaller than the limits of photolithography; at present,

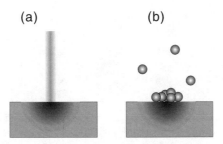

Figure 7.2 A schematic showing the electrodeless assembly method of Fudouzi et al.[3] (a) An electron beam is used to inject charge into an insulating substrate. (b) When conducting colloids in solution are placed over the pattern, the potential and corresponding electric field due to the charge causes particles to be attracted from solution onto the written pattern by dielectrophoresis.

many masks for photolithographic processes are written using an electron beam (e-beam) machine, often with a resolution of 30 nm or less.

Results using similar ideas — the use of electrostatic forces for the assembly of nanoscale wires — but different processes were reported in *Science*[4] in 2001. By applying a potential across a 5-mm interelectrode gap containing a solution of gold colloids of 15–30 nm diameter, suspended in an aqueous NaCl solution, it was found that colloidal "wires" would spontaneously form in solution. Notably, the colloid concentration is relatively high (upward of 0.13% by volume). It was observed that, when a low-frequency (between 10 Hz and 150 Hz) field of the order of 10^4 V m^{-1} was applied to this solution, the colloids spontaneously formed wire shapes that sprouted form the electrodes and crossed the chamber to the opposing electrode. Where the voltage was lower, these nanowires would sprout branches heading in different directions; higher fields produced straighter wires. Speed of wire growth could be as fast as 0.5 mm sec^{-1}, a considerable speed when considering the distances over which the technique might wire up. This is illustrated in Figure 7.3a. If latex colloids are added to the solution, they also aggregate with the wire but do so in such a way as to form an outer sheath surrounding the gold core. Furthermore, the wire (under certain conditions) can spontaneously grow multiple offshoots to form fractal patterns; since this is such a new field of research, no reasons for this behavior (or the extent to which it might be harnessed) have yet been explored.

The effect occurs due to the high local field deformations caused by the curvature of the wire — effectively one colloidal diameter thick at the tip — distorting the electric field sufficiently to cause particles in the immediate area to be attracted to the tip, thereby extending the wire and moving the point of collection forward. The initial collection process begins by induced dipole–induced dipole interactions, or pearl chaining — except that the pearl chains form across relatively large distances. Furthermore, as long as the particles remain in solution, the wire is self-healing; in the event of disturbances breaking the wire, regrowth to completion occurs almost immediately.

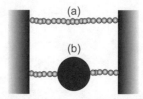

Figure 7.3 Fabrication of colloidal nanowires, as demonstrated by Hermanson et al.[4] (a) When conducting colloid solutions of fairly high concentration are placed between electrodes, they spontaneously form single-thickness pearl chains across the inter-electrode gap. (b) Where there is a conducting island within the interelectrode gap, the wire forms connections to the island preferentially over cross-chamber wires. This is also true for more complex structures involving multiple islands.

Significantly, nanowire growth is usually directed toward any fixed bodies, such as conducting islands of material between the electrodes (but not at any fixed potential) or larger colloidal particles within the electric field volume, so that the nanowires automatically form connections between structures placed within the electrode chamber (see Figure 7.3b). Where there are many such features, complex shapes connecting them can be achieved. This has significant implications for the formation of self-assembled nano-devices; if the technology is to be harnessed for this purpose, then there is a requirement that the process should be able to self-guide between appropriate points, as discussed later.

7.5 Electronics with nanotubes, nanowires, and carbon-60

As described in the previous chapter, there is a group of complex molecular, or molecular-scale, objects receiving great attention as a potential foundation for a new class of semiconductors. The most well known of these are the forms of carbon known as fullerenes and nanotubes. Together, these have been described as forming the basis of "carbon electronics," seen by many as the successor to silicon devices. However, it is important to also include nanowires in this group; these are solid cylinders of material consisting of many molecules but having similar proportions to nanotubes.

As with the materials described earlier in the chapter, the principal consideration for turning them into viable electronic components is the development of a means to position them within an electrode array such that the devices form the appropriate contacts to other devices, in order to make the multipart structures needed for viable electronic components.

In order to understand this, we should consider what structures we need in order to construct viable electronic components. Chief among these is the transistor, the cornerstone of the vast majority of electronic equipment, both analog and digital. In its digital role, the transistor can be considered to be a switching device; it allows current to flow between two terminals (the *source* and the *drain*) provided that a sufficient potential is available at the controlling input (the *gate*). By varying the gate voltage we can switch the transistor on or off. A simpler, but similar, device is the diode, effectively an electronic valve that only allows current to flow along it in one direction; a variant is the light-emitting diode or LED, which glows when a voltage is applied.

By combining many transistors together, we can form *combinational logic* circuits, which form the basis of computers. These are arrangements of transistors whose output state is based on the state of the inputs; common *logic gates* (the name given to circuits performing simple logic operations) include AND (the output current flows if the voltage is high at inputs A *AND* B) and OR (the output current flows if the voltage is high in either input A *OR* input B). A third common logic gate, the NOT gate, produces no output when the input is present and vice versa. Combinations of these produce logic gates such as the NOR gate, active when neither A *NOR* B are active.

In order to fabricate devices with these materials, the most difficult problem is the positioning of materials with respect to both the energizing electrodes and to other materials. For many years, the principal method of performing this operation was to make electrode connections retrospectively — that is, to randomly scatter nanotubes (for example) across a substrate, and then pattern electrodes over them. Obviously this method is of limited use for component assembly, particularly since the combination of multiple nanotubes would happen purely by chance. Methods such as dielectrophoresis had been suggested for dielectrophoretic manipulation of nanotubes in 1998,[5] but it was not until 2000 that the first demonstration of electrostatic manipulation for electronic component assembly was demonstrated by Erdman et al.;[6] by applying an electric field to two electrodes, a 20-μm-diameter LED was positioned from low-conductivity solution by positive dielectrophoresis. The solution was then drained and flux applied to render the connection between LED and electrodes permanent. The force used in the manufacture process is described as electrophoretic but may in fact be dielectrophoretic or, more likely, a combination of the two.

The field of electrostatic component assembly moved into the realm of nanotechnology early in 2001, when Xiangfeng Duan and other workers in the laboratory of Charles M. Leiber at Harvard University reported the fabrication of an LED using electrostatically positioned nanowires in the journal *Nature*.[7] In this paper, an assembly method was described whereby a crossed electrode array was covered with a solution containing only *p*-doped indium phosphide nanowires. This was introduced over a quadrupolar array containing thin, pointed electrodes with interelectrode gaps of about 10 μm across the center of the chamber. An electric field was applied to one opposing pair of electrodes, causing a single nanowire to be attracted to the electrodes; after one was trapped, the solution was removed and a second introduced, containing only *n*-doped nanowires. The second pair of electrodes was energized, allowing a second nanowire to be trapped and creating a *p-n* junction: a diode. This was subsequently found to have light-emitting properties, demonstrating that the method was useful for

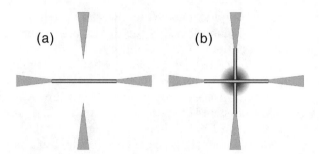

Figure 7.4 A schematic showing the assembly methods of Duan et al.[7] (a) The solution over needle-shaped quadrupolar electrodes contains *p*-doped nanowires. One opposing pair of electrodes is energized with a DC potential, attracting a nanowire across the interelectrode gap. (b) The solution is removed and replaced with another containing only *n*-doped nanowires. The remaining electrodes are energized, causing the trapping of a single nanowire. The solution is removed. The remaining device has a *p-n* (diode) junction remaining, which is observed to be a light-emitting diode.

construction of viable nanoscale components. The actual diode, formed on the junction of the two wires, was a mere 26 nm across. Figure 7.4 shows a schematic of this fabrication procedure.

This method was developed by the Harvard group in conjunction with another method, that of fluidic assembly, which allows assembly of oriented patterns of nanowires, but still requires electrodes to be deposited to order after the nanowires are deposited. Using this method, the group performed a second coup for nanoscale electronics by producing the first functional logic gates with nanowires, this time in the journal *Science*.[8] Once again these consisted of crossed nanowires, with orthogonal patterns of *p*- and *n*-type semiconductor nanowires running between electrode structures (shown in Figure 7.5). These structures — *OR, AND,* and *NOR* gates — were combined to produce a functional, more complex logic device called a half-adder, demonstrating the feasibility of using nanowires as the basis for more complex fabrication technology.

In the same issue of *Science*, another paper[9] demonstrated the possibility of constructing logic gates using carbon nanotubes instead of nanowires. Carbon nanotubes, while possessing many features that make them an obvious material for electronic assembly, are possessed of drawbacks not present in nanowire research. First, their length is not controllable; they must be fabricated in an ensemble of random lengths. Second, their electrical properties depend on the arrangement of carbon atoms in the nanotube wall, which again cannot be predetermined, and a mixture of electrical and semiconducting nanotubes is obtained during preparation. Separation of nanotubes according to electrical properties is a laborious task, requiring individual sorting to take place. Finally, the majority of nanotubes are MWNTs; these consist of many concentric nanotubes, each of which may be either semiconducting or conducting; MWNTs with six or more walls

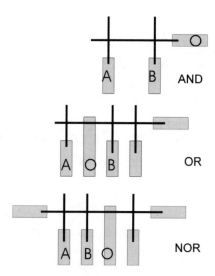

Figure 7.5 A schematic illustrating the method of assembling logic gates from nanowires.[8] By organizing nanowires between electrodes (with *p*-type running vertically and *n* type running horizontally), it is possible to fabricate *AND, OR,* and *NOR* gates between the connecting electrodes. A and B indicate the input lines of the devices, and O indicates the output electrode.

are almost certain to contain at least one conducting layer and hence be considered to be effectively a conducting nanotube. Methods have been suggested for overcoming this problem, such as the passing of a high current through MWNTs: this causes the conducting walls to break up, leaving only the semiconducting parts.[10] However, no sorting procedure has been developed yet. We may speculate that an appropriate method of sorting, by either functionality or length, might be to develop dielectrophoretic separation techniques such as those developed by Washizu et al. for DNA studies, described in Chapter 6,[11] or that such technology may be employed to arrange nanotubes for electronic devices such as those described for nanowires.[8] The experimental results[5] demonstrate that such a task would be feasible, particularly in the light of successes by Duan,[7] but as yet all nanotube fabrication is performed by depositing electrodes to connect to scattered nanotubes, where the electrode design is performed to order.

Another possibility for carbon electronics is the carbon-60 transistor; this was reported in 2000 by Park and colleagues[12] and consisted of two needle electrodes approximately 1 nm apart into which a fullerene was deposited by random scattering in a toluene solution. The fullerene, being slightly smaller than the interelectrode gap (about 0.7 nm), is retained at one electrode by van der Waals forces, but the addition of a single electron gives it sufficient energy to transfer to the other electrode. When it arrives at the second electrode, it receives enough energy to move back to the first, where the process is repeated. This vibration essentially means that the device acts as a single-electron transistor (the lowest power electronic device producible). The rate at which this occurs is governed by the gate voltage. This device represents the near-limit of molecular electronics, where the size of the device is of the order of a single nanometer. If this technology is to be adopted more widely, then we may speculate that dielectrophoretic assembly of the devices — in the manner used by Bezryadin et al.[13] to trap colloids — would be the most effective form of device manufacture.

7.6   Putting it all together: the potential for dielectrophoretic nanoassembly

If nanotube (or nanowire) technology is to become viable as a replacement for silicon, then methods will be required for the semiautomation of the fabrication process. The Intel® Pentium® III microprocessor required 37 separate lithography steps to produce transistors, so there is an extent to which any replacement technology may also involve many steps without being regarded as impractical. On the other hand, those 37 fabrication steps produced up to 44 million transistors. Considering the fact that the advantage of nanoscale components is that they offer the possibility of packing many more components onto a single chip than silicon technology, we may expect that by the time such devices come onto the market, the demands may be for hundreds of millions of transistors on a chip. As such, in order to fabricate useful devices, semiautonomous processes for the insertion of millions of subcomponents at a time will need to be developed.

We may speculate, based on the research already performed and described both in this chapter and in earlier ones, that electrokinetic assembly may play a role in at least some of these steps. Consider that we have already seen that particles in solution form chains between conducting "islands" within an alternating electric field. If we were to place such islands at the contact points where we wish our nanodevices to be positioned to apply an electric field to them, it may be the case that the distortion in the electric field may be sufficient to locate particles between those conducting islands. Applying the field with a solution of only one component type — a single fabrication step — would allow the positioning of many components required to be in one orientation across the whole chip surface. We may then remove the field and apply it in a different direction, or change the solution to introduce a second type of component such as a different variety of nanoparticle. Repeating this would allow a layer-by-layer assembly of the device; at the end of fabrication, a fixative — perhaps a polymer of some kind — could be introduced across the whole array to hold everything in place. This would truly meet the conditions for Drexler's[14] bottom-up approach to fabrication. The devices would be self-assembled from nanometer-scale components without the use of large-scale equipment to control them, with the exception of the introduction of the conducting points between which the particles are connected. Even these might in fact be deposited using the Fudouzi method of electron implantation.

One of the most exciting possibilities with nanotube fabrication is that fabrication does not need to be limited to a single dimension. We may envisage a scenario where devices are assembled in three dimensions, crisscrossing each other like the pipes that fill a ship's engine room. At this point we really do begin to enter the realm of science fiction, but such a move may ultimately be the best way to move computing forward. If a technology is able to pack one transistor into every square micrometer of chip real estate, then a 1 cm × 1 cm array will hold one hundred million transistors. However, if the same device could be constructed in three dimensions, then even a transistor that required a cube of volume 10 μm along a side would produce a billion transistors in a cube 1 cm down a side — and if the transistors could be packed into the space of a cube 1 μm down a side, that value becomes one trillion.

7.7   Dielectrophoresis and materials science

The final aspect of the application of dielectrophoresis to molecular-scale engineering is its role — demonstrated and potential — in materials science and engineering. Materials science can in many ways lay claim to being the original nanotechnology, being the study and design of new materials on the molecular scale and upward. Such feats of engineering have typically been brought about by means of chemistry, but electrokinetics has a role to play in the deposition of materials in an ordered manner.

7.7.1   Deposition of coatings

The simplest application of dielectrophoresis to materials science is that of forming coatings. The engineering of a material's surface layer is of great importance where that material is going to be immersed in hostile conditions, such as corrosive liquids. Alternatively, in electronics applications such as the development of low-power, high-resolution, flat-panel displays there exists a need to increase the emission of electrons from the surface. In order to achieve the latter aim, several groups[15–18] have investigated the use of dielectrophoresis to coat surfaces with so-called *nanodiamond* — essentially a diamond powder, consisting of diamond fragments with diameters on the 10 nm scale. The application is the enhancement of cold cathode emitters, which are arrays of microengineered spikes protruding vertically from a surface to a height of 50 μm or more, typically spaced 10 μm apart in regular square formation. The tips narrow to a sharp point (perhaps 30 nm across), and produce a stream of electrons when a voltage is applied between the needle array and a plane set above it in parallel. This charge can be focused to a display, in the same way that three electron beams can be used to generate a television image; in this instance, each pixel would use its own electron beam.

Such needle shapes, when exposed to an electric field, are sufficiently nonuniform to distort the electric field, and this distortion is significant enough for dielectrophoresis of nanodiamond powder to take place as shown in Figure 7.6. This usually takes place in nonaqueous solvents (such as

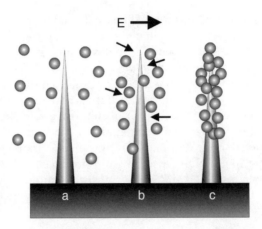

Figure 7.6 Coating of emitter tips with nanodiamond. Silicon emitter tips are immersed in a solution containing nanodiamond (a). When an electric field is applied across the tips, the field around the tips is sufficiently distorted for dielectrophoresis to take place, and the nanodiamond is attracted to the tip surfaces (b). When the field is removed and the solution is dried away, the coating remains on the tip surface (c)

acetone or isopropanol), for ease of particle suspension and drying and avoidance of hydrolysis effects in very high field strengths. The applied potentials are usually of a DC rather than AC nature.

Work on dielectrophoretic deposition by Alimova et al.[15] showed that altering a number of factors gives a degree of control over both the thickness and morphology of the deposited nanodiamond films. They considered deposition in a number of different solvents, electric field polarity (since the applied electric field is DC, there will be both dielectrophoretic and electrophoretic processes acting on the powder in solution), applied voltage magnitude, deposition time, and type of powder used. It was found that the maximum film size that could be attained was about 1 µm, though much thinner films cold be formed. Interestingly, the morphology of the deposited film was found to vary according to the properties of the type of nanodiamond powder used. When nanodiamond powder is suspended in aqueous solutions, it alters the pH of the solution in different ways according to the type of powder used; some types will decrease the pH, others increase it. It was found[15] that low-pH powders form films of uniform thickness across the surface of the tip, whereas high-pH powders selectively collect at the end of the needle tip. No explanation has been given for this, though it may be due to the electrophoretic interaction between the low-pH forms of nanodiamond and any charge injection that may occur at the electrode tip during the application of the DC potential.

7.7.2   Three-dimensional material structuring

Another advantage of dielectrophoretic force for materials science applications is its applicability of the ordering of polarizable particles within a

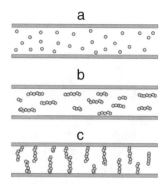

Figure 7.7 When fabricating composite materials, it advantageous to control the arrangement of particles within a matrix. By placing a mixture of particles and curable material within an electrode structure, it is possible to use pearl chaining to align the particles along the field lines. The resultant material has different electrical properties according to whether the material is aligned along or across the field lines. Where rods or permanent chains (rather than induced-dipole chains) are used, this orientation can be frequency controlled.

three-dimensional volume, particularly in combination with other, related forces such as electro-orientation. An important aspect of materials science is the control of the nanostructure of materials, which can have significant impact on the mechanical and electrical properties of such materials.

For example, a number of workers (e.g., Randall et al.,[18] Bowen et al.,[19] Rao and Satyam,[20] and Hase et al.[21]) have used dielectrophoretic forces to align particle dispersions in fluid matter. Randall et al.[18] used mutual dielectrophoresis to cause pearl chaining of 100-nm-diameter $BaTiO_3$ powder dispersed through uncured silicone elastomer. These pearl chains formed continuous lines through the material, which remained there while the elastomer was cured, as shown in Figure 7.7.

Further work by the same group[19] improved the technique, which they referred to as "tunable electric field processing" of materials. Using a range of elastomer and epoxy materials to form the matrix, a wide range of filler particles was aligned including insulators (such as dioxides of zinc, titanium, and silicon), semiconductors (including graphite and zinc oxide), and conductors (aluminum and silver-coated resin). It was demonstrated that any material whose dielectric constant is greater than that of the matrix material may be used. While AC or DC fields may be used, AC fields suppress any electrophoretic interaction between the field and the matrix polymers (which are sill in a fluid state). However, low frequencies proved to have the strongest interparticle binding force, with 10 Hz being an optimum in silicone but 650 Hz being the optimum in epoxy. This was determined by measuring the shear stress of the still-fluid material while the electric field was applied. Other materials may also be used. As described in Chapter 6, similar materials processing has been described for nanotubes, which aside from their electrical properties have notable mechanical characteristics.

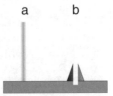

a   b

Figure 7.8 A method for constructing microstructures on a conducting surface. A laser is used to drill or evaporate holes in the surface of the conductor. After the drilling is complete, conical structures around the hole remain and can be used as "primers" for controlled pearl chaining.

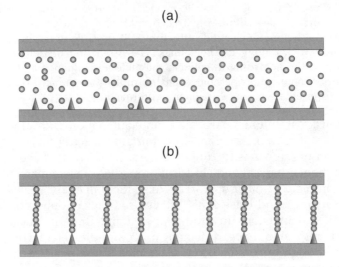

(a)

(b)

Figure 7.9 Arrangement of pearl-chained particles within a linear, organized matrix using the field-deforming electrode structures illustrated in Figure 7.8.

More recently, an electrode array has been devised by a group in Japan for optimizing the growth of particle columns.[21] When a laser was used to evaporate small areas of the surface of a 1-mm thick copper electrode, projections of melted copper formed around the holes as shown in Figure 7.8. These conical formations, dubbed micropillars, were typically up to 40 μm tall. By applying a second (untreated) copper electrode and applying a potential across a suspension containing latex beads, columns of particles could be formed. Unlike the methods used by Randall's group, however, these formations were a single bead in width and formed on the surface of the laser-treated electrode and extended in a straight line across the chamber, as shown in Figure 7.9. Since the positions of the laser sites could be controlled with high accuracy, so could the positions of the bead towers.

A number of applications for dielectrophoretically processed materials have been suggested, which have advantages due to their electrical, mechanical, and thermal properties. Devices include direction-sensitive pressure sensors (where the sensitivity is limited to the direction in which the lines of material have been dielectrophoretically assembled) for acoustic detection in aqueous environments, such as underwater sensors and implantable biosensors for blood flow measurement, and thermistors that conduct electricity through the assembled powder lines at low temperatures, but when the temperature is increased, the matrix expands, increasing the gap between adjacent colloids and increasing the resistance. Material applications include the engineering of composites with electrical and thermal conduction capabilities and structural applications for nanoscale composites with rods (such as nanotubes) forming the equivalent of nanometer-scale GRP (glass-reinforced plastic, or fiberglass). Most exciting is the idea that materials could be tuned *in real time*, with the mechanical or electrical properties of a material being altered by application of an electric field to reorder colloids within a fluid-phase gel, allowing, for example, the possibility of changing its mechanical resonance by the addition of stiffening due to the presence of colloidal rods across the material. Where the structure of the aligned chains can be controlled accurately — as in the example of the laser-drilled microtowers, there are still broader possibilities. For example, it was suggested that, along with applications for sensors such as those described above, this technology could also be used for structuring devices for photonics applications; by missing certain paths through the material matrix, flexible waveguides could be produced.

It has even been suggested[20] that dielectrophoretic force may be responsible for the reordering of conducting graphite particles in thick-film polymer resistors. These resistors are trimmed to appropriate lengths by the application of a high voltage (AC or DC) pulse during fabrication, but it has been observed that after the application of such pulses, the electrical properties of the material change. It was suggested by Rao and Satyam[20] that this may be due to the dielectric force acting on the conducting grains embedded in the polymer during the application of the high-voltage pulses, causing them to concentrate at the resistor edges and depleting them from the center.

7.7.3 *Dewatering*

Finally, we will consider an example of electrokinetic effects (including dielectrophoresis, electrophoresis, and electro-osmosis) on a much larger scale (that of meters) that still concerns the manipulation of molecules, in this case water. Research has been ongoing for some decades on the use of dielectrophoresis for the removal of water from clay. This work, pioneered in the 1960s by the U.S. military, is concerned with the removal of water from muddy areas to allow soldiers and equipment to pass and has also been investigated for the mining industry. The process works by inserting perforated steel pipes 5 cm in diameter at regular intervals across the area being drained and an iron mesh laid across the top. A DC potential with magnitude in the kV range is then applied between the rods and the grid. The interaction of the field with the electrical double-layer surrounding the charged clay particles causes the movement, over a period of many days, of the water in that double layer toward the region of high electric field (the pipe), where it seeps through the perforations and is collected. Over a period of 28 days, significant dewatering of the soil can be achieved. In order to determine the contribution of dielectrophoresis to these effects, Lo et al.[22] examined an AC case (where electrophoresis and electro-osmosis contributions are minimized) where two rod electrodes were placed approximately 30 cm apart and significant strengthening of clay was achieved.

While this application is significantly outside the scope of the rest of this book, it is worthwhile to consider that although the effects we consider here are important only on the nanometer scale, given a sufficiently high electric field strength and a sufficiently long period of time, much larger effects can be achieved!

7.8   Nanoelectromechanical systems

As a final thought, it is worthwhile considering that electromechanical forces may have an important role to play in the development of actual mechanical devices — machines — on the nanometer scale. For example, by using electrostatic forces, it is possible to attract carbon nanotube contacts across a gap to close a switch,[23] forming a nanomechanical relay. Such devices might have great potential for nanocircuitry, forming hybrid electromechanical computing devices. Indeed, in his great thought experiment, Drexler[14] considered an entirely mechanically based computing paradigm for avoiding the issue of connecting electrical currents to nanodevices altogether; however, such speculations are decades from realization, if they are ever realized at all.

Beyond the field of computing, other ideas for nanoelectromechanical devices (NEMS) are being considered, often using nature (which is excellent at producing protein machines on the nanometer scale) as a guide. One such development is the study of the potential for mechanical devices such as electric motors on the nanoscale. Rotary and linear protein motors exist in nature for processing of molecules, for transport around the cell interior, and most importantly for motility. Muscle fibers use linear protein motors to contract; bacterial flagella use rotary protein motors to turn corkscrew-like flagella the way a ship's engine uses a propeller. Many of these machines use variations on the thermal ratchet process — described in the next chapter — to effect transport, while others use mechanisms related to charge movement or proteins changing conformation; this is a process we can replicate using dielectrophoresis, opening up new avenues for the development of artificial nanomotors. For example, in order to produce a rotary dielectric motor, we can use dielectrophoresis and electrorotation to provide stabilizing force and torque; this idea can even be mathematically modeled to determine its effectiveness. Such a thought experiment is presented, along with suggestions for possible applications, in Appendix A.

References

1. Xia, Y., Rogers, J.J.A., Paul, K.E., and Whitesides, G.M., Unconventional methods for fabricating and patterning nanostructures, *Chem. Rev.*, 99, 1823, 1999.
2. Velev, O.D. and Kaler, E.W., In situ assembly of colloidal particles into miniaturised biosensors, *Langmuir*, 15, 3693, 1999.
3. Fudouzi, H., Kobayashi, M., and Shinya, N., Assembling 100 nm scale particles by and electrostatic potential field, *J. Nanoparticle Res.*, 3, 193, 2001.
4. Hermanson, K.K., Lumsdon, S.O., Williams, J.J.P., Kaler, E.W., and Velev, O.D., Dielectrophoretic assembly of electrically functional microwires from nanoparticle suspensions, *Science*, 294, 1082, 2001.
5. Yamamoto, K., Akita, S., and Nakayama, Y., Orientation and purification of carbon nanotubes using AC electrophoresis, *J. Phys. D: Appl. Phys.*, 31, L34, 1998.
6. Erdman, C.F., Swint, R.B., Gurtner, C., Formosa, R.E., Roh, S.D., Lee, K.E., Swanson, P.D., Ackley, D.E., Coleman, J.J., and Heller, M.J., Electric field directed assembly of an InGaAs LED onto silicon circuitry, *IEEE Photon. Technol. Lett.*, 12, 1198, 2000.
7. Duan, X., Huang, Y., Cui, Y., Wang, J., and Lieber, C.M., Indium phosphide nanowires as building blocks for nanoscale electronic and optoelectronic devices, *Nature*, 409, 66, 2001.
8. Huang, Y., Duan, X., Cui, Y., Lauhon, L.J., Kim, K.H., and Lieber, C.M., Logic gates and computation from assembled nanowires building blocks, *Science*, 294, 1313, 2001.
9. Bachtold, A., Hadley, P., Nakanishi, T., and Dekker, C., Logic circuits with carbon nanotube transistors, *Science*, 294, 1317, 2001.
10. Avouris, P., Collins, P.G., and Arnold, M.S., Engineering carbon nanotubes and nanotube circuits using electrical breakdown, *Science*, 292, 706, 2001.
11. Washizu, M. and Kurosawa, O., Electrostatic manipulation of DNA in microfabricated structures, *IEEE Trans. Ind. Appl.*, 26, 1165, 1990.
12. Park, H., Park, J., Lim, A.K.L., Anderson, E.H., Alivisatos, A.P., and McEuen, P.L., Nanomechanical oscillations in a single C_{60} transistor, *Nature*, 407, 57, 2000.
13. Bezryadin, A. and Dekker, C., Nanofabrication of electrodes with sub-5 nm spacing for transport experiments on single molecules and metal clusters, *J. Vac. Sci. Technol. B*, 15, 793, 1997.
14. Drexler, K.E., *Nanosystems: Molecular Machinery, Manufacturing and Computation*, Wiley, New York, 1992.
15. Alimova, A.N., Chubun, N.N., Belobrov, P.I., Ya Detkov, P., and Zhirnov, V.V., Electrophoresis of nanodiamond powder for cold cathode fabrication, *J. Vac. Sci. Technol.*, 17, 715, 1999.
16. Xu, N.S., Chen, J., Feng, Y.T., and Deng, S.Z., Nanostructured diamond film on etched silicon and its field emission behavior, *J. Vac. Sci. Technol.*, 18, 1048, 2000.
17. Choi, W.B., Cuomo, J.J., Zhirnov, V.V., Myers, A.F., and Hren, J.J., Field emission from silicon and molybdenum tips coated with diamond powder by dielectrophoresis, *Appl. Phys. Lett.*, 68, 720, 1996.
18. Randall, C.A., Miyazaki, S., More, K.L., Bhalla, A.S., and Newnham, R.E., Structural-property relationships in dielectrophoretically assembled $BaTiO_3$ nanocomposites, *Materials Lett.*, 15, 26, 1992.

19. Bowen, C.P., Shrout, T.R., Newnham, R.E., and Randall, C.A., Tunable electric field processing of composite materials, *J. Intel. Mat. Syst. Struct.*, 6, 159, 1995.
20. Rao, Y.S. and Satyam, M., Dielectrophoretic model of conductance increase in PVC-graphite thick film during electrical breakdown, *Int. J. Microcircuits and Electronic Packaging*, 20, 57, 1997.
21. Hase, M., Egashira, M., and Shinya, N., Development of a novel method to create three-dimensional arrangements of particles using dielectrophoresis in artificially nonuniform electric field, *J. Intel. Mat. Sys. Struct.*, 10, 508, 1999.
22. Lo, K.Y., Shang, J.Q., and Inculet, I.I., Electrical strengthening of clays by dielectrophoresis, *Can. Geotech. J.*, 31, 192, 1994.
23. Dequesnes, M., Rotkin, S.V., and Aluru, N.R., Calculation of pull-in voltages for carbon-nanotube-based nanoelectromechanical switches, *Nanotechnology*, 13, 120, 2002.

chapter eight

Practical dielectrophoretic separation

8.1 Limitations on dielectrophoretic separation

As we have seen in the preceding chapters, dielectrophoresis can be used to separate different particles from a mixture into homogeneous groups on an electrode array. However, in the examples described, the number of particles involved in the separation process is very small. A polynomial electrode array such as the ones described in Chapters 4 and 5 will generate sufficient force to trap particles within perhaps 100 μm of the electrodes (or more for larger particles) but will be unable to trap particles beyond that limit because the field gradient diminishes rapidly from the electrode edges. This means that the array is only capable of separating particles within a volume of the order of picoliters in size. This might be sufficient if we are using dielectrophoresis as an investigative tool, such as to study the effects of drugs on virus particles. However, many applications exist where dielectrophoretic selection could be used to isolate particles from much larger samples — for example, the isolation of viruses from a groundwater or blood sample. In such cases, samples of microliters or even milliliters need to be processed for the device to be useful.

In order to meet this increase in sample volume, the electrode array must be sufficiently large for the whole sample to be within reach of the dielectrophoretic field, and the sample must be passed across an electrode array to an output either by an external force (such as a pump) or by the action of the electrodes themselves. The principal methods for so-called bulk dielectrophoretic separation employ conventional dielectrophoresis to achieve these goals; another uses a different electrokinetic phenomenon, called traveling-wave dielectrophoresis. These methods either perform straightforward *binary separation* — where a heterogeneous mixture is split into two subpopulations that then appear in different outputs at different locations, or one population is expelled while another is retained — or *fractionation*, whereby many subpopulations are separated into groups or fractions that appear at the output as homogeneous groups that are output in sequence, fastest first.

8.2   Flow separation

The simplest method of separating large numbers of particles is to scale up the principle used for separating latex beads, viruses, and DNA described in previous chapters. An electrode array is energized, and particles are either attracted to it or repelled into the bulk medium according to their dielectric properties.[1] Separation can be performed on the basis of different polarities of force (one attracted, the other repelled) or different magnitudes of force (one attracted strongly and held at the electrodes, the other attracted weakly and carried away by an applied flow). This method has been used for some decades and is described in Pohl's book[2] for the separation of larger particles of polymers and minerals (of the order of hundreds of micrometers in diameter), as well as biological samples such as cells. In order to overcome the problem of the limited effective range of dielectrophoretic force from the electrode and to provide a separation process that not only separates the particles on an array, but also provides outlets that produce refined sources of both particle types, the sample is pumped across the electrodes. Pumping forces are usually provided by a device external to the electrode array, such as a peristaltic pump that feeds an inlet, and excess fluid is removed via one or more outlets. In order to maximize the dielectrophoretic trapping, the electrode array covers as large a volume as possible. For microengineered arrays, this means covering an area as large as possible with active electrode sites, for example by using a geometry such as the castellated array described in Chapter 4, which is in turn similar to that used by Benguigui and Lin.[3]

The simplest method of separation is to select a field frequency and suspending medium such that one type of particle experiences positive dielectrophoresis, and the other negative dielectrophoresis as with previous examples involving latex beads, viruses, and so forth. The mixture is passed over the array and is separated according to the polarity of the induced force. Those experiencing positive dielectrophoresis are trapped on the electrode array, while those experiencing negative dielectrophoresis continue to flow across the array and are removed at the outlet as a homogeneous population. When all the particles have been passed over the array and sorted, the particles that have been held by positive dielectrophoresis can be released and are collected separately, either at the same outlet (the previous particle type having been removed) or at a different outlet, being pumped from another source in a different direction (e.g., Becker et al.[4] and Washizu et al.[5]). This is shown schematically in Figure 8.1. If many types of particle must be separated, then multiple stages of filtering can be used to remove unwanted cells at different frequencies. Alternatively, repeated separation of a sample will refine the filtration process where the dielectric properties of the particles to be separated are similar.[6]

The electrodes employed in such systems are required to induce electric fields over an area as large as possible, to maximize the chances of capturing particles passing overhead. However, they are also required to allow particles experiencing either no or negative dielectrophoretic forces to pass over

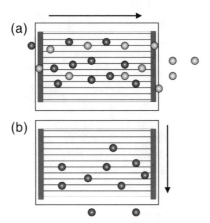

Figure 8.1 A dielectrophoretic flow separator. (a) A mixture of particles flows onto an electrode array from left to right. The dark particles are trapped by positive dielectrophoresis, while the light ones are not trapped and leave the array at the right where they are collected. (b) When all light particles have been recovered, the electric field is removed and a second flow pushes them to the bottom of the array where they are collected separately.

the array without becoming trapped in a field cage. Early systems of this type used a planar electrode array with sufficient height of chamber that repelled particles could move over the regions where force was high in order to get to the other side of the chamber. Such a strategy has limited effectiveness because particles flowing in the upper part of the chamber are too far from the electrodes to experience either positive or negative dielectrophoresis and will therefore not be selected but will pass through to the output. This is particularly important for nanometer-scale particles; since dielectrophoretic force is proportional to volume, small particles require large electric field gradients that can only be sustained a small distance from the electrodes. In order to overcome this problem, the height of the chamber is limited by placing a lid a defined distance above the electrode array. In some cases the lid can have an extra set of electrodes written onto it, allowing a doubling of the effective range of the field. However, in this case it is fundamentally important that repelled particles have a route by which to travel over the electrodes through the regions of high electric field strength if they are to get to the outlet. Electrode arrays such as the castellated geometry can be interdigitated — that is, large numbers of castellated electrodes can be powered from two power rails running along the side of the array, with the electrodes interlocked like fingers in two clasped hands. The direction of flow is orthogonal to the main interdigitations, so that when particles flow across the electrodes, some will be attracted by the high field and will be immobilized at the points of closest interelectrode approach. Particles experiencing negative dielectrophoresis pass along the bays, where the dielectrophoretic force is too weak to resist the motion of the particle due to the solution being pumped; repelled particles are herded into lines that pass

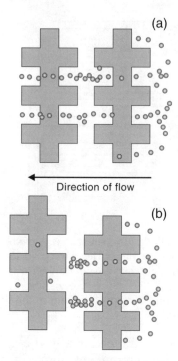

Figure 8.2 In flow separation, it is important that particles flowing over electrodes while being repelled by negative dielectrophoresis have a clear path through the device (a), or pockets of local minima may cause particles to be trapped on the array (b).

over the electrodes and are collected at the outlet (as shown in Figure 8.2). Such an electrode array was successfully used by Washizu and colleagues to perform dielectrophoretic chromatography on protein molecules in solution,[5] as described in Chapter 6. By flowing protein solutions across interdigitated electrodes, the trapping force attracted larger protein molecules in preference to small ones (due to the fact that dielectrophoretic force scales by volume), enabling purification to take place.

A more advanced version of this design uses electrodes slanted in one direction, so that the lines of particles experiencing negative dielectrophoresis are herded in one direction in preference to another. This is advantageous because a single inflow source can provide the impetus for one population (that trapped by positive dielectrophoresis and then released) to be collected at an outflow directly across the electrodes from the inflow, while a second outflow, offset from the inflow, can be used to collect the second population.[7]

A more complex method of separation using dielectrophoresis was introduced by Markx and Pethig[8] and illustrated in Figure 8.3. Demonstrated for bacteria, yeast, and plant cells, the device uses two ports that may act either as inlets or outlets, arranged at either end of an interdigitated array similar to that shown in Figure 8.1. Particles are introduced into the center of the array by flowing them from one of the ports (the other acting as outlet). A

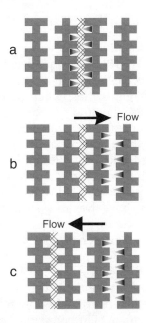

Figure 8.3 A dielectrophoretic separator using castellated electrodes. (a) A sample of two types of particle, experiencing positive (hatched) and negative (shaded) dielectrophoresis, respectively, are separated on the electrode array. (b) A flow is introduced across the array; the loosely bound particles collected by negative dielectrophoresis move to the next interelectrode gap. (c) All particles are released and a flow in the opposite direction pushes them back up the chamber, after which steps (b) and (c) are repeated to increase the separation distance. Eventually the two populations reach opposite ends of the array.

frequency is selected at which one population experiences positive dielectrophoresis while the other experiences negative dielectrophoresis, causing separation. The pumps then push the weakly held repelled particles into the next interdigitation. The particles are then released by switching off the field, and the pumps operate in the reverse direction to move all the particles back so that the population experiencing positive dielectrophoresis has net movement in the opposite direction to those experiencing negative dielectrophoresis. By the careful use of two inlet/outlet pipes at each end of the array, it is possible to "shuffle" the populations apart, with each population being moved toward a different end of the array. While this method of separation is inherently slow and complex, the repeated action of moving and trapping particles increases the likelihood of achieving 100% separation efficiency (i.e., all particles are retrieved in the appropriate homogeneous group). Such a method could in theory be used to separate many populations of particles, provided appropriate frequencies can be found; particles can be separated according to large differences in dielectric properties, and when these have been separated into two populations, subpopulations can be identified by finding appropriate frequencies to separate within the subpopulation.

8.3   Field flow fractionation

A second method of fractionation employing dielectrophoresis is its use in field-flow fractionation, FFF. This technique has been used since the 1960s for separating macromolecules and colloids up to about the micrometer scale. The operating principle is thus: when fluid travels along a tube or channel, viscous forces cause the fluid traveling near the surface of the tube to travel more slowly than the fluid flowing near the center of the tube (this effect is visible in rivers, where the flow at the center is much faster than at the banks). If particles are suspended within this fluid, then they will travel at different speeds according to their distance from the surface. FFF exploits this by adding an additional force field, which restricts different types of particles to specific heights above the surface of a thin, flat channel according to how the particle responds to the imposed field. Particles traveling at different heights will travel at different velocities due to the different rates of flow and, consequently, if those particles are presented at the inlet to the chamber simultaneously, they will emerge at the output at different times according to their responses to the field. The combination of the field and the flow enables fractionation, hence field-flow fractionation. Typical fields include gravity (sedimentation FFF), temperature gradient (thermal FFF), and viscous properties of the particle in a crossflow (flow FFF).[9]

Dielectrophoresis is used in this technique to provide the force field — that is, the means by which particles travel at specific heights above the electrodes. While positive dielectrophoresis could be used to drag particles to the edge of the tube, it might also cause particles to be trapped and become a flow separator in line with the previous description. The use of negative dielectrophoresis provides an ideal force field because it forces the particles away from the tube sides and into the medium, and the distance to which the particles are repelled is proportional to the magnitude of the force exerted on the particles.

A particle passing over an electrode array will experience forces due to both buoyancy (related in turn to gravity) and dielectrophoresis. When the forces are in equilibrium, the particle remains at a stable position so that on average the total force in the z direction is zero (as shown in Figure 8.4), i.e.,

$$F_{TOTAL,z} = F_{DEP} + F_{BUOYANCY} = 0 \qquad (8.1)$$

i.e.,

$$\mathrm{Re}[K(\omega)]2\pi r^3 \varepsilon_m \nabla E^2 = \frac{4}{3}\pi r^3 (\rho_p - \rho_m)g \qquad (8.2)$$

where ρ indicates density, g acceleration under gravity, and p and m refer to particle and medium, respectively. Rearranging, we obtain

$$\mathrm{Re}[K(\omega)]\nabla E^2 = \frac{2(\rho_p - \rho_m)g}{3\varepsilon_m} \qquad (8.3)$$

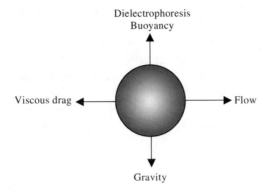

Figure 8.4 A schematic illustrating the forces experienced by a particle in a field-flow fractionation system.

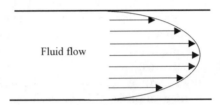

Figure 8.5 A schematic of the variation in flow rate across an enclosed capillary. In an FFF separation system described here, the electrodes are arrayed across the lower surface of the capillary.

Therefore, the particle will be levitated at a constant height according to its density and the dielectrophoretic force it experiences. If a mixture of particles flows along a tube and over such electrodes, the particles will fraction at different heights according to their density and dielectrophoretic response to the electrodes at the bottom of the chamber. When these particles reach equilibrium at different heights, the effects of viscous drag between the sides of the channel and the fluid cause them to be retarded at different rates, since the flow at the center is much greater than the flow at the edges. A typical velocity distribution in a channel is as shown in Figure 8.5. Since particles are levitated to different heights, they consequently travel at different velocities down the separator. If all particles start at one end of the separator at the same time, then by the time the particles have traveled to the end of the separator, they will have fractioned into different bands according to how fast they have traveled, and those bands will leave the separator at different times, which allows them to be separated.

The application of dielectrophoresis as a means of providing a force field was simultaneously developed by Markx et al.,[10,11] Huang et al.,[12,14] and Wang et al.[13] This type of device has been demonstrated to be effective for the fractionation of a number of bioparticles including latex beads, bacteria, and cells. The electrode geometry used to impart the negative dielectrophoretic force is essentially the same as the electrode array described previously, being a planar electrode array (usually of an interdigitated type) fabricated onto a glass surface and sealed with a coverslip or similar, at a fixed height above the array. Recent developments by Müller et al.[15] have described a novel approach wherein electrodes are used to confine particles to one side of the channel by use of a negative dielectrophoretic funnel, then spacing the particles through the channel by use of a bow electrode. A single inlet and outlet are required for the provision of flow (provided by pumping external to the array) and to allow the entry and egress of the particles.

8.4   Thermal ratchets

The next separation technique discussed here differs from the first two presented in that it is only truly useable for nanometer-scale particles. It exploits a key property in the definition of a colloid: that Brownian motion plays a significant part in the position of the particle. The concept of thermal ratchets (or Brownian ratchets, or forced thermal ratchets) was originally voiced by Pierre Curie in the late 19th century,[16] when he described motion provided by ratchet mechanisms. The concept is as follows: if particles could be trapped in an asymmetric force field such that they were more likely to move in one direction than another, then a cycle of applying and then removing the field would cause particles to move, then disperse, then move again. Careful field design would allow the development of a ratchet mechanism to allow particles to be moved along the field; furthermore, the system could be used to separate particles either on the basis of their response to the field or their rate of diffusion when the field is turned off. The idea was revisited by Richard Feynman in his famous Lectures on Physics,[17] in which he described how two boxes — one containing a weather vane, another containing a ratchet-and-pawl mechanism — could be used to generate movement (such as lifting a very small object) by rectifying the random motions of the gas against the vane by using the ratchet and pawl. Since then, the field of using ratchet mechanisms has grown, and for a summary of the field the reader is directed toward the comprehensive review on the subject by Reimann.[18] The underlying mechanism has been attributed to the manner in which many biological functions occur on the molecular scale, including the function of muscle,[19] molecular motors,[20] and molecular pumps.[21]

    The concept of the dielectrophoretic thermal ratchet — using dielectrophoresis as a means of rectifying Brownian (also known as thermal) motion to move particles around an electrode array — was first described by Ajdari and Prost[22] who considered the forces acting on a particle in suspension and subject to Brownian motion. A particle exposed to a potential with sawtooth variation in space, when repeatedly applied and removed for finite periods of time, will theoretically show a biased overall motion along the direction in which potential increases for the longest physical distance, leading Chauwin et al. to the assertion that this principle provided *mouvement sans force*.[23] Analysis by Magnasco,[24] and subsequently by Astumian and Bier,[25] illustrated that models of this nature could be devised to explain the motion of proteins along biopolymer chains using thermal noise to advance the smaller molecules through a series of potential "ratchets," or the mechanism by which molecular motors such as those found in flagellate bacteria might work.[26] The practical application of this principle, using dielectrophoresis to provide the necessary potential gradient, was first proposed by Ajdari and Prost[22] and subsequently demonstrated experimentally by Rousselet et al.[27] Rousselet and co-workers used latex spheres of varying diameters to attain particle motion of 0.2 µm s^{-1}, with diffusion rates of particles advancing from one ratchet to the next at 40% per step for significant times of zero applied field. Ajdari and Prost[22] also proposed that this method has applications in the separation of particles according to their relative sizes. The method has since been enhanced so as to separate particles in a continuous manner according to their relative dielectric properties.[28] Furthermore, under the correct conditions it is possible to drive particles of specific dielectric properties *backward* through the ratchet system while other particles are simultaneously being driven *forward* in the manner described previously, enhancing spatial separation.

A dielectrophoretic ratchet system operates thus. Consider a quantity of colloidal particles, of greater polarizability than the surrounding medium and uniformly dispersed through the volume immediately over an appropriately shaped electrode array. The volume under study is exposed to an imposed potential energy profile — such as an electric field gradient imposing a positive dielectrophoretic force — which has an asymmetric pattern of tips arranged in a sawtooth pattern, with a distance d between successive tips. When the field is activated, the particles will move to collect at the highest points on that potential energy profile. They will do this by moving up the field gradient, thereby also moving in space. As can be seen in Figure 8.6, many more particles — those suspended in the volume of length d_2 — will move to the right of the picture than those in the volume of length d_1, which will move to the left. For spherical particles of radius r, the dielectrophoretic force is calculated using the now well-established equation:

$$\mathbf{F}_{\text{DEP}} = 2\pi\varepsilon_m r^3 \, \text{Re}\big[K(\omega)\big]\nabla E^2 \qquad (8.4)$$

Owing to the ∇E^2 term, dielectrophoretic motion will be directed along the path of increasing the local electric field gradient. Owing to the asymmetric design of the electrodes, the field gradient is biased such that a greater proportion of the space between successive electrode tips generates dielectrophoretic motion to the right of the diagram than to the left. Hence, the majority of particles will translate toward the right. After collection, the concentration profile of the particles resembles that shown in Figure 8.6, concentration A. Following the collection of particles over a period τ_{ON}, the potential difference across the electrodes is removed. Under Brownian motion, the particles will then drift from the electrode tips over a period of time.

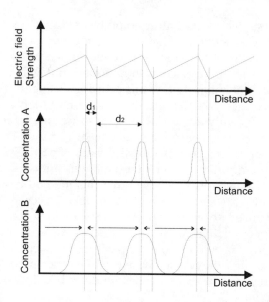

Figure 8.6 A demonstration of the function of thermal ratchets. (top) A nonsymmetrical electric field is imposed, consisting of units of length d made from ascending regions of potential energy of length d_2 and descending regions of length d_1, viewed left to right. The value of the potential energy alternates between an on state and an off state. When activated, particles move up the field gradient and accumulate at the point of highest energy, resulting in a narrow concentration distribution (center). When the field is released, the particles diffuse out in bell-shaped formations (bottom). When the field is reapplied, particles that have diffused more than d_1 to the right will move forward to the next point of high field; the remaining particles will return to the same collection point to which they were attracted. In this manner, a net left-to-right motion occurs.

After sufficient time, some particles will drift a distance greater than d_1, which will place those translating to the right within the dielectrophoretic capture zone of the next (right) electrode. After a period of time τ_{OFF}, the electric field is reapplied. At that point, the distribution of particles will resemble concentration B in Figure 8.6. Assuming particle dispersion has taken place at an approximately equal rate, all particles except those that have traveled a distance greater than d_1 will be attracted to the same electrode tip. However, those that have moved greater than d_1 to the right will be trapped by the next electrode to the *right*. Provided no particles have moved a distance greater than d_2 to the left, thereby entering the capture zone of the *previous* electrode, there is a net motion of particles to the right.

Particle diffusion can be modeled using Fokker–Planck equations,[29] which define the change in probability of a particle traveling a distance along axis x as a function of time t

$$\frac{\partial P}{\partial t}(x, t) = -\text{div } J(x, t) \tag{8.5}$$

where J represents the probability current, which can be expressed within Equation 8.5, thus,

$$\frac{\partial \mathbf{P}}{\partial t}(\mathbf{x}, \mathbf{t}) = -\text{div}\left(\frac{D}{kT} P(x,t)\mathbf{F}(x,t) - D\nabla P(x,t)\right) \tag{8.6}$$

where P indicates the probability density function of the particle location, D is the diffusion coefficient, k is Boltzmann's constant, T the temperature, and $\mathbf{F}$ is the force due to an imposed field. The first term of the equation in the brackets corresponds to drift due to an externally applied force field; the second part corresponds to particle diffusion. It has been shown[22] that for optimum transport this diffusion will be bounded by the lower diffusion limiting case of the above expressions, where the diffusion rate is small enough to prevent particles passing beyond a single repeating electrode unit (that is, diffusing a distance greater than d_2) in a single time interval τ_{OFP}. This limiting case is given by the expression:

$$\tau_{\text{OFF}} \ll \frac{(d_2 - d_1)^2}{D} \tag{8.7}$$

At this stage, we must consider the electrode geometry required in order to generate our asymmetric field. The most common geometry in use is that of the so-called *Christmas tree* electrodes such as those shown in Figure 8.7. These generate a force in the right direction covering a larger area than those forcing particles to the left, which is similar to the potential profile in Figure 8.6 along the center of the interelectrode gap; they are also easily fabricated using conventional electrode manufacturing techniques. The divisions between capture zones for each electrode are shown as a curved dotted line between the corners of adjacent electrodes.

For an ideal case (where the condition in Equation 8.7 is met), a two-dimensional isotropic diffusion (i.e., diffusion at equal rates in all directions in a plane) of particles has a probability of crossing a semicircular boundary of radius d_1 in time τ_{OFP} and thus the fraction of particles having crossed that boundary at that time is given by the expression[27]

$$P = \tfrac{1}{2}\exp\left(\frac{-d_1^2}{4D\tau_{\text{OFF}}}\right) \tag{8.8}$$

where the 1/2-factor indicates isotropic diffusion. However, Rousselet and co-workers determined through experimentation that in practice the boundary between capture zones is only approximately semicircular, but is not exactly so (as can be seen in Figure 8.7) and proposed a more accurate empirical model based on their experimental observations:

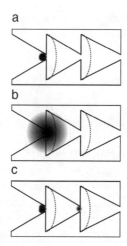

Figure 8.7 (a) Particles are collected at the point of closest approach between two Christmas tree electrodes. The particles are attracted from within the region defined by the two vertical, curved lines that define the capture zone for the electrode pair between the lines. (b) The electric field is switched off and the particles diffuse in all directions. Those diffusing over the boundary to the right (indicated by the horizontal shading) enter the capture zone for the next set of electrode tips; if the field is reapplied, they will move forward while the remaining particles will be reattracted to the tips in the center of the picture. (c) If the diffusion process is left for too long, some particles will enter the capture zone for the previous tips (to the left, with vertical shading), reducing the efficiency of the device.

$$P = 0.9 \exp\left(\frac{-d_1'^2}{4D\tau_{\text{OFF}}}\right) \tag{8.9}$$

where d_1' is a radius variable chosen to fit the equation from experimental data. The coefficient 0.9 indicates that the diffusion is nonisotropic, with particles in experiments tending not to mount the electrode surfaces and thus being more likely to diffuse forward from the collection point.

After the optimum time τ_{OFF} some particles, i.e., those that have traveled a distance $d_2 - d_1$ forward (toward the right in Figure 8.7), will be captured by the next electrode on reapplication of the electric field. If τ_{OFF} is greater than this, some particles will be captured by the previous (left) electrode and the efficiency of the ratchet will be reduced. Thus, maximum velocity has a defined maximum period time $(\tau_{\text{OFF}} + \tau_{\text{ON}})$ as shown in the studies of Prost et al.[30] However, the longer this time period lasts, the lower the net velocity V will be. V cannot exceed a value given by

$$V = \frac{d_2 - d_1}{\tau_{\text{ON}} + \tau_{\text{OFF}}} \tag{8.10}$$

in the direction in which the ratchets point.

Since the boundary between capture zones does not follow a semicircular pattern, it is difficult to evaluate the efficiency of a geometry in terms of proportion of particles migrating forward within the time τ_{OFF}. In order to address this, Ajdari and Prost[22] proposed a dimensionless factor x as a ratio based on the distance d_1 as a proportion of the total distance d. A similar measure of this ratio, Λ, can be used for the comparison of different electrode geometries. Λ is expressed as the ratio of the distance along the axis through the center of the gaps between the electrodes along which particles are attracted to the next electrode tip at time τ_{OFF}, as a proportion of the distance along this axis where electrodes do not change electrode tip:

$$\Lambda = \frac{d_2 - d_1}{d_2 + d_1} \tag{8.11}$$

The value Λ may be interpreted as a measure of asymmetry and may take values from 0 (a symmetrical electrode assembly) to 1 (complete asymmetry). In practice $\Lambda = 1$ is unattainable, but by maximizing Λ ratchet performance may be measured and improved.

8.5   Separation strategies using dielectrophoretic ratchets

The mechanism by which thermal ratchets operate, as described above, has obvious applications in the moving of particles across an electrode array. However, the concept is easily adapted for use in particle separation. There are many ways in which this can be achieved, which are based on either the physical or electrical properties of the particles, or both.[31–33] This has potential applications similar to methods of continuous dielectrophoretic flow separation, but appropriate to situations where it is impractical to provide a fluid flow through the particle chamber or where smaller numbers of particles need to be separated.

The most widespread approach to particle separation by dielectrophoretic ratchets is for the fractionation of a mixture of colloidal particles according to particle size (or another factor that affects the diffusion of particles through the suspending medium). The principle of operation is as follows. Particles trapped by dielectrophoretic ratchets, and subsequently released, diffuse from the point of collection (as demonstrated in Figure 8.7) until they fall into the capture zone of the next pair of electrode tips along the ratchet. If the field is reapplied before particles have diffused that far, then no net movement is achieved; the longer the diffusion period after that initial barrier is overcome, the more particles will diffuse into the next capture zone and the greater the likelihood of particles achieving movement of one step forward along the ratchet becomes. Fractionation of particles using dielectrophoretic ratchets by this method was first demonstrated by Faucheux and Libchaber,[32] who fractioned particles with radii 1.5 μm and 2.5 μm.

If a mixture of two particles having different values of diffusion constant D (for example, particles having different radii but similar electrical properties) is placed in a dielectrophoretic ratchet system, then the aggregate net forward movement of the particle populations along the array will be different according to the time interval between the application of the field, in accordance with Equation 8.9. So, if the mixture were placed at one end of the array, then the populations would arrive at the other end of the array at different times according to the factors chosen in Equation 8.9, with those particles that diffuse fastest reaching the end of the array first, and those diffusing less quickly taking longer to reach the end. This can be extended to a mixture of many types of particles. Since the different populations diffuse at different rates, then if they are all present at one end of the array at the start of the procedure, they will have diffused into different *bands* or homogeneous groups on the array after a number of on/off cycles, in a similar manner to the fractionation of different color dye molecules on filter paper dipped into a dye solution. The use of thermal ratchets for particle fractionation[33] was modeled by simulation by Schnelle and colleagues[33] in 2000 by examining the fractionation of five different sizes of latex bead on a ratchet array after a period of 5 h. Furthermore, the fractionation can be combined with the variation in the dielectric properties of the particles, as used for the basis of separation; for example, if particles experience forces of different intensities, then they will require different lengths of time τ_{ON} to collect at electrode tips. Therefore, by removing the field after one population has fully collected but before the other has reached this state, further purification can be achieved.

An alternative method makes use of the fact that since the dielectrophoretic force attracting the particles between electrode tips is related to the AC dielectric properties of the particles, particles with differing dielectric properties will respond differently when subjected to electric fields of a given frequency. Nonpolar particles, or polarizable particles exposed to fields of frequency equal to their crossover frequency, will not respond to dielectrophoretic forces and therefore a suspension of polarized and nonpolarized particles will separate, the former being drawn out of the population while the latter remains in place. This is, however, of limited efficiency since the nonattracted component will exist in some quantity at the exit of the device, without being repelled from it. This method can, however, be extended by the application of negative dielectrophoresis to the second population (the first experiencing positive dielectrophoretic collection). It has been shown[31] that if conditions are such that particles on a Christmas tree electrode are repelled by negative dielectrophoresis, then they collect at the point on the interelectrode space where the potential energy is at a minimum; that is, they collect *along the boundary* between two adjacent capture zones (along the dotted line in Figure 8.7). If they are then released and recaptured in the same manner as described above, particles will move in exactly the same manner as those propelled by positive dielectrophoretic ratcheting, but in the *opposite* direction across the electrode array. So for the electrodes shown in Figure 8.7, all particles diffusing to the left of the electrode tips will, on reapplication of the field, be pushed to the next intercapture zone boundary, and so forth. Therefore, with judicious choice of operating conditions, mixtures of particles where two subpopulations have different crossover frequencies can be simultaneously driven in opposite directions across the array by applying an electric field of frequency in the window between the crossover frequencies of the two particle species.

8.6   Stacked ratcheting mechanisms

The primary drawback with using forced thermal ratchets as a practical method of particle separation is the reliance on Brownian motion to provide a means of driving particles from the collection points. This restriction limits the use of such a separator to applications involving small particles with correspondingly large diffusion constants. Practical diffusion rates for micron-sized particles[27] are approximately 120 sec for 40% of particles to pass one unit forward on an array with 50 μm between pairs of electrode tips. This may be improved by optimizing electrode design, but while the fractionation of particles may be useful where physical removal of particles is not necessary (and therefore the separation between the subpopulations does not need to be large), the reliance on diffusion constrains the ability of dielectrophoretic ratchets to attain particle velocities required for practical continuous particle separation. However, there is a method of using dielectrophoretic separation by ratchet mechanisms that eliminates the dependence of the system on Brownian motion to provide the method of particle dispersion.

Consider the electrode assembly shown in Figure 8.8. The assembly is composed of two pairs of ratchet electrodes, with one pair suspended above

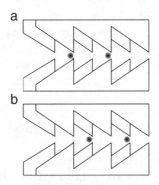

Figure 8.8 By stacking two pairs of dielectrophoretic ratchet electrodes, it is possible to eliminate the need for Brownian motion to provide forward motion. If the tips of the top and bottom pair are equally spaced as shown above, then for both positive and negative dielectrophoretic collection, the particles collected on one pair are all in the trapping zone of the next electrode pair on the other layer, and so by energizing top and bottom electrodes alternately, motion is imparted.

the other by a distance equal to the distance between capture zones (or adjacent tip pairs) and displaced along the main axis by half of the same distance. Potentials are applied to only one pair of electrodes at any given time. If a mixture of particles is suspended between the electrode levels and potentials are applied to the lower pair of electrodes, particles undergo dielectrophoresis and are attracted to, or repelled from, the electrode tips toward their respective collection points. After time τ_{ON} collection has taken place and the cells have aggregated. At this time, the potentials on the lower pair of electrodes are removed and the top pair of electrodes is activated. Those particles undergoing positive dielectrophoresis, located at the previous collection point and therefore within the capture zone of the *new* collection point, will be attracted forward to that point. Similarly, particles undergoing negative dielectrophoresis will be repelled to the collection point of the previous capture zone. The repulsion of the two stacked electrodes in this displaced form ensures that particles undergoing negative dielectrophoretic collection will not settle over the electrode surfaces. After a further time τ_{ON} the potential reverts to the lower set of electrodes and the cycle is repeated.

This process offers many advantages over the single-ratchet mechanism. The types of particles that may be used in this manner are limited according to their dielectrophoretic response rather than their size, and thus much larger particles may be used than is possible using a single pair of electrodes. More significantly, the cycle has a duration of $2\tau_{ON}$ rather than $\tau_{ON} + \tau_{OFF}$ in the single-pair case. Experiments[27] have shown that for particles of approximately 1 μm diameter, τ_{ON} has value 30 sec or less, compared to τ_{OFF} of approximately 120 sec for diffusion-based methods. Thus the stacked-ratchet method offers approximately twice the efficiency of the first method, a value that increases with increasing particle size. This is further enhanced by the percentage of particles being drawn forward per cycle approaching 100% due to the concentration of all particles beyond the crossover limit of the electrodes at the switch between pairs. However, the limitation of the stacked-ratchet device compared to the single ratchet version is that the only criterion for particle separation is the difference in the dielectric properties of the two particle types, rather than also allowing separation on the basis of diffusion constant and hence size. However, since (as we discovered in Chapter 4) the electrical properties are affected by the particle size (with all other factors equal), this can be overcome relatively easily.

The stacking of two dielectrophoretic ratchets on top of one another was first described theoretically by Chauwin et al. in 1994,[34] and was subsequently demonstrated by Gorre-Talini et al. in 1998,[35] who demonstrated the separation of latex beads using stacked ratchets. Stacked ratchet electrode arrays have potential applications similar to methods of continuous dielectrophoretic cell separation described earlier in this chapter, but are particularly appropriate to situations where it is impractical to provide a fluid flow through the particle chamber or where smaller numbers of particles need be separated.

8.7 Traveling wave dielectrophoresis

Unlike the preceding separation methods, the final separation technique discussed here does not rely on conventional dielectrophoresis to separate particles. Instead, it relies on a fundamentally different method of inducing translational forces, an effect known as *traveling wave* dielectrophoresis. This shares with dielectrophoresis the phenomenon of translational induced motion, but rather than acting toward a specific point (that of the highest electric field strength), the force acts to move particles *along* an electrode array in the manner of an electrostatic conveyor belt. The phenomenon was first discovered by Batchelder[36] in the early 1980s, but was not adopted by the wider dielectrophoresis community until the work of Masuda et al.,[37,38] Fuhr and coworkers,[39, 41,42] and Hagedorn et al.[40] essentially rediscovered the phenomenon nearly a decade later.

Consider a particle in a sinusoidal electric field that *travels* — that is, rather than merely changing magnitude, the field maxima and minima move through space, like waves on the surface of water. These waves cross a particle, and a dipole is induced by the field. If the speed at which the field crosses the particle is great enough, then there will be a time lag between the induced dipole and the electric field, in much the same way as there is an angular lag in a rotating field that causes electrorotation. This physical lag between dipole and field induces a force on the particle, resulting in induced motion; the degree of lag, related to the velocity of the wave, will dictate the speed and direction of any motion induced in the particle. The underlying principle is closely related to electrorotation; it could be argued that the name *traveling wave dielectrophoresis* is misleading because the origin of the effect is not dielectrophoretic, that is, it does not involve the interaction of dipole and field magnitude gradient. Instead, the technique is a linear analogue of electrorotation, in a similar manner to the relationship between rotary electric motors and the linear electric motors used to power magnetically levitated trains. As with the rotation of particles, the movement is asynchronous with the moving field, with rates of movement of 100 μm sec^{-1} being reported.

Since it is difficult to create and control a moving, perfectly sinusoidal, electric field, we employ electrode structures that are energized with sinusoidal signals, each of which has the same frequency but a different phase as its neighbors, with the difference in phase between phases being regular. The net effect of the phase change along the electrodes is to provide a sampled version of the sine wave, with the potentials describing a wave with a wavelength equal to the distance between electrodes sharing the same phase, assuming that the phase difference is an integer fraction of 360°. Typically, electrodes are energized by either three or four phases, with the same equipment being used to generate the electric fields. Electrodes are typically 10-μm wide, with interelectrode gaps of about the same size. The width of the electrodes creates distortions in the field that will be described later, but at distances of 5–10 μm from the electrodes, a sufficiently sinusoidal electric field is generated. In order to maintain a steady rate of transport, we

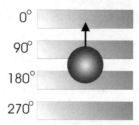

Figure 8.9 A schematic of the action of traveling wave dielectrophoresis. A traveling electric field is generated by applying phased AC signals. This induces a moving dipole that, if the electric field is moving fast enough, is physically displaced from the electric field peak. The interaction between field and induced dipole indicates a force in the direction of the field travel. The electrodes are a sequence of parallel bars on a substrate.

require a field that is as uniform as possible. We therefore do not require high field gradients and thus we do not require the use of ultrasmall electrodes in order to manipulate ultrasmall particles. An example of electrodes with 90° phase shifting is shown in Figure 8.9. Electrodes are typically arranged as an array of parallel bars, forming tracks along which particles travel in a perpendicular direction to the arrangement of the electrodes.

As we have learned in Chapter 2, we can express the full term of the dielectrophoretic force thus:

$$\mathbf{F} = 2\pi\varepsilon_m r^3 \, \mathrm{Re}\big[K(\omega)\big]\nabla E^2 + \mathrm{Im}\big[K(\omega)\big]\big(E_{x0}^2\nabla\varphi_x + E_{y0}^2\nabla\varphi_y + E_{z0}^2\nabla\varphi_z\big) \quad (8.12)$$

In fact, we can split this into two separate expressions:

$$\mathbf{F}_{DEP} = 2\pi\varepsilon_m r^3 \, \mathrm{Re}\big[K(\omega)\big]\nabla E^2$$

$$\mathbf{F}_{TWD} = 2\pi\varepsilon_m r^3 \, \mathrm{Re}\big[K(\omega)\big]\big(E_{x0}^2\nabla\varphi_x + E_{y0}^2\nabla\varphi_y + E_{z0}^2\nabla\varphi_z\big)$$

$$(8.13)$$

The first expression, for a force dependent on the electric field magnitude gradient, is the conventional dielectrophoretic force equation. The second part gives the force due to an electric field phase gradient; this gives us the force in a traveling electric field, which can be described as a field with a linearly varying phase relationship and therefore a phase gradient. We can take the second expression and consider its value for a situation where the phase gradient of the traveling electric field is incremental in only one direction (the direction of travel). For example, let us examine an electric field traveling in the *x* direction. First, this means that the *y* and *z* terms disappear since there is no phase gradient in these directions. Furthermore, let us postulate that the traveling field is ideal — that is, there is a constant, linear phase gradient along the path the particle is traveling. If the wave has wavelength (defined by the distance between electrodes carrying the same phase) λ and corresponding phase gradient $-\lambda/2\pi$,

$$\mathbf{F}_{TWD} = \frac{-4\pi^2 \varepsilon_m r^3 \, \text{Im}\big[K(\omega)\big] E^2}{\lambda} \tag{8.17}$$

Compare this with the equation for electrorotational torque; the force is equal to the expression for torque, multiplied by π divided by the wavelength of the wave. Since the force is governed by the imaginary part of the Clausius–Mossotti factor, any dielectrophoretic forces induced act independently of, and in addition to, the traveling wave force, and can be attractive or repulsive according to the properties of the particles. The interactions between conventional and traveling wave dielectrophoretic forces govern the net motion of the particles.

This interaction is further complicated by the fact that the electric field geometry around interdigitated electric fields is considerably more complicated than might first appear from the simplicity of the electrodes themselves, and we have the additional factor that we are concerned not only with local variations in magnitudes, but also with the *phase* of the local field with respect to adjacent points. Traveling wave dielectrophoresis is dependent on an electric field phase gradient to give it direction, in a similar manner to the way in which dielectrophoresis is dependent on the gradient of the magnitude of the electric field in order to give it direction. However, the magnitude of the traveling wave dielectrophoretic force is also dependent on the magnitude of the electric field.

In order to study the variation of these, we can use dynamic (time variant) electric field simulation, as described in more detail in Chapter 10. The simulations here are a variant of the standard traveling wave array, in that they consist of a channel running orthogonal to the electrode array. The potentials applied to the electrodes on opposing sides of the channel are in antiphase. This array was developed by Huang et al.[43] in order to provide a region along which the traveling wave is near ideal; that is, with almost constant magnitude and with a phase that varies linearly with distance (providing a constant force in accordance with Equation 8.16), in a position in which particles could be observed moving at a constant height — that is, along the bottom of the electrode chamber. A schematic of the area of simulation and of the electrode configuration is shown in Figure 8.10.

The magnitudes and phases of the three coordinate components of the electric field in a plane 3 μm above the plane containing the electrodes are shown in Figure 8.11; the x and y planes describe the electrode plane, with the field traveling in the x direction. If we examine the electric field magnitudes first, we find that they are as we might expect: the field is largest in the x direction between adjacent electrodes (since the electrodes run in the y direction, the interelectrode gaps are arrayed across the x direction); the field is largest in the y direction across the channel; and, in the z direction,

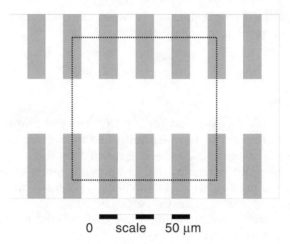

0   scale   50 µm

Figure 8.10 The electrode array simulated using methods described in Chapter 10, together with the area shown by the simulations. The electrodes are 10-µm wide and have 10-µm spacing between them. There is a 30-µm channel cutting across the electrode array. The electrodes are energized by sinusoidal potentials with 90° phase shift between each electrode, so that the simulation area represents one complete cycle of the sinusoid. The horizontal simulation area is 3 µm above the electrode plane.

the maxima are found across the upper surface of the electrodes. This is much as we would expect, since it indicates that the field gradient causes dielectrophoresis to move particles toward the electrodes, and in particular to the electrode edges.

If we then examine the phase distribution, we begin to observe features that we would not have expected. We would anticipate that the phase distribution would have a constant gradient, covering a full 360° sweep for every four electrodes covered. This is in fact what we observe in the y-directional phase. Across the electrodes there is a near-constant phase gradient, interrupted only at the midpoint of the electrodes beyond which the field points in the opposite direction, equivalent to a 180° phase change. Note that this does not change the direction of the traveling wave force, since the direction of the *gradient* does not change. Examining the phase changes in the x and z directions shows two interesting effects. The first is that there is a phase change at the midpoint of the channel (which is to be expected since there are opposing phases on opposite sides of the channel), and the second is that there is a stepped appearance to the phase distribution across the upper surfaces of the electrodes. Closer study of these effects shows that, across the electrode surfaces, the phase of the electric field does not just level out (which is to say, the electric field changes simultaneously across the electrode surface), but the phase gradient actually moves *in the opposite direction* to the traveling field. This is actually due to the distorting effect of the flat electrodes; the ideal array would feature a continuous sine wave, but we are using four flat planes, each with constant potential across its surface, to

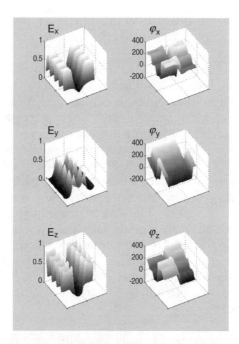

Figure 8.11 The variation in electric field peak magnitude (left column) and phase shift (right column) for electric fields in the x, y, and z directions (top, middle, and bottom rows, respectively), for the area of the electrode array indicated in Figure 8.10.

mimic this. As a result, the sine wave is distorted with flat regions across the electrode surface and much steeper regions in the interelectrode gaps; at short distances above the electrode edges, the compressed, steeper sections dominate over the flat regions and this results in a backward step in the electric field. The electric field over the electrode surface is observed to travel in the opposite direction to the electric field in the channel, and at greater distances over the electrode edges.

If we view the effects of this distribution on the forces generated in the vertical plane, then a clearer interpretation of the implications of this phase change can be made. Figure 8.12 shows the magnitude and direction of

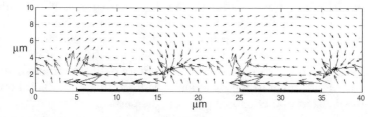

Figure 8.12 The traveling-wave forces generated by a sequence of electrodes with 90° phase shifts, two of which are shown. As can be seen, the direction of the forces changes within about 5 µm of the electrodes.

traveling wave force for a particle with arbitrary properties, as determined using a simulation similar to that described above, but in a vertical plane intersecting two electrodes. The force has been determined using Equations 8.13–8.15, and is in arbitrary units. As can be seen, above about 5 μm from the electrode surface, the force acts uniformly in one direction. Nearer the electrode than this, the force distribution is considerably more complex, and the force does not become approximately constant along a path until the distance about the electrodes reaches toward 8 μm. We can generalize this to say that the force field is approximately uniform at heights equal to the electrode width, and the reversal effect occurs at approximately half that distance.[44,45]

If we now consider that the force experienced by particles in the traveling field is the combination of conventional and traveling wave dielectrophoretic forces, in proportions dictated by the real and imaginary parts of the Clausius–Mossotti factor, then the behavior of such particles becomes even more complex. Workers in the field (e.g., Fuhr et al.,[39] Huang et al.,[43] Morgan et al.,[44] and Hughes et al.[45]) have identified a number of regimes according to the proportions of the two forces experienced by a particle. The four notable particle behaviors, three of which are illustrated in Figure 8.13, can be summarized as follows:

(i)   If the real part of the Clausius–Mossotti factor is both positive and greater than or equal to the imaginary part, then the positive dielectrophoretic trapping force overcomes any induced motion due to the traveling wave, and particles become trapped at the electrode edges.

(ii) If the particles experience a negative dielectrophoretic force but no significant traveling field force, then they will be repelled from the electrodes into the solution. Provided this force is sufficiently large to propel the particles into the solution rather than trapping them in clusters in between the electrodes, then particles will be prevented from entering the region directly above the electrodes. If there is a channel in the center of the electrode array, the particles may collect there.

(iii) If both the real and imaginary parts of the Clausius–Mossotti factor are significant and $\text{Re}[K(\omega)]$ is negative, then the particles will be levitated over the array to where the traveling force dominates and particles will travel along the electrode structures. If the value of $\text{Im}[K(\omega)]$ is positive, then the direction of travel will be in the opposite direction to the traveling field; if negative, the direction will be with the field. As described above, distortions in the phase of the wave mean that the forces within close proximity of the electrodes direct the particle to travel in the direction *opposite* to that described; however, the dielectrophoretic force prevents the particles entering this region.

(iv) Finally, if the trapping and translational forces are of approximately equal magnitude — that is, when $|\text{Im}[K(\omega)]| > 4\text{Re}[K(\omega)]$ and $\text{Re}[K(\omega)] > 0$ — then the force exerted on the particles is dependent on whichever force (and hence, local gradient, be it in field magnitude

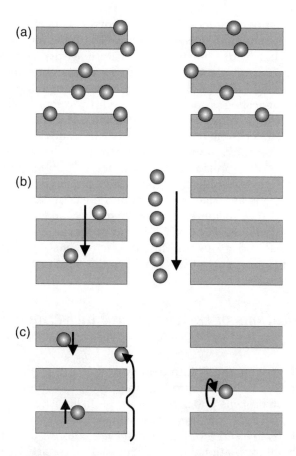

Figure 8.13 A schematic indicating some of the different behaviors demonstrated by particles in traveling electric field structures. (a) Where the real component of the Clausius–Mossotti factor is large and positive, particles will experience positive dielectrophoresis and be attracted to the electrode edges. (b) Where the real part is negative and the imaginary part is large, particles are repelled from the electrode array and travel along it. (c) When the real part is small and the imaginary part is large, particles exhibit seemingly random behavior including erratic motion, changing direction, and rotating at electrode edges.

or phase) is dominant at any particular point in space. Since the traveling wave force is highly dependent on spatial position near the electrodes, this means the motion of particles in this regime can be extremely chaotic and inherently unpredictable. Particles are observed to move between electrodes, then remain at an electrode edge for a few seconds, then return in the direction from which they came. Particles also exhibit a tumbling motion due to the localized phase changes in electric field causing the induction of localized electrorotational torque. This mode of behavior was termed the fundamentally unstable (FUN) regime by Huang et al.[43] who first reported the behavior.

Since the first two regimes described above are effectively reproducing the action of conventional dielectrophoresis electrodes, and the fourth is so unstable as to have little practical value, the majority of traveling wave dielectrophoresis manipulation is concerned with the third regime wherein particles are repelled from the electrode and travel along the array. Regarding this regime, an interesting proposition was made by Morgan and co-workers,[44] who analyzed the distribution of forces generated in planar arrays using analytical Fourier series methods. Their work indicates that, since both the height of levitation above the array and the traveling wave–imparted velocity are related to the applied electric field, the velocity of particle travel is actually independent of voltage. As the applied potential (and hence electric field), the particle is levitated further from the electrode surface, to where the electric field due to the electrodes is smaller, and hence the particle motion remains the same. However, this only occurs under conditions where the electric field, and Clausius–Mossotti factor, cause the particle to be repelled to a height greater than 1.5 times the width of the electric-field generating structures.

8.8. Applications of traveling wave dielectrophoresis

8.8.1 Manipulation

Masuda et al.[37,38] were the first to demonstrate that traveling electric fields could be used to induce controlled translational motion of bioparticles including red blood cells and lycopodium particles. These traveling fields were generated by applying three-phase voltages, of frequency ranging from 0.1 Hz to 100 Hz, to a series of bar-shaped electrodes. At these low frequencies the dominant translational forces acting on the bioparticles were electrophoretic in origin, and Masuda et al.[38] proposed that such traveling fields could eventually find application in the separation of particles according to their size or electrical charge. The first demonstrations of the application of asynchronous traveling fields of frequency between 10 kHz and 30 MHz were later shown by Fuhr and coworkers.[39] These were used to impart linear motion to pollen and cellulose particles, and Huang et al.[43] later showed that traveling fields of frequency between 1 kHz and 10 MHz can be used to manipulate yeast cells and, by altering the frequency, could alternate between dielectrophoretic and traveling-wave behaviors. As with ratchet devices, traveling-wave dielectrophoresis offers a method of pumping particles through the suspending solution, without the need to actually move the solution itself — particles move while the medium itself remains still. Pumping of the fluid can be achieved, since water is a polar liquid, an effect described by Fuhr et al.[42] However, this requires a relatively high medium conductivity and achieves fluid pumping phenomena similar to those described in Chapter 3. Most traveling wave manipulation occurs in lower conductivity media where fluid motion can be disregarded.

8.8.2   Separation

Since traveling waves can be used to manipulate particles according to their dielectric properties, it follows from work discussed previously that the force exerted on different populations of particles with different dielectric characteristics will have different magnitudes. Given the array of different behaviors described previously that can be generated by dielectrophoresis, it is possible to separate two groups of particles with different characteristics and transport them to different ends of a traveling wave electrode array, without any requirement for external fluid pumping and with the only control required being that of the applied electric field.

This was first demonstrated by Huang et al.[43] who showed that yeast cells could be propelled along an electrode array while bacteria in the same sample were trapped at the electrodes by positive dielectrophoresis. This method was considerably enhanced by Talary et al.[46] using viable and nonviable yeast cells. When a 35-kHz wave is applied, the nonviable cells are attracted to the electrodes by positive dielectrophoresis, while viable cells are both repelled by negative dielectrophoresis and transported along the array by traveling wave dielectrophoresis. When those cells have all reached the end of the array, the applied field frequency is changed to 4 MHz, at which frequency the viable cells are trapped at the end of the array by positive dielectrophoresis while the nonviable cells are repelled and travel *in the opposite direction* along the array, as the value of Im[$K(\omega)$] for nonviable cells at that frequency has a sign opposite to that for viable cells at the lower frequency. At the end of the procedure the viable cells are collected at one end of the array, the nonviable cells collected at the other.

8.8.3   Fractionation

An alternative method of performing particle separation using traveling wave dielectrophoresis is by using an electrode array to fractionate particles in a method similar to that described previously for dielectrophoretic ratchets. Particles belonging to different subpopulations are induced to move in the same direction, but the speed at which they move across the array is dictated by the properties of the particles. In the case of traveling wave dielectrophoresis, the factor responsible for differentiating between the particles is the different induced force due to the different dielectric properties of the particles. However, unlike the previous methods described in this chapter, the discrimination between particles is based on the *imaginary* part of the Clausius–Mossotti factor, which is responsible for dictating velocity over a large distance, and as such is potentially much more sensitive.

A traveling wave fractionation array has been demonstrated by Morgan and co-workers[47] and Green et al.[48] who used an array with 1000 electrodes with total width and length 2 cm; the array was demonstrated by separating different types of blood cells. By starting the cells at one end of the electrode array, the different populations of cells move across the array at slightly different velocities, so that after 80 sec the two types of cells studied (red blood cells and white blood cells) would be physically separated by 1 mm; studies have indicated that particles that differ in their dielectric properties by 0.2% would be separated by 100 μm by the end of the 2 cm array. Methods for ensuring both particle types start across the array at the same time are discussed in the next chapter.

An alternative strategy for traveling-wave fractionation has been suggested by Jan Gimsa,[49] in which the traveling wave force is effectively applied as a force field in the form of field-flow fractionation. The system operates by providing a flow along a chamber in one direction (let us say the x-axis), across a series of traveling-wave electrodes arranged parallel to the direction of the flow. When energized, these impart a force field that makes the particles travel in the direction orthogonal to that of the imparted flow. When the particles reach the opposing end of the electrode chamber, they will have fractioned across the plane at the end of the chamber (the y-axis). If the particles consisted of well defined and predictable mixtures of particles, outlets could be provided at the appropriate points at the end of the chamber; alternatively, the particles appearing across the entire end of the chamber could be analyzed in the manner of the output from an electrophoresis gel. Furthermore, by tuning the system appropriately, it is theoretically possible to use this technique to perform conventional dielectrophoresis-based field flow fractionation *at the same time*. Such a system would rely on fractionation according to the distance and direction the particle had traveled up the y-axis at time of exit from the chamber, and also the time taken to traverse the distance from one end of the chamber to the other. A schematic of such a separation device is shown in Figure 8.14.

8.8.4   Concentration

In the previous examples, traveling wave dielectrophoresis has been applied in a linear fashion — that is, the particles are input to the array at one end and proceed in a linear fashion to an output, or fraction along the array to allow the total distance traveled to be measured and population content to be determined. However, since the force can be applied in any direction across a surface on which electrodes can be fabricated, it can also be used to direct particles to travel in different directions, either to direct particles to a single point on an array, or to manipulate them around a laboratory on a chip as described in the next chapter.

Considering the first application, it is possible to fabricate so-called *spiral electrode arrays* formed by looping the four electrodes carrying the phased signals in an ever-decreasing spiral, such as those shown in Figure 8.15. After each group of four electrodes the pattern is repeated, so that there is an appearance of a sequence of electrodes with 90° shift along the direction toward the center of the array, and therefore there is a traveling field imposed that travels from the *outside* of the array to the *center*. Particles that experience

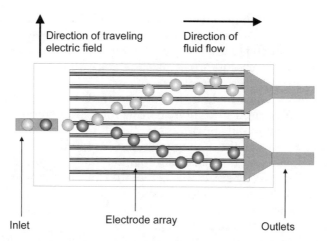

Figure 8.14 A schematic of the traveling-wave separator proposed by Jan Gimsa. An imposed flow imparts movement of particles along the horizontal axis, while traveling-wave electrode structures impart a motion at 90° to this. When arriving at the opposing end of the chamber, a particle's properties can be determined by its position along the *y*-axis, or (as here) separate outlets could be used for fractionation. Furthermore, the system could be combined with dielectrophoretic field-flow separation (as shown in Figure 8.4) to provide two-dimensional separation.

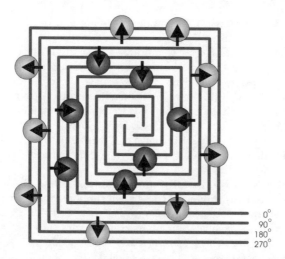

Figure 8.15 A schematic showing the function of spiral electrode arrays. The four electrodes are arranged in a spiral such that a continuing traveling wave is established between the center of the array and the edge. Particles will move to the center of the array or away from it according the their dielectric response.

traveling wave dielectrophoresis will either be conveyed into the center of the array or will be propelled off the array. This type of array has a number of advantages, principally in ease of fabrication. By allowing the four tracks to form an interlocking spiral, a very large area can be covered using just four electrodes on a single-layer electrode array. Normal traveling wave electrode arrays are restricted in total size by the necessity to connect each electrode to an off-chip power supply independently, or else use multi-layered fabrication techniques, which adds considerably to the cost and complexity of manufacture.

Another advantage of spiral electrode arrays is that they can easily be adapted so as to be used in combination with electrorotation, since it was demonstrated by Fuhr and co-workers[50,51] that at the center of the array, the four electrodes bearing the four phases can be terminated in electrodes arranged in a square, creating an electrorotation chamber for particle assessment. This can be directly applicable to the study of submicrometer particles through the use of an optical detection and measurement system such as the one described in Chapter 5, which could allow the measurement of rotation rates of particles concentrated at the center of the array. A number of subsequent studies have been made of spiral electrodes consisting of either square-shaped spirals such as those seen in Figure 8.15 or circles with ever-decreasing radius.[52,53] Spiral electrode arrays can be used for another application — that of manipulating or steering particles around laboratories on a chip, which is described in more detail in Chapter 9.

References

1. Pohl, H.A. and Hawk, I., Separation of living and dead cells by dielectrophoresis, *Science*, 152, 647, 1966.
2. Pohl, H.A., *Dielectrophoresis*, Cambridge University Press, Cambridge, 1978.
3. Benguigui, L. and Lin, I.J., Phenomenological aspect of particle trapping by dielectrophoresis, *J. Appl. Phys.*, 56, 3294, 1984.
4. Becker, F.F., Wang, X.B., Huang, Y., Pethig, R., Vykoukal, J., and Gascoyne, P.R., The removal of human leukemia-cells from blood using interdigitated microelectrodes, *J. Phys. D: Appl. Phys.*, 27, 2659, 1994.
5. Washizu, M., Suzuki, S., Kurosawa, O., Nishizaka, T., and Shinohara, T., Molecular dielectrophoresis of biopolymers, *IEEE Trans. Ind. Appl.*, 30, 835, 1994.
6. Talary, M.S., Mills, K.I., Hoy, T., Burnett, A.K., and Pethig, R., Dielectrophoretic separation and enrichment of CD34+ cell subpopulation from bone-marrow and peripheral-blood stem-cells, *Med. Biol. Eng. Comput.*, 33, 235, 1995.
7. Ertaz, D., Lateral separation of macromolecules and polyelectropytes in microlithographic arrays, *Phys. Rev. Lett.*, 80, 7, 1998.
8. Markx, G.H. and Pethig, R., Dielectrophoretic separation of cells — continuous separation, *Biotechnol. Bioeng.*, 45, 337, 1995.
9. Giddings, J.C., Field-flow fractionation: separation and characterization of macromolecular, colloidal, and particulate materials, *Science*, 260, 1456, 1993.
10. Markx, G.H., Rousselet, J., and Pethig, R., DEP-FFF: Field-flow fractionation using non-uniform electric fields, *J. Liq. Chromatogr. Relat. Technol.*, 20, 2857, 1997.
11. Markx, G.H., Pethig, R., and Rousselet, J., The dielectrophoretic levitation of latex beads, with reference to field-flow fractionation, *J. Phys. D: Appl. Phys.*, 30, 2470, 1997.

12. Huang, Y., Wang, W.B., Becker, F.F., and Gascoyne, P.R., Introducing dielectro-phoresis as a new force field for field-flow fractionation, *Biophys. J.*, 73, 1118, 1997.
13. Wang, X.-B., Vykoukal, J., Becker, F.F., and Gascoyne, P.R., Separation of poly-styrene microbeads using dielectrophoretic/gravitational field-flow-fraction-ation, *Biophys. J.*, 74, 2689, 1998.
14. Huang, J., Yang, J., Wang, X.B., Becker, F.F., and Gascoyne, P.R., The removal of human breast cancer cells from hematopoietic CD34(+) stem cells by dielectro-phoretic field-flow-fractionation *J. Hematother. Stem Cell Res.*, 8, 481, 1999.
15. Müller, T., Schnelle, T., Gradl, G., Shirley, S.G., and Fuhr, G., Microdevice for cell and particle separation using dielectrophoretic field-flow fractionation, *J. Liq. Chromatogr. Relat. Technol.*, 23, 47, 2000.
16. Curie, P., Sur la symétrie dans les phénomènes physiques, *J. Phys.* 383, 111, 1894.
17. Feynman, R.P., Leighton, R.B., and Sands, M., *The Feynman Lectures on Physics Vol. 1*, Addison-Wesley, Reading, MA, 1963.
18. Reimann, P., Brownian motors: noisy transport far from equilibrium, *Phys. Rep.*, 361, 57, 2002.
19. Huxley, A.F., Muscle structure and theories of contraction, *Prog. Biophys.*, 7, 255, 1957.
20. Cordova, N.J., Ermentrout, B., and Oster, G.F., Dynamics of single-motor molecules: the thermal ratchet model, *Proc. Natl. Acad. Sci. U.S.A.*, 89, 339, 1992.
21. Westerhoff, H.V., Tsong, T.Y., Chock, P.B., Chen, Y., and Astumian, R.D., How enzymes can capture and transmit free energy from an oscillating electric field, *Proc. Natl. Acad. Sci. U.S.A.*, 83, 4734, 1986.
22. Ajdari, A. and Prost, J., Mouvement induit par un potentiel périodique de basse symétrie: diélectrophorese pulsée, *C.R. Acad. Sci. Paris*, 315, 1635, 1992.
23. Chauwin, J.-F., Ajdari, A., and Prost, J., Mouvement sans force, *SFP 4émes Journées de la Matire Condensèe*, 2, 669, 1994.
24. Magnasco, M.O., Forced thermal ratchets, *Phys. Rev. Lett.*, 71, 1477, 1993.
25. Astumian, R.D. and Bier, M., Fluctuation driven ratchets: molecular motors, *Phys. Rev. Lett.*, 72, 1766, 1994.
26. Elston, T.C. and Peskin, C. S., The role of protein flexibility in molecular motor function: coupled diffusion in a tilted periodic potential, *Siam J. Appl. Math.*, 60, 842, 2000.
27. Rousselet, J. et al., Directional motion of Brownian particles induced by a periodic asymmetric potential, *Nature*, 370, 446, 1994.
28. Doering, C.R., Horsthemke, W., and Riordan, J., Nonequilibrium fluctuation-induced transport, *Phys. Rev. Lett.*, 72, 2984, 1994.
29. Risken, H., *The Fokker-Planck Equation: Methods of Solution and Applications*, Springer-Verlag, Berlin, 1984.
30. Prost, J., Chauwin, J.F., Peliti, L., and Ajdari, A., Asymmetric pumping of particles, *Phys. Rev. Lett.*, 72, 2652, 1994.
31. Gorre-Talini, L., Jeanjean, S., and Silberzan, P., Sorting of Brownian particles by the pulsed application of an asymmetric potential, *Phys. Rev. E*, 56, 2025, 1997.
32. Faucheux, L.P. and Libchaber, A., Selection of Brownian particles, *J. Chem. Soc. Faraday Trans.*, 91, 3163, 1995.
33. Schnelle, T., Müller, T., Gradl, G., Shirley, S.G., and Fuhr, G., Dielectrophoretic manipulation of suspended submicron particles, *Electrophoresis*, 21, 66, 2000.
34. Chauwin, J.F., Ajdari, A., and Prost, J., Force-free motion in asymmetric structures — a mechanism without diffusive steps, *Europhys. Lett.*, 27, 421, 1994.

35. Gorre-Talini, L., Spatz, J.P., and Silberzan, P., Dielectrophoretic ratchets, *Chaos*, 8, 650, 1998.
36. Batchelder, J.S., Dielectrophoretic manipulator, *Rev. Sci. Instrum.*, 54, 300, 1983.
37. Masuda, S., Washizu, M., and Iwadare, M., Separation of small particles suspended in liquid by nonuniform traveling field, *IEEE Trans. Ind. Appl.*, 23, 474, 1987.
38. Masuda, S., Washizu, M., and Kawabata, I., Movement of blood-cells in liquid by nonuniform traveling field, *IEEE Trans. Ind. Appl.*, 24, 217, 1988.
39. Fuhr, G., Hagedorn, R., Müller, T., Benecke, W., Wagner, B., and Gimsa, J., Asynchronous traveling-wave induced linear motion of living cells, *Studia Biophysica*, 140, 79, 1991.
40. Hagedorn, R., Fuhr, G., Müller, T., and Gimsa, J., Traveling wave dielectrophoresis of microparticles, *Electrophoresis*, 13, 49, 1992.
41. Fuhr, G., Über die Rotation dielektrischer Körper in rotierenden Feldern, dissertation, Humboldt-Universität, Berlin, 1985.
42. Fuhr, G., Schnelle, T., and Wagner, B., Travelling-wave driven microfabricated electrohydrodynamic pumps for liquids, *J. Micromech. Microeng.*, 4, 217, 1994.
43. Huang, Y., Wang, X.B., Tame, J., and Pethig, R., Electrokinetic behaviour of colloidal particles in travelling electric fields: studies using yeast cells, *J. Phys. D: Appl. Phys.*, 26, 312, 1993.
44. Morgan, H., Izquierdo, A.G., Bakewell, D., Green, N.G., and Ramos, A., The dielectrophoretic and travelling wave forces generated by interdigitated electrode arrays: analytical solution using Fourier series, *J. Phys. D: Appl. Phys*, 34, 1553, 2001.
45. Hughes, M.P., Pethig, R., and Wang, X.-B., Forces on particles in travelling electric fields: computer-aided simulations, *J. Phys. D: Appl. Phys.*, 29, 474, 1996.
46. Talary, M.S., Burt, J.P.H., Tame, J.A., and Pethig, R., Electromanipulation and separation of cells using travelling electric fields, *J. Phys. D: Appl. Phys.*, 29, 2198, 1996.
47. Morgan, H., Green, N.G., Hughes, M.P., Monaghan, W., and Tan, T.C., Large-area travelling-wave dielectrophoresis particle separator, *J. Micromech. Microeng.*, 7, 65, 1997.
48. Green, N.G., Hughes, M.P., Monaghan, W., and Morgan, H., Large area multi-layered electrode arrays for dielectrophoretic fractionation, *Microelectronic Eng.*, 35, 421, 1997.
49. Gimsa, J., unpublished work, 2001.
50. Fuhr, G., Fiedler, S., Müller, T., Schnelle, T., Glasser, H., Lisec, T., and Wagner, B., Particle micromanipulator consisting of two orthogonal channels with traveling-wave electrode structures, *Sensors and Actuators A*, 41, 230, 1994.
51. Fuhr, G., Müller, T., Schnelle, T., Hagedorn, R., Voigt, A., Fiedler, S., Arnold, W.M., Zimmermann, U., Wagner, B., and Heuberger, A., Radiofrequency microtools for particle and live cell manipulation, *Naturwissenschaften*, 81, 528, 1994.
52. Wang, X.-B., Huang, Y., Wang, X., Becker, F.F., and Gascoyne, P.R.C., Dielectrophoretic manipulation of cells with spiral electrodes, *Biophys. J.*, 72, 1887, 1997.
53. Goater, A.D., Burt, J.P.H., and Pethig, R., A combined travelling wave dielectrophoresis and electrorotation device: applied to the concentration and viability determination of Cryptosporidium, *J. Phys. D: Appl. Phys.*, 30, L65, 1997.

chapter nine

Electrode structures

9.1 Microengineering

As has been described in previous chapters, one of the great advantages of dielectrophoretic manipulation of particles is that the size of particles manipulated is not dependent on the size of electrodes used, but on the inhomogeneity of the electric fields produced by those electrodes. This is fortunate, as the minimum size of electrodes that can be easily manufactured today is still significantly larger than the smallest particles that can be manipulated by dielectrophoretic methods. In fact, most of the electrode arrays used to describe much of the material in this book are somewhat larger than the smallest devices possible, although electrodes with dimensions at the limits of fabrication technology were described for the trapping of single conducting colloids[1] and DNA molecules.[2]

Microengineered electrodes for dielectrophoretic applications first appeared in the late 1980s (e.g., Price et al.[3]) and have in the intervening years almost completely superseded the previous methods of generating nonuniform electric fields through the use of machined electrodes such as needles, planes, and curves (some of which are described in the next chapter). In particular, the development of microengineered electrode structures has meant that the dielectrophoretic behavior of nanoparticles can be observed in ways that would not be achievable by other methods.

One of the reasons that microengineered electrodes are so useful for dielectrophoresis is actually a byproduct of their shape. While microengineers will design suitable geometries for determining the electric field (by controlling the distance between neighboring electrodes) and locating electric field maxima and minima, the dielectrophoretic forces generated are greatly enhanced by the *planar* nature of the electrode structures. Microengineered electrodes are produced from thin metal film deposited across flat surfaces, so that when these are patterned into electrode geometries such as those seen in the preceding chapters, the electrode edges (viewed in cross section) are like sharp needles or razor blades pointing toward one another. Even a simple array, such as two straight and parallel electrodes (which, if extended to three dimensions, would produce a uniform electric field) produces a highly inhomogeneous electric field; this type of array is commonly used in the study of electrohydrodynamic fluid motion as described in Chapter 3.

Thus far, we have examined the use of microelectrode structures — quadrupoles, castellated electrodes, interdigitated traveling wave arrays, dielectrophoretic ratchets — and their applications for the manipulation of nanoparticles. In this chapter we will examine the processes by which such structures are made and consider how such electrodes can be used to perform a range of tasks beyond what we have already seen. We will also examine the concept of the laboratory on a chip, where all the processes we have discussed so far (particle trapping, detection, analysis, and separation) are combined into a single microfabricated device.

9.2   Electrode fabrication techniques

The development of dielectrophoresis was, for the first few decades, dominated by the use of machined, three-dimensional shapes such as rods, pins, and planes to generate nonuniform electric field shapes. While these geometries were eminently suitable for dielectrophoresis of cell-sized particles, they did have disadvantages — principally that trapping particles on the nanoscale required high voltages to generate sufficient nonuniformities (with corresponding problems with fluid heating, as discussed in Chapter 3) and shapes on this scale did not facilitate the building of enclosed field minima for easy observation of negative dielectrophoresis. It was with the adoption of techniques originated in the electronics industry for the construction of semiconductor devices (silicon chips) that electrode construction was revolutionized. Since the end of the 1980s, almost all the electrode structures used for dielectrophoresis have been microfabricated. It is worth noting that the term *fabricated* is used almost exclusively to describe electrode construction, a term adopted from the microelectronics industry. It is occasionally abbreviated to *fab*, hence the description of semiconductor fabrication facilities as *fab labs*. A range of microfabrication techniques exists for the construction of electrode structures, but the principal method employed is that of photolithography. Other methods exist for particularly fine electrode structures, the principal method being direct-write methods outlined in Section 9.2.5.

9.2.1   Photolithography

The term *photolithography* derives from Greek roots and means "to write on stone with light," an apt description when one considers that the process was developed for optically patterning silicon, a common constituent of minerals such as granite. The process has its origins in photography and shares many of photography's characteristics; fundamentally, the process operates by using an equivalent to a photographic negative to expose a light-sensitive surface to produce a required pattern.

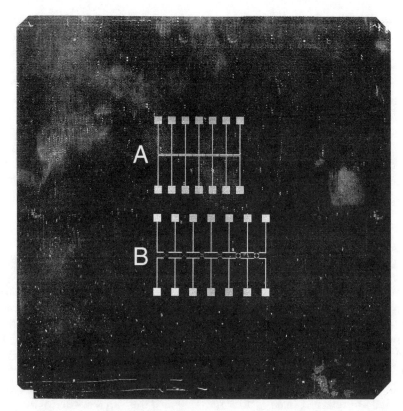

Figure 9.1 A life-sized image of a photomask for two dielectrophoresis electrode arrays. The photomask is a negative, with the dark region covered by chrome, and the electrode patterns transparent. A positive photomask would reverse these, with a chrome pattern in a largely transparent mask.

The equivalent to the photographic negative in microfabrication is the *photomask*, also called a *mask plate* or *mask*. This can be constructed of any clear material (capable of passing ultraviolet light), partially covered in an opaque patterned layer (capable of blocking ultraviolet). Ordinary photographic masks have been used for the process, particularly for on–off productions where resolution is limited to the grain size of the film; however, more common is the use of the standard photomask of the electronic industry, consisting of a quartz sheet with a thin film of chrome. This layer of chrome is patterned with the design of the electrode, usually using a technique called e-beam lithography (described in Section 9.2.5). The pattern can be represented in two ways, which will again be familiar to those with basic knowledge of photography; the pattern can be either a positive image or a negative image of the required structures, as shown in Figure 9.1. However, it is very important to note that the reference to positive and negative can refer to the mask, the processes involved in fabrication, or the output, quite independently.

When using a mask, we can either have a mostly clear mask with chrome only present where we want electrode structures to be on our finished device (again, the finished product, for example consisting of electrode structures on a substrate of glass, is called a *device* from semiconductor terminology). The process of producing this device is called a *positive process* since the device resembles the mask. The converse is a *negative process*, where we would use a mask that was mostly chrome, which is only transparent in those areas we wish to have material (such as a conductor) present on our final device. This can be compared to a photographic negative where white and black on the negative correspond to the reverse on the final product. Where things can become confusing is that mask production facilities refer to the means by which masks are produced in order to describe them. Masks are produced from quartz sheets (usually square with side dimensions measured in whole inches) completely covered in chrome. This chrome is then removed to produce the mask using electron beams, allowing very high definition (typically the smallest feature that can be written is a spot 30 nm in diameter, though 3 nm is achievable). Where the mask is exposed to the beam, the chrome can be removed. In order to produce a mask for a negative photolithographic process, the mask is only written in the areas to be exposed; that is, the areas that will be clear on the mask and filled in on the final devices. Where a mask is required for positive photolithography, the mask must be written in all the areas *except* those where the pattern is, so that only the chrome electrode pattern remains. For this reason, masks required to be a *negative* of the finished device (i.e., exposed where you want material) are referred to as *positive* masks (since they are exposed where you *don't* want chrome). Similarly, masks for positive photolithography are referred to as negative masks, at least by the people who make them. This contradiction is rarely explicit and can lead to expensive mistakes (which the author can report from bitter experience!); however, it is possible to make so-called *reverse polarity* (a positive mask from a negative, or vice versa) copies with reasonable accuracy (typically ±1 μm) for much less than the cost of a new mask. At time of writing, a written 4-inch mask plate costs about $1000, with a reverse polarity copy costing approximately one-third of this.

9.2.2   Wet etching

The fabrication process can be described as the process of selectively adding and removing material in specified ways until a final product is achieved. The first of these processes is deposition, the adding of a thin layer of material to the surface. The second process — and arguably the most important — is etching; the selective removal of material from a surface. The means by which the material is etched can vary, but the two most important are wet etching and dry etching; the former involves the use of liquid solvents and acids to remove material, whereas the latter process uses reactive ionized gases. For the majority of cases where a single electrode layer is required,

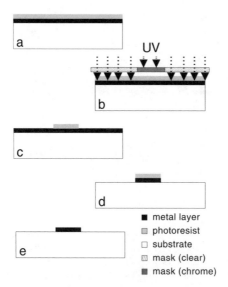

Figure 9.2 A schematic showing the fabrication process for positive photolitho-graphy. (a) The substrate, precoated in metal, is spin coated with photoresist. (b) The photoresist is exposed to UV via a photomask. (c) The photoresist is developed. (d) The excess metal is removed by acid treatment. (e) The remaining photoresist is removed, leaving the finished device.

wet etching is the simplest and most effective procedure. The procedures used for positive and negative photolithography by wet etching vary slightly, but both are used equally and will be described here.

Most etching processes involve the use of a material called *photoresist*. This is a polymer that can, under appropriate conditions, resist the action of acids used to remove electrode material; those conditions vary according to whether it is being used for positive photolithography (termed positive photoresist) or negative photolithography (negative photoresist).

We can examine the use of photoresist by considering a *positive* photo-lithography process as an example (shown schematically in Figure 9.2). In this example we will consider the fabrication of a single-layer device consisting of metal electrodes on a glass substrate (note that the substrate does not usually count as a layer) such as a microscope slide, patterned into the electrode shapes we have seen in previous chapters. In this procedure, we start with a slide completely covered in metal. Typically this metal will be gold; however, gold does not adhere to glass very well, and in order to secure it another layer (an *interlayer*) of another metal (such as titanium) that does adhere to glass is added first. Typically, the interlayer only needs to be thick enough to attach gold to glass (a layer about 10-nm thick will suffice), whereas the gold needs to be thick enough to carry a reasonable electrical current, and is typically of the order of 100 nm thick. There are two ways of depositing these gold films, called *evaporation* and *sputtering*. The former method involves heating a piece of the material to be deposited in a vacuum,

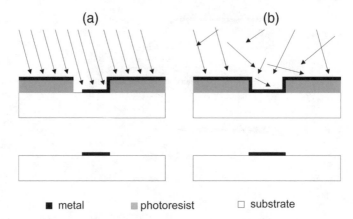

■ metal   ■ photoresist   □ substrate

Figure 9.3 Metal films can be deposited by evaporation (where metal is evaporated in a vacuum, causing atoms to scatter linearly) or sputtering (where a heated cloud of ion atoms is generated by heating). The latter method overcomes the problem of *shadowing*, where regions not covered by photoresist may not be exposed because of shadows cast by adjacent photoresist, because the atom cloud acts along the entire exposed surface.

so that when the material sublimates it literally evaporates, sending molecules in straight lines from the source to the sample (devices are often called samples during fabrication, and devices when complete) where they adhere. Sputtering is similar, but involves heating a piece of material (referred to as a target) with an electron beam until it produces an ionized gas; since the ions move freely in all directions in the gas, they contact the sample from all directions. This overcomes a problem with evaporation called shadowing, which occurs when structures on the surface of the sample block the ions and cast shadows on the surface of the material where no metal is deposited, as seen in Figure 9.3. Where multiple layers of metal are to be deposited (for example, when using a titanium interlayer), deposition should take place without breaking the vacuum present during deposition by either method; this prevents contaminants from interfering with the metal. For example, titanium readily oxidizes in air to form a layer to which gold will not attach so readily.

With our slide now covered in metal (the term is *metallized*, the process *metallization*), we can begin to form our patterns. First, we coat the slide in a layer of photoresist. This is done by *spin coating* or *spinning*. The slide is placed on a spinner, which is a device consisting of an upturned motor with a vacuum chuck at the center. The sample is held in place by the vacuum and spun at a regulated speed. A drop of the photoresist is applied to the sample prior to spinning, and when the sample is spun, most of the photoresist flies off the edge of the sample. However, a thin, uniform layer of photoresist is left across the surface of the sample; the thickness of the layer is determined by the viscosity of the photoresist and the speed and duration of rotation. Spin speeds are typically between 1000 and 10,000 rpm to produce resists of thickness 500–1000 nm. After the resist has been spun for an appropriate length of time, it is baked until hard.

The sample is now ready for the key part of the photolithography process, and the one that gives the process its name. In order to transfer the pattern on the photomask to the photoresist, the photoresist must be exposed to ultraviolet light. Many fab labs use devices called *mask aligners*; these large and expensive pieces of equipment use a microscope to align the sample with the mask. They are principally designed for use with complex, multi-layer electrode structures and are not strictly speaking necessary for single-layer devices where exposure could be performed by bringing mask and sample into direct contact (called contact exposure). However, doing this does reduce the lifetime of the mask, since any damage (such as scratches) caused to the mask plate will appear on all future devices exposed using that mask. When the mask and sample are brought together by which-ever means, the sample is then exposed to UV light through the photomask. Where the UV contacts the photoresist, it causes damage to the polymer; where it is blocked by the chrome layer on the mask, no damage is caused. The determination of exposure time and intensity is something of an art in itself — there are standard guidelines, but experience is invaluable, particu-larly where a design mixes areas with fine detail with other areas covering large spaces. Both overexposure (causing the undercutting of the mask) and underexposure (meaning the photoresist is not sufficiently damaged to be removed at the next stage) are to be avoided.

When the sample has been exposed, it is *developed*; as with photography, this is used to turn the image into something visible. In this case, the developer removes the material that has been exposed to UV, leaving the metal below it exposed; the remainder of the material is still covered by baked photoresist. The exposed material is that which we wish to remove to obtain our pattern. This is where we select wet etching; this basically means the removal of the metal using acids. Combinations of acids are used to etch specific metals; hydrofluoric acid (HF) is used for etching glass. Occasionally a two-step etching process may be required where an acid will remove one layer (e.g., gold) but not the interlayer below (e.g., titanium). Throughout the process, the remaining resist protects the areas where we wish to have metal in the final structure. Once the surplus metal has been etched away, the remaining resist can be removed. This leaves the substrate with metal present where the baked resist was, and the device can now be mounted for use. Typically, this means connecting the device to a structure with simple con-nections to external power supplies, such as a printed circuit board of some kind; connections between circuit board and device can be made by soldering directly to the device, but this is unreliable and can destroy a device. A more gentle method is to connect between bonding pads and wire using conduct-ing paint. This is most effective if the wire is fairly rigid (e.g., single core cable), and the bonding pad on the device should be sufficiently large to allow the user to apply the paint without spilling onto neighboring bonding pads — a 2 mm distance between neighboring bonding pads is usually

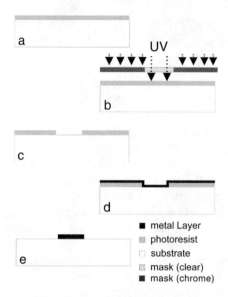

Figure 9.4 The lift-off fabrication technique. (a) The bare substrate is spin coated with photoresist. (b) The sample is exposed to UV via a photomask. (c) The photoresist is developed, leaving exposed substrate in the shape of the required pattern. (d) The sample is sputtered with the required metal. (e) The remaining resist is removed; as it dissolves, the remaining metal *lifts off* the surface, leaving the required pattern.

sufficient, although where pushed for space, smaller distances can be used if application is performed while viewed through a microscope.

Taking this procedure as a starting point, we can examine the differences involved in using other fabrication protocols. First of these is the use of a negative rather than positive mask, or negative rather than positive resist. A negative resist acts as the chemical reverse of positive resist; it hardens when exposed to UV light. This means that baked negative photoresist will be removed where it has *not* been exposed to UV. However, using a negative photoresist with a negative photomask would allow the same procedure outlined above to be repeated. Where a negative mask is used with positive photoresist, or vice versa, another procedure, known as *lift off*, is used. This procedure, summarized in Figure 9.4, begins with a bare substrate, to which the photoresist is applied, baked, exposed, and developed, leaving gaps to the bare substrate where the final design requires metal to be. The sample is then inserted into an evaporator or sputter coater and completely covered in metal. When the sample is immersed in solvent, the resist *underneath* the metal is dissolved, causing the metal above it to *lift off* (which gives the process its name). Lift off is in some ways a more complicated process than conventional wet etching, since it requires metallization midfabrication; it is possible to buy samples premetallized from a variety of sources, but sending samples coated with photoresist is not standard practice, and makes the process time consuming and expensive unless access to an evaporator or

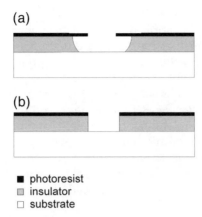

(a)

(b)

■ photoresist
▢ insulator
☐ substrate

Figure 9.5 Most wet chemical etching processes are isotropic; that is, material is etched away at the same rate in all directions, such as example (a) above. This can lead to undercutting of holes. Where more vertical sides are required, an *anisotropic* etching process such as reactive ion etching (RIE) is required.

sputter coater is available. However, lift off does have the advantage of being a more accurate process and is widely used in the semiconductor industry, since it avoids the problem of *undercutting* as described below.

9.2.3   Dry etching

There are circumstances, particularly in the fabrication of multilayered devices, where wet etching is not ideal for the formation of required structures. The main reason for this is that most wet etching processes are *isotropic* — when material is etched, the etching occurs in all directions simultaneously and at the same rate. While this is not a problem where the material to be etched is effectively two-dimensional (such as our thin films of metal), where an insulator is used to form a channel or through hole between different layers, it may be extremely important to etch the material with sides as vertical as possible (the term is *a high degree of verticality*). The reason for this is shown in Figure 9.5; if a thick layer of material is covered in resist with a window through as shown, then an isotropic etch would act to undercut the resist. This is not a problem with thin films. In order to completely etch through a 100 nm film, a device may suffer an undercut of 100 nm at the edge, but this is unlikely to be noticed. However, a film 1 μm thick or more may be significantly impaired if undercut by the same distance in each direction.

Dry etching acts by bombarding the surface of the material with reactive plasmas; inert gases such as O_2 or more reactive species such as Cl_2 are ionized by glow discharge. These gases are then exposed to the surface at a controlled pressure. These ions react with the exposed material that is stripped off (plasma stripping) or burnt (plasma ashing). The significant difference between wet and dry etching comes about because the ions are charged, and hence respond to an applied electric field. When a radio-frequency electric field is applied across the chamber in one direction (that is, the direction in which etching is desired), the ions bombard the surface preferentially in the direction of the field (along the field lines) while little etching takes place at right angles to the field. Hence, the etching only takes place in one direction — it is *anisotropic*. This form of dry etching, technically known as *reactive-ion etching* (RIE), is widely used to achieve vertical structures several hundred micrometers in height, and as such is ideal for the fabrication of channels for microfluidics applications (see Section 9.2.7) and for via holes in multilayered systems (see Section 9.2.6). However, recent developments in photoresist technology have meant that RIE may be supplanted to a degree by wet-etch methods for producing such devices.

9.2.4   Laser ablation

Another anisotropic etching technique, which can be applied equally for removal of metal for electrode fabrication, and to other materials for channel or via hole formation, is laser ablation (also called excimer laser fabrication). This uses a focused laser beam to vaporize (ablate) material, to a depth controlled by the intensity and duration of the beam. The beam is focused through the mask and onto the sample, allowing the removal of material with high precision (although not as precise as photolithography). There are other advantages to laser ablation, perhaps the most important being in its usefulness in prototyping. The mask and sample are held within the ablation system, and can be positioned with a quoted accuracy of 100 nm. Since the mask can be repositioned during the fabrication procedure, the mask need not represent the complete final design; it is possible for the mask to contain modules representing unique patterns, which can then be copied and pasted across the sample. Where two designs are to meet, the modules can be stitched together. However, there are complications such as the ablated material causing damage to nearby parts of the sample, requiring predeposition of a layer of polyimide to protect the sample. The polyimide must be removed after each layer is processed. The depth of the material removed is determined by the energy delivered with each laser pulse, according to the damage threshold of the material.

Since the material is removed along the path of the laser beam, excimer techniques are capable of producing highly vertical structures. However, there are circumstances where a nonvertical profile (as achieved by isotropic etching) is required. In such circumstances, the same effect can be obtained by etching shallow concentric troughs of decreasing area to produce a stepped gradient. The applications of laser ablation techniques have been discussed in detail by Pethig et al.[4] with particular reference to laboratory on a chip devices; however, other workers in the field have examined the application of the technique both for forming electrode structures[5] and for modification of existing structures.[6] Laser ablation devices have great potential for microengineering of devices, but their use has so far been restricted to a few workers due to the very high initial cost of purchasing the laser ablation system.

9.2.5 Direct-write e-beam structures

The primary disadvantage with optical methods of photolithography — that is, their use of light (in the ultraviolet waveband) to define the features of the pattern — is that from the laws of physics, the smallest feature size that can be defined is limited by the factor of the wavelength of light used with the mask. In theory, features smaller than the wavelength of light used cannot be reproduced, meaning that the limit of definition of conventional UV systems is of the order of 300–400 nm; smaller features can be defined using complex methods such as phase-shift exposure, capable of producing features as small as about 100 nm. While conventional photolithography is generally adequate for the task at hand, there are applications where a higher definition is required. For these applications, the best alternative is e-beam lithography.

Electron beam lithography is similar in principle to photolithography but uses a focused stream of electrons and specific photoresists (most common of which is poly methyl methacrylate, or PMMA, also known by the tradenames Perspex™ and Plexiglass™). Where the electron beam impacts the surface, the PMMA is chemically damaged and can be removed. The size of the beam depends on its energy, but it is commonly possible to focus it to a spot 30 nm across, with spot sizes as small as 3 nm being reported. The process of forming the shape of the intended device by eliminating a series of spots is referred to as *writing*; patterns are written using spots with a range of different sizes, with large spots used to fill in large, open areas and small dots used to fill in detail such as edges. The electron beam can only move within a limited range across the surface; defining spot size (by focusing) and positioning (by beam deflection) are achieved electromagnetically. In order to write across a greater area than the limited deflection of the beam can allow, the stage holding the sample can move between defined positions allowing the design to be written in chunks that are stitched together at the edges.

E-beam lithography is the method of choice for the production of photomasks; the chrome layer is covered in PMMA and then written, and the exposed areas removed by etching. As well as the production of masks, e-beam lithography can be used for direct exposure of devices, bypassing the mask stage altogether. By coating a substrate with PMMA and writing a pattern on it with e-beam, it is possible to use lift-off processes outlined at the beginning of this section to produce electrode shapes directly. This process has allowed the fabrication of highly detailed electrode structures, such as clearly defined castellated electrode shapes with dimensions of 1 μm along each face, finishing in sharp corners[7] or well-defined lines of interdigitated electrodes 250-nm wide.[8] However, the down side to e-beam direct-write structures is their cost in terms of time — each sample must be loaded into the beamwriter and written separately, whereas conventional exposure allows many samples to be processed one after another — and corresponding cost in terms of money. E-beam structures are expensive, but they are the only way to make structures with nanometer-scale precision.

9.2.6   Multilayered planar construction

As indicated in previous sections, it is sometimes necessary to form complete devices from more than one layer of material. There are two common reasons for this practice; either a protective layer is placed across the entire device except for an active area where the particles are to collect, electrically insulating the medium from the electrodes; or the electrode array is sufficiently complex to require a separate arrangement of electrodes and power rails instead of a straightforward connection from electrode to bonding pad. In both cases, there is a requirement for the addition of an insulating material, which is then etched.

Insulating material can be added in a number of ways. The simplest is to spin material onto the surface, in the same way that photoresist is added. This is mainly used for polymers such as polyamide and polyimide. However, where this is not possible, the principal alternative is the process of chemical vapor deposition (CVD). In a CVD process, chemical vapors are introduced over a heated sample; at the surface of the sample they react to form a solid product. It is possible to use glow discharge (similar to that used in RIE) to heat the gases instead, meaning that the sample does not need to be heated; this is referred to as plasma-enhanced CVD or PECVD. Common films deposited by this method include silicon oxide and silicon nitride; these are referred to as *passivation layers* since they protect the reactive surface in silicon manufacture. Once deposited, an insulating film can be etched using the methods described previously.

Where there is to be a connection between two conducting layers — for example, connecting an electrode array to a power rail — the means of interconnection must be considered. In any interconnect, the two layers are connected via a hole etched in the insulating layer. Where the gradient into the hole is shallow and the thickness of the insulating layer is small (perhaps a few hundred nanometers), it may be possible to simply deposit the top layer directly over the hole;[4,9] the top layer simply runs into the hole and forms an electrical connection at the bottom. However, where the insulator is thicker and the hole is anisotopically etched with a high degree of verticality, it is not straightforward to deposit material up the vertical walls on the side of the hole. In such circumstances, it is possible to use electroplating. By immersing the sample in an electroplating solution with the electrodes powered via the interconnects, gold (or some other suitable material) can be grown into the holes, until the deposited metal forms studs that pass all the way from the bottom layer to the top. The top layer of the conductor can then be plated over the studs, making electrical contact. This method was used with some success by Green et al.[9] but problems were encountered where the area of the electrodes, and number of holes to be plated, is large. Figure 9.6 shows how these protocols differ.

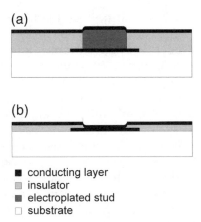

(a)

(b)

- ■ conducting layer
- ▨ insulator
- ■ electroplated stud
- □ substrate

Figure 9.6 Schematics of structures consisting of two conducting layers (an electrode layer and a power-rail layer), separated by a layer of insulator. If the insulating layer is thick (a), then electroplating can be used to add metal into the holes from the lower layer, forming studs across which the top layer can be deposited. If the insulating layer is thin (b), then metal can be deposited straight into the hole.

One important feature of mask design for multilayered electrodes is the presence of *registration marks*. When exposing resist for the upper layers while constructing the device, it is necessary to position the mask exactly over the lower layers. Since the top layer needs to be positioned over the bottom layer with micrometer-scale precision or smaller, potentially over a very large area, it can be a difficult and time-consuming task to adjust the mask (remember this can be adjusted for position *and* rotation). The task can be made easier through the addition of marks on each layer, whose positions directly correspond. If these are as far apart as possible on the mask, then aligning them will mean that the whole mask should also be aligned. For aligning masks, it is important to use a mask aligner; while this is not vital for single-layer devices, for multiple layers it is essential for this reason.

9.2.7   Microfluidics

Thus far, we have considered only the processes whereby we can fabricate planar electrode structures; even where supporting parts such as insulating layers and multilayer conducting paths are used, the total thickness of the device is rarely more than 1 μm. In many circumstances, planar devices will suffice; once such devices are complete and bonded up, they can be wired to a voltage supply, coated across the entire surface via a micropipette with a solution containing particles, put under a microscope, and used. However, while this is appropriate for simple laboratory experiments, more complex devices such as separators requiring extraction of the separated particles require closed paths for fluids to travel. Similarly, for some experiments involving particle separation, distinct outlet flows are required. Where some form of flow-through system is required, for example, where external pumping must be used, some means of constraining the flow is required once onto the device itself. This can be achieved by microfabricating channels across the surface of the electrodes, effectively forming a network of pipes across the electrode array. The term used to describe such microfabricated devices for controlling the delivery of fluids is *microfluidics*; one can compare this term to the word microelectronics (the controlled flow of electrons around a device) to see its origins. Microfluidics, and in particular the study of the flow of liquids around such miniscule capillaries, is a complex field that could easily fill a book in its own right (see the supplementary reading list at the end of the chapter for examples); however, we can examine the simple use of channels for liquid management and how to fabricate them.

There are three main methods of fabricating channels that will be discussed here; depositing and etching channels onto the substrate, etching them for some form of cover, and molding them. Considering these methods in order, the simplest way to form channels in the light of the fabrication methods so far discussed is to deposit a layer of material to the depth required, then etch it anisotropically to form the channel. Until recently it has been difficult to deposit stable layers of material more than a few micrometers thick; the material usually becomes stressed and flakes off when layers of more than about 10 μm are deposited. However, this limit has recently been broken by the development of Epon SU-8™ photoresist (available from a number of suppliers); this material can be spin coated onto the sample, then exposed as with any other material. Photolithography with SU-8 can form structures with very high verticality, enabling the formation of channels out of the resist itself; it has been shown to be very stable and can be used as part of the final structure. For example, SU-8 has been used in the manufacture of channels for traveling-wave dielectrophoresis,[10] where fluidic channels of depth 100 μm were fabricated and used to hold a solution by having a lid placed across the whole assembly; no leakage was observed after a liquid sample was stored in the device for several days.

The second method of constructing channels is to etch them into a *lid* that is attached to the planar device. There are two ways of achieving this. The first is to use a glass lid and etch the channels from the glass. This can be achieved using wet-etch methods as described earlier; glass can be etched isotropically using HF, producing a channel of semicircular cross section from a narrow gap in photoresist. Etching of channels is achieved by exposing those areas to be etched by HF through the photoresist. For isotropic etching, there will be as much undercutting on all sides of the gap in the photoresist as there is increasing of channel height. The advantage with etching onto glass is that it is straightforward to bond glass structures together, either by anodic bonding (passing a current through the structure and heating it), or by direct bonding by baking the two glass components at ~800°C; the lid and substrate carrying the electrodes can thus be joined for a permanent seal.

An alternative to glass is PMMA, which can be exposed and etched using a process called LIGA. LIGA (in German, *Lithographe, Galvanaformung,*

Abformung or lithography, electroplating, molding) is a process of using x-rays to expose PMMA, which can then be etched with a high degree of verticality (when defining shapes that are not vertical, the term used is *high aspect ratio*) and electroplated with nickel or nickel-cobalt to produce a master. This can then be used as a mold for other materials. One such material is poly-di-methyl-siloxane (PDMS), a two-part curing silane elastomer. This can be poured into the PMMA mold, acquire the required shape, and then be removed and applied to the substrate (alternatively, an intermediate stage using electroplating can be used to form a negative). PDMS has many advantages; it can be used to mold three-dimensional shapes with micrometer-scale resolution, it is optically transparent, flexible, and tough, and it is biocompatible. However, LIGA is an expensive process since it requires access to a synchrotron in order to produce the x-rays. Studies are ongoing into the possibility of using SU-8 as a molding material instead, with good success rates reported so far.

In addition to these methods, there exist a wide number of methods for pumping, separation, and the production of valves and so forth for microfluidics, usually etched into silicon.

9.2.8   Other fabrication techniques

In addition to the methods outlined, there are many other techniques that can be applied to the construction of electrodes; many are specialized, or are still in development. One example of each of these are the processes for constructing very small (nanometer-scale) interelectrode gaps, and embossing technology.

Thus far we have encountered two examples of nanoscale electrode gaps. One was discussed in Chapters 4 and 6 for the trapping of conducting colloids and DNA; the other was for trapping fullerene molecules for molecular transistors in Chapter 7. The techniques for fabricating these structures is different, although the end result is similar. Taking the latter example first,[11] the simplest method of construction (shown in Figure 9.7) is to begin with an e-beam-written structure consisting of two touching triangles. Passing an electrical current through the structure will cause heating at the narrowest point (100 nm) in the structure, where the current density will be highest; if the current is sufficiently high, the metal will burn out at that point. At the moment that the material is burnt through, the electrical current is open-circuited and no more damage is done. With this technique, interelectrode gaps as small as 1 nm have been reported.

The alternative method, developed by Bezryadin and Dekker,[1,12] and Porath et al.[2] uses two electrodes suspended over a gap in silicon, etched away isotropically to allow any material attracted to the electrodes to attach directly to the electrodes without the substrate getting in the way. Two methods are presented for this; in one, conventional e-beam lithography is used to form two protrusions across the interelectrode channel, which when sputtered accumulate thickness (and hence get closer), until the tips are

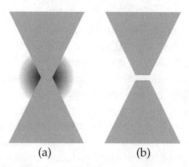

(a)   (b)

Figure 9.7 Very small interelectrode junctions can be fabricated by forming an interlocking-triangle structure using e-beam lithography, such that the waist of the device is of the order of 100 nm. (a) By passing a current through the device, the current density at the waist of the device is sufficient to burn it out. (b) The resultant interelectrode gap is about 1 nm across.

20 nm apart. Even smaller gaps can be achieved using a technique called *contamination lithography.* By focusing an electron beam onto organic material, carbon rods can be grown in directions defined by the position of the beam. By growing such rods from the electrode edges using a scanning electron microscope and then sputter coating these rods, electrodes with sub-5-nm gaps can be formed.

There are a number of other methods for device construction in development — including dielectrophoresis itself, as outlined in Chapter 7. Other techniques include *embossing* — using a molded pattern to stamp polymer/ plastic substrates, or stamping a shape into another material (one example of this method is the production of compact disks), which has been demonstrated to be effective for fabrication down to tens of nanometers,[13] as well as the use of scanning probes and optical tweezers to manipulate components into place. A review of methods being explored is available,[13] though this field is highly dynamic (since it underlies the progress in the speed of computer processors, itself a massive industry) and the reader is advised to seek the most up-to-date material available on this rapidly changing field.

9.3   Laboratories on a chip

Throughout this book so far, we have been largely concerned with four basic electrode geometries: quadrupolar arrays, castellated electrodes, thermal ratchets, and interdigitated arrays for traveling-wave dielectrophoresis. While these arrays are capable of trapping, analyzing, or separating particles on the array itself, they are generally designed to perform one process. If a device is to be useful beyond the laboratory, then it will be required to (a) interface with other components within a system and (b) perform a number of operations in order to achieve a single function. In the 1990s[14] a term was invented to describe devices that do perform multiple operations on a sample in order to perform a complete and useful function. This term represents the *ethos* of using miniaturization to replicate the equipment one normally finds in the laboratory — the devices were called *laboratories on a chip.*

Laboratories on a chip are devices wherein the components of a modern laboratory — fluid handling, reactors, heaters, pumps, separators, and sensors — are integrated in miniature onto a single device, typically of a size between that of a postage stamp and a credit card. The original labs on chips were miniaturized chemistry devices such as capillary electrophoresis columns and liquid chromatographs, but since then the term has expanded to include a wide range of devices from DNA microarrays to miniaturized chemical assays to dielectrophoretic particle sorters. Implementations vary, but a common presentation is to enclose the lab on a chip in a cartridge format that is inserted into another unit containing the ancillary electronics (power supplies, signal generation, optics for analysis, and so forth) and liquid handling (reservoirs and pumps). The liquid sample to be analyzed is inserted either into the control unit or into the cartridge itself, where it is processed and analyzed. From a commercial perspective, the reduction in size, with benefits for mass production, transportability, and ease of use, means that such devices have great potential for point-of-care diagnostics, portable water screening equipment, and rapid cell, protein, and DNA analysis for rapid drug discovery. Such devices could, for example, potentially allow the testing of blood samples at the patient's bedside, or at least provide the potential for cheap analysis of blood or urine samples on a hospital-by-hospital basis. Another application is for the identification of pathogens such as toxic viruses used for biological warfare. The integrated nature of these devices, combining many analysis methods, lends itself to another general term — Micro-total analytical systems, abbreviated MicroTAS or μTAS.

Since cells are easily manipulated using dielectrophoresis and we are able to discriminate between cell types on the basis of both surface and interior properties, much of the work on dielectrophoretic laboratory-on-a-chip devices has been aimed at the development of particle detectors for medical applications. However, most lab-on-a-chip devices have features in common, the principal feature being that they are fabricated using semiconductor methods and operate on small samples (of the order of microlitres) using channels etched into glass, photoresist, or some other polymer, through which material is pumped from an external source.

There have been a number of approaches to the application of dielectrophoretic techniques to the laboratory on a chip concept. Perhaps the first example of the use of microengineering techniques to construct a dielectrophoretic laboratory on a chip is the "dielectrophoretic fluid integrated circuit" described by Washizu, Nanba, and Masuda;[15,16] this device was able to move cells around microfabricated channels and sort them into different outlets. Some researchers have described attempts to perform a range of functions using dielectrophoresis, from separation to trapping and analysis (e.g., Pethig et al.[4]); others have used dielectrophoresis as part of a broader system including electroporation or biochemical methods for cell detection, still others have used dielectrophoresis primarily as a method for isolating specific cells at a preliminary stage (e.g., Arai et al.[17]). In this section, we will examine how dielectrophoresis can provide additional functionality for lab-on-a-chip systems, how many activities can be combined into a single analysis system, and how laboratories on a chip can be interfaced with the outside world for analysis and control.

9.3.1   Steering particles around electrode structures

An advantage of miniaturizing laboratory components to the micrometer scale is that it allows manipulation of small amounts of material — even single particles. We examined in Chapter 4 the possibility of trapping single particles, but once held these particles are static, held within small and immobile field cages. We have also examined methods of dielectrophoretic propulsion, such as by the use of dielectrophoretic ratchets or traveling-wave dielectrophoresis, but these operate on bulk material containing large numbers of particles. However, it is possible to selectively steer *single* particles around an electrode array, provided some form of monitoring (usually video-based) is available for tracking the particle's position.

Three methods for single particle manipulation are presented here. The first was presented in 1990 and represents an early formulation of the lab on a chip in the fluid integrated circuit (FIC) of Washizu et al.[16] By switching between microelectrodes in sequence (as illustrated in Figure 9.8), particles can be handed from one electrode to another. One of the innovations with the device is that it is not merely the electrodes that are used to attract the particles — the electrodes are largely flat and parallel, and do not cause significant field inhomogeneity — but the inclusion of field-constructing insulating protrusions between the electrodes that cause field distortion, with the effect being that particles are attracted to the tip of the insulator by positive dielectrophoresis. By switching between electrodes, particles are propelled between protrusions and hence along the array. When the particle reaches the end of the electrode array, it is possible to switch the particle into one of two outputs by energizing electrodes next to the appropriate output as shown in the figure.

A second method of particle manipulation, which shares some common features with the Washizu system, was presented by Suehiro and Pethig in 1998.[18] By employing two orthogonal sets of parallel electrodes across the top and bottom of an electrode chamber and selectively activating individual strips on the top and bottom of the chamber, a single particle can be trapped by positive dielectrophoresis. By then energizing the adjacent line with a potential at a frequency known to impart negative dielectrophoresis, the particle could be induced to move from one electrode intersection to the next. In order to observe the particle's position on the grid, the electrodes were constructed from indium tin oxide (ITO) rather than conventional gold. ITO has a higher resistance than gold, but benefits from a higher resistance to bioparticle adhesion,[19] and, most importantly, it is highly optically transparent. This enables direct observation of the particle even when suspended within a three-dimensional cage. This is illustrated in Figure 9.9.

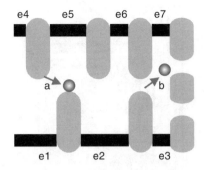

Figure 9.8 The "fluidic integrated circuit" of Washizu and co-workers. (a) By energizing adjacent electrodes e1 and e2 either side of an insulating protrusion (gray), particles can be attracted to the end of that protrusion. By switching between potentials on either side of the channel, the electrodes can pass the particle between protrusions. (b) By activating the electrodes e6 and e7 on one side of the arrangement, the particle can be guided into one of the two outputs.

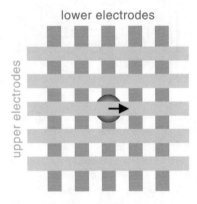

Figure 9.9 The single-particle handling system of Suehiro and Pethig. By switching applied potentials between adjacent electrodes on the top and bottom parallel electrode arrays, it is possible to steer a particle around the grid formed by the electrodes.

The final method of individual particle handling presented here is the only one of the three to employ traveling wave dielectrophoresis instead of conventional dielectrophoresis to manipulate single particles. By using a pair of spiral electrode arrays (similar to that shown in Chapter 8) next to one another, across which channels were deposited to limit the paths which the particle could take, Fuhr and co-workers[20] devised a system whereby single particles can be steered to a number of different outputs. The system, shown schematically in Figure 9.10, uses variations in the voltages applied to the traveling-wave structures to bias movement around the array. A particle entering one of the channels is transported to the center of the array by traveling wave dielectrophoresis. On reaching the center of the array, it falls into the field null at the center, since in order to travel effectively, the particle should experience both traveling wave dielectrophoresis and negative

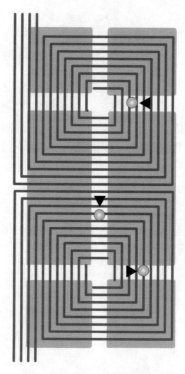

Figure 9.10 The orthogonal spiral electrode arrays used by Fuhr and co-workers. The spiral electrode arrays use traveling waves to move them toward, or away from, the center of the electrode arrays. Once at the center, the particle can rapidly be guided toward one of the outlets by lowering the potential to the electrode terminating at that outlet. This enables particles to be steered into and out of any of the channels.

dielectrophoresis (see Chapter 8). The phasing of the field is reversed (this can be achieved by reversing the phasing of two channels, as discussed in Appendix A) to propel the particle out of the field minimum. This acts along all directions leading from the center of the array, but the direction the particle moves in can be selected by lowering the potential to the electrode blocking the particle from moving in the desired direction. The negative dielectrophoretic forces on the remaining electrodes act to move the particle into the start of the channel, and from there, the traveling-wave dielectrophoretic motion acts as a conveyor to move the particle. The system as shown was capable of steering a particle from any inlet to any outlet in the six-port system illustrated. As before, human observation is required to determine the timing of the changing of the field. However, there are other methods of monitoring particles, as discussed below.

9.3.2   *Particle detection*

Ideally, any laboratory on a chip should be able to function without requiring external human control. The operator wishes to input a sample and obtain results without having to observe particle motion and determine the particle collection rate, observe frequency-dependent dielectrophoretic behavior, or instruct controlling electronics to move particles this way or that. If the device is to be automated, methods are required for automatically interrogating the sample to determine the position, behavior, and number of particles. Methods have been developed for assaying rotation rates of particles in electrorotation electrodes using image processing,[21,22] and similar image processing techniques have been used to detect particles undergoing positive and negative dielectrophoresis.[23] However, while these methods offer accurate detection methods for determining particle behavior, they do not sit comfortably with the idea of the simple, portable electronic devices.

A number of methods based on more simple means of observation have been explored. An early method of determining particle collection rates was devised by Price et al.; the particles are attracted to electrodes held within a chamber consisting of two patterned microscope slides, and the concentration of particles is determined by shining the laser through the slides and measuring the turbidity of the solution. A later version by Talary et al.[24] measured the particle density along the gap between the slides, thereby measuring both positive and negative dielectrophoresis. Laser interrogation can also be achieved on-chip. A recent development by Cui et al.[25] uses an optical fiber within the chip itself, with the fiber bisected by a microfabricated channel along which particles are propelled by traveling-wave dielectrophoresis; the track electrodes are across the bottom of the channel. Particles passing between the halves of the optical fiber are illuminated by the laser from one side of the channel. They are detected either by measuring the change in brightness at the opposing fiber or the fluorescence emission from labeled particles back along the original fiber. By using more than one sensor along the channel, the speed at which the particles are propelled by the traveling waves (and hence the polarizability) can be determined.

Finally, as described in Chapter 5, it is possible to measure the light intensity variations across an interference pattern with two crossed lasers of slightly different frequency in order to measure the velocity of unlabeled nanoparticles such as viruses.[26] Lasers can also be used to determine the rotation rate of particles. As described in Chapter 5, measuring the change in phase of a laser shining through a chamber where particles are rotating gives an indication of the rotation rate, a method effective even for nanoscale particles.[26] Another method has been used by Berry, where the changes in light intensity of a beam of light projected across the particle and centered on the middle of four light sensors are measured. As the particle (in this case a bacterium) rotates, the intensity on the detectors varies sinusoidally, and this variation can be used to determine the rate of rotation of the particle.

In addition to optical methods, electronic detectors have been developed. These methods measure the change in electrical impedance between two electrodes when particles are present compared to when they are absent.[27,28] When a particle passes between the two measurement electrodes, the impedance between the electrodes should change, particularly if the particle is markedly different in electrical characteristics from its surroundings and occupies a large portion of the channel. While this method is principally of use for larger particles such as cells, it can be used for nanoparticles after they have been attracted to the electrode edges by positive dielectrophoresis; particles can have a proportionally larger effect on the impedance at the electrode than in the bulk, and this method[28] has been shown to be effective for nanoparticles.

9.3.3   Integrating electrokinetic subsystems

A complete laboratory-on-a-chip device requires the integration of many subsystems in order to perform useful functions such as particle separation and analysis. A number of workers in the field have constructed such systems, either relying totally on dielectrophoretic methods of particle manipulation or integrating other methods. Both approaches have advantages; using only dielectrophoresis allows the entire device to be fabricated in a single operation, and since dielectrophoresis requires no extra material beyond what is used (except some ancillary devices for detection and fluid handling), operation of the device is easier. However, by restricting the device to all-dielectrophoretic methods, other possibilities of particle discrimination according to factors that do not affect electrical properties — such as the presence of a certain gene — cannot be used, whereas methods borrowed from molecular biology, such as flow cytometry and PCR, are sensitive to these types of factors.

Where the system relies largely on dielectrophoresis to provide many different functions, there are additional requirements that are placed on the dielectrophoretic system. For example, particles entering a separator should be guided so as to enter the device in an appropriate place (where differences in field are greatest) or time. An example of this was presented by Cui et al.[29] as part of a traveling wave dielectrophoresis-based particle separator. As described in Chapter 8, traveling wave methods offer the possibility of highly sensitive particle fractionation, but only if the particles start from one end of the array at the same time; if they drift onto the array at different times, fractionating different types of particles is impossible. In order to avoid this, Cui et al. employed an extra pair of individually addressable electrodes at the beginning of the array. By energizing these electrodes so as to attract all particles to them by positive dielectrophoresis, particles can be positioned in a "starting gate," as shown in Figure 9.11. When the traveling field is applied to the remainder of the electrodes, the phases applied to the trapping electrodes can be changed so that they become part of the traveling array, ensuring that all particles start at the same point. Once moving, particles can be detected using the optical methods described above.

Another proposal for a traveling-wave-based lab on a chip was made by Pethig et al. in 1998.[4] However, unlike the Cui design, traveling wave dielectrophoresis is employed principally as a means of transporting particles between inlets, analysis electrodes, and dielectrophoretic traps. However, Pethig's proposal contains a novel idea; by using a forked junction in the

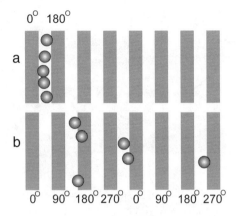

Figure 9.11 A schematic of the starting gate device used by Cui et al. (a) By collecting particles between the first two electrodes by dielectrophoresis (i.e., signals phased 180° apart), particles can start down the traveling wave array simultaneously, allowing them to be fractionated (b).

traveling wave conveyor belt, energized at two different frequencies, particles could be encouraged into one of the two forward paths according to their dielectrophoretic properties, allowing separation. These junctions could theoretically shift particles toward different dielectrophoretic traps, analysis electrodes, or outlets.

Other workers have used conventional dielectrophoresis as a means of sorting and analyzing particles as they are pumped around the microsystem by using a conventional external pump. In order to organize particles into a single-file sequence for analysis, devices such as dielectrophoretic funnels and concentrators were used.[30,31] Triangular electrode arrays placed across the flow are shown schematically in Figure 9.12. Electrodes are placed at the top and bottom of the array and energized so as to repel the oncoming particles. This forces them toward the constriction at the end of the array, where the gap between the electrodes is small enough to let only a single particle pass. The device was shown to be effective for flow rates up to 3.5 mm sec^{-1} allowing high-speed processing of cells to be achieved. Subsequent arrays provide dielectrophoretic trapping, analysis, and switching, the latter process operating by energizing one or two electrodes by negative dielectrophoresis to selectively move the particles toward one of two outlets (also shown in Figure 9.12).

Other workers in the field have sought to integrate dielectrophoretic manipulation with other methods. For example, U.S. company Nanogen has integrated dielectrophoretic trapping of cells with hybridization techniques for bacterial analysis.[32–34] Dielectrophoresis is used to provide initial cell sorting to isolate the bacteria from blood cells, then trap them on electrodes where a high voltage is used to break apart the bacteria. The DNA are then allowed to drift free, and PCR is used to identify the type of bacterium in order to identify it. Other workers have used dielectrophoresis with laser

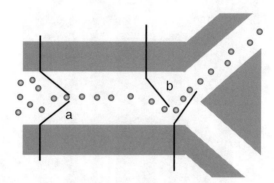

Figure 9.12 Components of the lab-on-a-chip system devised by Fiedler and colleagues. Particles are driven through these devices by an external pump at rates of up to 3.5 mm sec^{-1}. The electrode patterns are written on glass on both the top and the bottom of the chamber, separated by 25–40 µm; electrodes on the two planes have opposite phase, creating a repulsive dielectrophoretic curtain. (a) A concentrator or funnel, which guides incoming particles into single file. (b) A switch that allows the output channel to be selected. The leftmost electrode is always energized, guiding the particles to the bottom of the channel and hence out of the bottom channel; if the rightmost electrode is activated, the stream is diverted into the top channel.

trapping to sort cells, with unwanted cells being trapped by positive dielectrophoresis and an automated tracking system being used to identify bacteria and move them to an outlet port.[17]

9.3.4 Contact with the outside world

Since the laboratory on a chip is ultimately required to interface with a control unit of some kind, consideration must be given to the way in which the device interfaces with that control unit. For example, the choice of substrate will depend the way in which the device is connected to outboard electronics and sensors. The most common substrate for construction of microelectrodes for these applications is glass, since it is resistant to acids and solvents during the etching process, and since it is transparent, and thus it allows the use of various optical techniques as described above. Furthermore, precut glass of the appropriate size of such devices is readily available in the form of microscope slides, which form the basis of the majority of devices. On the other hand, silicon has been used in some systems (e.g., Docoslis et al.[35]) since it allows direct integration of electronics onto the device, which minimizes the amount of outboard equipment required by limiting the available sensing techniques. However, some sensors (impedance sensors or photodiodes, for example) can be built directly onto the chip. As ever, the choice must be dictated by the function of the device and the requirements placed upon it.

In order to impose an external flow, or to introduce or remove particles, it is necessary to interface the fluidic pathways through a lab on a chip with external sources and drains. The best method by which this is achieved varies

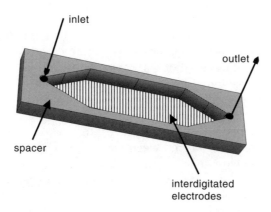

Figure 9.13 A fan-out profile inside a lab on a chip ensures an even flow across the channel.

according to the application. Interconnection between on-chip microfluidics and external pumps is usually performed via holes in the device, with a fan-out profile on the chip to ensure that input material is spread across the array evenly and without the formation of stagnant areas (e.g., Huang et al.[36] and Pethig[37]), as shown in Figure 9.13. Separation between top and bottom cover should be sufficient to ensure that the sample volume does not exceed the reach of the electrodes, or that where interaction between top and bottom electrodes is required (for example, in the field funnel described above), the electrodes are close enough for the force to be sufficient for particle manipulation; typical heights range between 25 μm[30] and 250 μm,[38] according to electrode geometries and applications.

In order to maintain a sealed fit to supply tubes, the tubes may either be glued to the device for a permanent connection[36,38] or through the attachment of capillaries and other connectors glued to the inlet and outlet holes, across which the joining tubes can be stretched.[29–34] The ideal format for a cartridge-based system, a press-fit separator device, has been demonstrated[35] but has not been widely adopted. Fluids are introduced via tubing (typically Teflon) driven from pumps capable of delivering suitable low flow rates (down to a few μl sec^{-1}); typically, peristaltic pumps and syringe pumps are used for these applications. Under many circumstances, more than a single microfluidic operation may be required. For example, one channel may be required to introduce a flow at right angles to a traveling wave electrode array. Then, after the particles have traveled along the array, a second outlet path is required to remove the particles, as demonstrated in the design of Cui et al.[29]

9.4   A note about patents

As can be seen from the previous section, the development of laboratory-on-a-chip systems represents a significant move toward the use of dielectrophoresis in commercial devices. However, as companies and researchers move into the commercial arena, there is an increased demand that investments of time and money should be protected; this has resulted in a dramatic increase in the patenting of dielectrophoretic devices and methods in recent years. Of the many aspects of dielectrophoresis that are novel and useful, those that are most amenable to patenting are the pieces of hardware, since they involve designs that are easy to define and can be applied to a wide variety of different particle types. The principles of dielectrophoresis were established in the 1950s, and any patents drawn up then have long since expired. However, the development of microengineered electrodes, laboratories on a chip, and so forth are all new developments, and as such the existing patent literature must be considered when developing commercial devices. Inspection of patent literature is also useful in that some ideas are patented without being published in the scientific literature.

Perhaps the earliest patent directly pertaining to the work considered here[39] was published by Batchelder, in which what amounts to a dielectrophoretic lab-on-a-chip system was described, using multiple electrodes on different planes for manipulating particles around an array. As with his early work with traveling-wave dielectrophoresis,[40] Batchelder's work was not followed up for many years.

The next nanoscale dielectrophoresis patent concerned the basic actions of particle trapping in planar microelectrodes, using both polynomial electrodes (see Chapter 10) and castellated electrodes to trap and separate particles suspended overhead.[41] Similar electrodes were used by Betts and Hawkes,[42] who patented a device using a similar electrode array with optical detection equipment to measure dielectrophoretic collection rate for particle characterization.

Many subsequent patents have been concerned with particle separation methods. The stepped dielectrophoretic separation process described in Section 8.2, where particles are stepped toward opposite ends of a castellated array, was patented by Pethig and Markx in 1994.[43] Betts and Hawkes[44] patented flow-through separation using interdigitated electrodes in 1996.[45] The use of field-flow fractionation and spiral electrodes, both described in Chapter 8, were patented by Huang and co-workers in 1997 and 1999, respectively.[45,46] Other cell sorters have been suggested by Crane,[47] using dielectrophoretic forces to array particles across the funnel, and Ager et al.,[48] who used traveling-wave dielectrophoresis to concentrate cells onto a spiral array. Although traveling-wave dielectrophoresis was published before patenting, other methods have been investigated, for example combining it with multiple electrode frequencies to improve separation effects.[50] Other methods seek to separate particles on an electrode array for the purposes of characterization; for example, Pethig and Markx patented a device wherein parallel electrodes are connected to separate signal sources so that each electrode along the array receives a frequency twice as high as its neighbor. Particles will only collect between the electrodes whose frequencies correspond to frequencies where the particles experience positive dielectrophoresis.[51]

More recently, components that form parts of labs on chips have been patented. For example, a number of patents have been filed by Hagedorn and co-workers[52] on such devices as funnels for focusing particles,[53] while Lock and Pethig have patented additional sensing electrodes in dielectrophoresis systems[54] and using curved interdigitated electrodes to achieve traffic control in traveling wave structures.[55] Similarly, Cheng et al. have patented a Y-shaped traveling wave device for cell separation[56] similar to the conveyor system described earlier in this chapter. A feature patented by Arnold[57] is the addition of heating to introduce thermal forces above the electrode chamber. Another recent trend in patent literature is the presence of companies who are primarily interested in the development of complete lab-on-a-chip systems; patents include those for the Nanogen system described previously[58,59] and another from an Italian company, Silicon Biosystems SRL, involving the use of microelectronics to specifically trap single-cell-sized particles.[60]

Finally, as new techniques are developed in nanotechnology using dielectrophoresis, they too are patented. Examples of these presented in this book are the colloid-based immunosensor developed by Velev and Kaler and presented in Chapter 7,[61] the method of dielectrophoretic assembly of devices also presented in Chapter 7,[62] the manipulation of DNA and other molecules in solution by dielectrophoresis as described in Chapter 6,[63,64] and the tunable construction of composite materials using dielectrophoresis also presented in Chapter 6.[65]

An examination of the dates of the patents listed here gives an indication of the pace at which dielectrophoresis is building commercial momentum; for the presented list (which is not exhaustive, since I have omitted patents where many similar applications are presented, identical versions of patents from different geographic regions, and patents involving electrode structures too large for the manipulation of nanoparticles, among others) eight patents were issued for material described in this book in the year before publication (2002), following a near-exponential trend. As the commercialization of dielectrophoresis increases, so will the restrictions on what can and cannot be used; however, as avenues become closed to commercial development, others are opened as researchers are challenged to find entirely new approaches to their work.

References

1. Bezryadin, A., Dekker, C., and Schmid, G., Electrostatic trapping of single conducting nanoparticles between nanoelectrodes, *Appl. Phys. Lett.*, 71, 1273, 1997.
2. Porath, D., Bezryadin, A., de Vries, S., and Dekker, C., Direct measurement of electrical transport through DNA molecules, *Nature*, 403, 635, 2000.
3. Price, J.A.R., Burt, J.P.H., and Pethig, R., Applications of a new optical technique for measuring the dielectrophoretic behavior of microorganisms, *Biochim. Biophys. Acta*, 964, 221, 1988.
4. Pethig, R., Burt, J.P.H., Parton, A., Rizvi, N., Talary, M.S., and Tame, J.A., Development of biofactory-on-a-chip technology using excimer laser micromachining, *J. Micromech. Microeng.*, 8, 57, 1998.

5. Schnelle, T., Müller, T., Gradl, G., Shirley, S.G., and Fuhr, G., Dielectrophoretic manipulation of suspended submicron particles, *Electrophoresis*, 21, 66, 2000.
6. Müller, T., Gerardino, A., Schnelle, T., Shirley, S.G., Bordoni, F., De Gasperis, G., Leoni, R., and Fuhr, G., Trapping of micrometer and sub-micrometer particles by high-frequency electric fields and hydrodynamic forces, *J. Phys. D: Appl. Phys.*, 29, 340, 1996.
7. Morgan, H. and Green, N.G., Dielectrophoretic manipulation of rod-shaped viral particles, *J. Electrostatics*, 42, 279, 1997.
8. Reimer, K., Köhler, C., Lisec, T., Schnakenberg, U., Fuhr, G., Hintsche, R., and Wagner, B., Fabrication of electrode arrays in the quarter micron regime for biotechnological applications, *Sensor. Actuators A*, 46, 66, 1995.
9. Green, N.G., Hughes, M.P., Monaghan, W., and Morgan, H., Large area multi-layered electrode arrays for dielectrophoretic fractionation, *Microelectron. Eng.*, 35, 421, 1997.
10. Cui, L. and Morgan, H., Design and fabrication of traveling wave dielectro-phoresis structures, *J. Micromech. Microeng.*, 10, 72, 2000.
11. Park, H., Park, J., Lim, A.K.L., Anderson, E.H., Alivisatos, A.P., and McEuen, P.L., Nanomechanical oscillations in a single C_{60} transistor, *Nature*, 407, 57, 2000.
12. Bezryadin, A. and Dekker, C., Nanofabrication of electrodes with sub-5nm spacing for transport experiments on single molecules and metal clusters, *J. Vac. Sci. Technol. B*, 15, 793, 1997.
13. Xia, Y., Rogers, J.A., Paul, K.E., and Whitesides, G.M., Unconventional methods for fabricating and patterning nanostructures, *Chem. Rev.*, 99, 1823, 1999.
14. Effenhauser, C.S. and Manz, A., Miniaturizing a whole analytical laboratory down to chip size, *Am. Lab.*, 26, 15, 1994.
15. Washizu, M., Nanba, T., and Masuda, S., Novel method of cell fusion in field constriction area in fluid integrated circuit, *IEEE Trans. Ind. Appl.*, 25, 732, 1989.
16. Washizu, M., Nanba, T., and Masuda, S., Handling biological cells using a fluid integrated circuit, *IEEE Trans. Ind. Appl.*, 26, 352, 1990.
17. Arai, F., Ichikawa, A., Ogawa, M., Fukuda, T., Horio, K., and Itoigawa, K., High-speed separation systems of randomly suspended single living cells by laser trap and dielectrophoresis, *Electrophoresis*, 22, 283, 2001.
18. Suehiro, J. and Pethig, R., The dielectrophoretic movement and positioning of a biological cell using a three-dimensional grid electrode system, *J. Phys. D: Appl. Phys.*, 31, 3298, 1998.
19. Selvakumaran, J., Hughes, M.P., Keddie, J.L., and Ewins, D.J., Assessing bio-compatibility of materials for implantable microelectrodes using cytotoxicity and protein adsorption studies, paper presented at the Proceedings of the 2nd IEEE – EMBS Conference on Microtechnologies in Medicine, Madison, 2002.
20. Fuhr, G., Fiedler, S., Muller, T., Schnelle, T., Glasser, H., Lisec, T., and Wagner, B., Particle micromanipulator consisting of two orthogonal channels with traveling-wave electrode structures, *Sensor. Actuators A*, 41, 230, 1994.
21. De Gasperis, G., Wang, X.B., Yang, J., Becker, F.F., and Gascoyne, P.R., Auto-mated electrorotation: dielectric characterization of living cells by real-time motion estimation, *Meas. Sci. Technol.*, 9, 518, 1998.
22. Zhou, X.F., Burt, J.P.H., and Pethig, R., Automatic cell electrorotation mea-surements: studies of the biological effects of low-frequency magnetic fields and of heat shock, *Phys. Med. Biol.*, 43, 1075, 1998.

23. Gascoyne, P.R.C., Huang, Y., Pethig, R., Vykoukal, J., and Becker, F.F., Dielectrophoretic separation of mammalian cells studied by computerized image analysis, *Meas. Sci. Technol.*, 3, 439, 1992.

24. Talary, M.S. and Pethig, R., Optical technique for measuring the positive and negative dielectrophoretic behavior of cells and colloidal suspensions, *IEEE Proc. Sci. Meas. Tech.*, 141, 395, 1994.

25. Cui, L., Zhang, T., and Morgan, H., Optical particle detection integrated in a dielectrophoretic lab-on-a-chip, *J. Micromech. Microeng.*, 12, 7, 2002.

26. Berry, R.M. and Berg, H.C., Absence of a barrier to backwards rotation of the bacterial flagellar motor demonstrated with optical tweezers, *Proc. Natl. Acad. Sci. U.S.A.*, 94, 14433, 1997.

27. Brown, A.P., Betts, W.B., Harrison, A.B., and O'Neill, J.G., Evaluation of a dielectrophoretic bacterial counting technique, *Biosensor. Bioelectron.*, 14, 341, 1999.

28. Milner, K.R., Brown, A.P., Allsopp, D.W.E., and Betts, W.B., Dielectrophoretic classification of bacteria using differential impedance measurements, *Electron. Lett.*, 34, 66, 1997.

29. Cui, L., Holmes, D., and Morgan, H., The dielectrophoretic levitation and separation of latex beads in microchips, *Electrophoresis*, 22, 3893, 2001.

30. Fiedler, S., Shirley, S.G., Schnelle, T., and Fuhr, G., Dielectrophoretic sorting of particles and cells in a microsystem, *Anal. Chem.*, 70, 1909, 1998.

31. Müller, T., Gradl, G., Howitz, S., Shirley, S., Schnelle, T., and Fuhr, G., A three-dimensional microelectrode system for handling and caging single cells and particles, *Biosens. Bioelectron.*, 14, 247, 1999.

32. Cheng, J., Sheldon, E.L., Wu, L., Uribe, A., Gerrue, L.O., Carrino, J., Heller, M.J., and O'Connell, J.P., Preparation and hybridization analysis of DNA/RNA from E-coli on microfabricated bioelectronic chips, *Nat. Biotechnol.*, 16, 541, 1998.

33. Cheng, J., Sheldon, E.L., Wu, L., Heller, M.J., and O'Connell, J.P., Isolation of cultured cervical carcinoma cells mixed with peripheral blood cells on a bioelectronic chip, *Anal. Chem.*, 70, 2321, 1998.

34. Huang, Y., Ewalt, K.L., Tirado, M., Haigis, R., Forster, A., Ackley, D., Heller, M.J., O'Connell, J.P., and Krihak, M., Electric manipulation of bioparticles and macromolecules on microfabricated electrodes, *Anal. Chem.*, 73, 1549, 2001.

35. Docoslis, A., Kalogerakis, N., Behie, L.A., and Kaler, K.V.I.S., A novel dielectrophoresis-based device for the selective retention of viable cells in cell culture media, *Biotechnol. Bioeng.*, 54, 239, 1997.

36. Huang, Y., Yang, Y., Wang, X.B., Becker, F.F., and Gascoyne, P.R.C., The removal of human breast cancer cells from hematopoietic CD34+ stem cells by dielectrophoretic field-flow fractionation, *J. Hemat. Stem Cell Res.*, 8, 481, 1999.

37. Pethig, R., Dielectrophoresis: using inhomogeneous AC electrical fields to separate and manipulate cells, *Crit. Rev. Biotechnol.*, 16, 331, 1996.

38. Wang, X.-B., Huang, Y., Wang, X., Becker, F.F., and Gascoyne, P.R.C., Dielectrophoretic manipulation of cells with spiral electrodes, *Biophys. J.*, 72, 1887, 1997.

39. Batchelder, J.S., Method and Apparatus for Dielectrophoretic Manipulation of Chemical Species, U.S. Patent 4,390,403, 1983.

40. Batchelder, J.S., Dielectrophoretic manipulator, *Rev. Sci. Instrum.*, 54, 300, 1983.

41. Pethig, R. and Burt, J.P.H., Manipulation of Solid, Semi-Solid or Liquid Materials, World Patent 9,111,262, 1991.

42. Betts, W.B. and Hawkes, J.J., Dielectrophoretic Characterization of Microorganisms and Other Particles, World Patent 91,082,484, 1991.

43. Pethig, R. and Markx, G.H., Apparatus for Separating by Dielectrophoresis, World Patent 92,422,583, 1994.

44. Betts, W.B. and Hawkes, J.J., Apparatus for Separating a Mixture, World Patent 9,320,927, 1996.

45. Huang, Y., Wang, X.B., Becker, F.F., and Gascoyne, P.R.C., Fractionation Using Dielectrophoresis and Field Flow Fractionation, World Patent 9,727,933, 1997.

46. Huang, Y., Wang, X.B., Becker, F.F., and Gascoyne, P.R.C., Method and Apparatus for Fractionation Using Spiral Electrodes, U.S. Patent 5,858,192, 1999.

47. Crane, S., Dielectrophoretic Cell Stream Sorter, U.S. Patent 5,489,506, 1996.

48. Ager, C.D., Dames, A.N., Purvis, D.R., and Safford, N.A., Methods of Analysis/Separation, World Patent 9,810,869, 2001.

49. Lock, G.M. and Pethig, R., Travelling Wave Dielectrophoretic Apparatus and Method, World Patent 0,105,514, 2001.

50. Pethig, R. and Markx, G.H., Apparatus and Method for Testing Using Dielectrophoresis, World Patent 9,804,355, 1998.

51. Kaler, K.V.I.S., Behie, L.A., Kalogerakis, N., and Docoslis, A., Filter for Perfusion Cultures of Animal Cells and the Like, U.S. Patent 5,626,734, 1997.

52. Hagedorn, R., Fuhr, G., Müller, T., and Schnelle, T., Electrode Arrangement for the Dielectrophoretic Diversion of Particles, World Patent 0,000,292, 2000.

53. Hagedorn, R., Fuhr, G., Müller, T., and Schnelle, T., Electrode Arrangement for Generating Functional Field Barriers in Microsystems, World Patent 0,000,293, 2000.

54. Lock, G.M. and Pethig, R., Electrodes for Generating and Analysing Dielectrophoresis, World Patent 0,105,511, 2001.

55. Lock, G.M. and Pethig, R., Dielectrophoretic Apparatus and Method, World Patent 0,105,512, 2001.

56. Cheng, J., Wang, X.B., Wu, L., Xu, J., and Yang, W., Apparatus for Switching and Manipulating Particles and Method of Use Thereof, World Patent 0,227,909, 2002.

57. Arnold, W.M., Method and Apparatus for Concentrating and/or Positioning Particles or Cells, World Patent 9,962,622, 1999.

58. O'Connell, J.P., Cheng, J., Wu, L., and Sheldon, E.L., Channel-less Separation of Bioparticles on a Bioelectronic Chip by Dielectrophoresis, World Patent 9,938,612, 1999.

59. O'Connell, J.P., Cheng, J., Diver, J., Heller, M., Jalali, S., Sheldon, E., Smolko, D., Wu, L., and Willoughby, D., Integrated Portable Biological Detection System, World Patent 0,037,163, 2000.

60. Medoro, G., Method and Apparatus for the Manipulation of Particles by Means of Dielectrophoresis, World Patent 0,069,565, 2000.

61. Velev, O.D. and Kaler, E.W., Miniaturized Immunosensor Assembled from Colloidal Particles between Micropatterned Electrodes, U.S. Patent 6,333,200, 2001.

62. Gurtner, C., Edman, C.F., Formosa, R., and Heller, M.J., Methods and Apparatus for the Electronic, Homogeneous Assembly and Fabrication of Devices, World Patent 0,134,765, 2001.

63. Kawabata, T. and Washizu, M., Method for Separating Substances Using Dielectrophoretic Forces, European Patent 1,088,592, 2001.

64. Miles, R.R., Balch, J.W., Wang, X.B., Gascoyne, P.R.C., Krulevitch, P.A., and Mariella, R.P., Microfluidic DNA Sample Preparation Method and Device, U.S. Patent 6,352,838, 2002.
65. Wilson, S.A. and Whatmore, R.W., Electric-Field Structuring of Composite Materials, World Patent 0,176,852, 2001.

Supplementary reading

Chang, C.Y. and Sze, S.M., *ULSI Technology*, McGraw Hill, New York, 1996.
Gad-el-Hak, M. *The MEMS Handbook*, CRC Press, Boca Raton, FL, 2001.
Gardner, J.W., Varadan, V.K., and Awadelkarim, O.O., *Microsystems, MEMS and Smart Devices*, Wiley, Chichester, 2001.
Koch, M., Evans, A., and Brunnschweiler, A., *Microfluidic Technology and Applications*, Research Studies Press, Baldock, U.K., 2000.

chapter ten

Computer applications in electromechanics

10.1 The need for simulation

Throughout this book, we have considered the forces generated on particles in nonuniform fields. We have established also that the magnitude of that force is dependent on the gradient of the electric field squared and on the gradient of the phase change where we are using traveling wave dielectrophoresis. However, if we are to understand the electric field for a specific electrode geometry or to calculate the magnitude of forces applied to particles within a nonuniform field, then it is necessary to determine the actual values of the electric field, in the volumes occupied by the particles, throughout the time they occupy them.

This can be achieved in a number of ways. For the simplest geometries, it is possible to use basic principles found in Chapter 2 to describe the electric field at all points between the electrodes. However, this is not possible for more complex shapes — in order to have any understanding of the field geometry, simplifying approximations must be made. However, by using those approximations we can predict the shape of the electric field (that is, the electric field *morphology*) of almost any electrode configuration, and hence calculate the dielectrophoretic force induced in particles around those electrodes. Simple postprocessing of the electric field morphology allows us to determine, for example, the electrode regions where we are most likely to observe positive or negative dielectrophoresis.

10.2 Principles of electric field simulation

We can divide the approaches to the determination of the electric field in a required volume (which we shall call the *solution space*) into two different categories. The first category contains those methods that use a series of equations to calculate the exact value of the electric field at all points within the solution space; this category is known as *analytical modeling*. The second category contains methods in which the solution space is not considered to be continuous but is divided either into a series of points or partitioned into adjoining elements so that simplifying approximations can be used to determine the electric field at those points, or across those partitions. This approach is known as *numerical modeling*.

Regardless of the approach used, the analysis of electric fields generally begins with a defining rule, such as Poisson's equation:

$$\nabla E = -\frac{\rho}{\varepsilon} \qquad (10.1)$$

where ρ is the charge density, ε is the permittivity, E is the electric field, and ∇ is the del vector operator. There are a limited number of electrode geometries where the potential distribution (and hence the electric field) may be determined directly from algebraic equations relating to Poisson's equation (e.g., Pohl[1] and Huang and Pethig[2]). However, there are many more configurations that are not calculable directly from algebra to provide an exact form of the electric field, and the solution set can only be determined by generating numerical answers at specific points based on an approximate mathematical model. These methods employ entirely different approaches to the analysis and design, and we will now examine them separately.

10.3   Analytical methods

10.3.1   Electrode geometries with analytical solutions to their electric fields

Some of the longest established electrode designs for dielectrophoretic applications are those whose electric field morphologies can be calculated easily using analytical methods. Many of these take advantage of the principal simplification we can make in analytical modeling, that of *dimensionality*. If we have electrode structures that have a nonuniform cross section, but that are infinitely long, then we can treat the problem as if it were two-dimensional rather than three-dimensional, which greatly simplifies the problem at hand. This still works if our electrode structures are not infinitely long, but considerably longer than the dimensions of the features being simulated. However, in that case, we must be aware of the fact that we are approximating an answer, rather than determining the real thing.

Early electrodes used field geometries that could easily be derived from Gauss's law, such as the example shown in Figure 10.1. For the case of a single conducting wire suspended coaxially along a conducting tube, the electric field as a function of the distance r_E from the common axis is given by

$$\nabla E^2 = \frac{2V^2 r}{r_E^3 \left(\log_e \left(\dfrac{r_1}{r_2} \right) \right)^2} \qquad (10.2)$$

Figure 10.1 A schematic of the coaxial electrode geometry, for which there exists an analytical expression for the field gradient squared.

where $\mathbf{r}$ represents the unit vector pointing toward the common axis, r_1 and r_2 represent the radii of the inner and outer tubes surfaces, respectively, and V denotes the potential difference between the inner and outer electrodes. However, while such electrode geometries are useful when analyzing the dielectrophoretic responses of cells, constructing such geometries on the scales required for nanometer-scale particles is not feasible.

One problem with the coaxial geometry is that the electric field changes at a rapid rate with respect to position. This means that the force at distances away from the central electrode is so small that complete particle separations cannot easily be achieved. In order to overcome this problem, a geometry was devised[1] in which the gradient of the electric field varied in a more modest manner over a much greater cross-sectional area. This *isomotive* geometry was devised to provide a force that was equal at all points within the cross-sectional area along the main axis of the electrode system. The isomotive electrode array is shown in Figure 10.2. Two curved electrodes carry opposing potentials, while a third, triangular conductor is grounded. By considering the dielectrophoretic force equation and Laplace's equation, Pohl showed[1] that the dielectrophoretic force increases linearly along the axis (shown by the arrow in the figure) where the electrodes are defined in cylindrical coordinates with an origin at the tip of the triangular electrode. The curved lines of the electrodes are defined by the equation for the equipotentials they define:

$$V = \frac{2}{3}\left(\frac{2F}{v\alpha}\right)^{\frac{1}{2}} r^{\frac{3}{2}} \sin\left(\frac{3\theta}{2}\right) \tag{10.3}$$

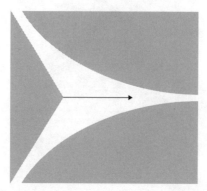

Figure 10.2 The *isomotive* electrode geometry. When AC potentials are applied between the two curved electrodes, the electric field gradient lies along the arrowed line. The triangular electrode at left is grounded, since it occupies the area where the equation defining the isomotive geometry does not apply.

where V is the potential, F is the magnitude of the constant force, v is the particle volume, α is the polarizability, and r and θ are the distance along the axis and the angle from it, respectively.

This electrode geometry proved very effective for the manipulation of particles by positive and negative dielectrophoresis, for example, for cell separation.[1] However, the repelled particles do not collect in well-defined regions, so that separation of particles into two homogeneous groups is difficult. One geometry that was designed to overcome this limitation is known as the *polynomial* electrode geometry. This electrode design forms the basis of the quadrupolar electrodes described in preceding chapters, although the principle underlying the design can be adapted for numbers of electrodes other than four. The design was originally developed by Huang and Pethig[2] and Wang et al.[3] for negative dielectrophoresis applications and was the first design presented to combine both analytical field distribution with defined regions of negative dielectrophoresis, generating electrode geometries that (like the isomotive geometry) can be described as uniformly nonuniform; that is, the electric field is defined by a polynomial expression, and that expression is used to define the electrode geometry. In that work, it was found that such electrodes can be defined by lines of a form such that, for 2N electrodes, the lines delineating the electrode edges in two dimensions must meet a polynomial expression of order N:

$$f_N(x,y) = af_{na} + bf_{nb} \qquad (10.4)$$

where a and b are constants and f_{na} and f_{nb} are independent functions that vary according to the order of the polynomial being used to construct the electrode geometry.[2] Taking the most common electrode geometry of $N = 2$ (four electrodes), the time-averaged force magnitude at a point x, y in the interelectrode gap (indicated in Figure 10.3) is given by the expression:

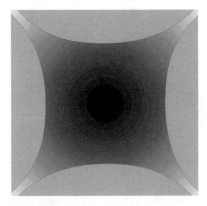

Figure 10.3 The *polynomial* electrode geometry. The electrodes are indicated by the gray area and extend away from the area illustrated with ever-decreasing interelectrode gaps. The shaded region indicates the magnitude of the electric field gradient squared, with a minimum value at the center of the array and an increasing value in concentric circles emanating from that point.

$$|\mathbf{F}| = -2\pi\varepsilon_m r^3 \,\mathrm{Re}[K(\omega)]\frac{V^2}{d^4}\left(x^2 + y^2\right) \tag{10.5}$$

where V is the root mean square (rms) value of the applied potential applied to the electrodes and d is the distance from the center of the electrode array to the electrode tips. As can be seen in Figure 10.3, which shows the variation in $|\mathbf{F}|$ across the interelectrode gap in accordance with Equation 10.5, the increments in force (indicated as the gradient from black to white) show a steady rise from zero at the center of the electrode array to a much higher value at the interelectrode gaps. Moreover, there are distinct regions of high and low force, where particles can collect by positive and negative dielectrophoresis, respectively. This geometry has since been widely adopted — for example, it forms the basis of the quadrupolar electrodes used in most of the work in Chapters 4 and 5. Note however that those electrodes are *approximations* to the one described above, since they are planar in nature rather than being cross sections of an elongated (approximately infinite) structure.

Electrode arrays with electrode geometries such than $N \neq 2$ have not been widely explored; the variation in ∇E^2 across an array was found by Huang and Pethig to vary as a function or distance from the center of the array r such that

$$\nabla E^2 \propto r^{(2N-3)} \tag{10.6}$$

for cases where $N > 1$. The simplest case of $N = 1$ (two parallel electrodes) would generate no dielectrophoretic force (though in planar form, it is used for the field-flow work described in Chapter 3). Orders of N greater than 2 result in field gradients with much steeper sides, resulting in broad regions

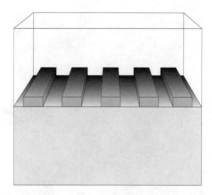

Figure 10.4 A section of interdigitated traveling-wave electrodes. Since the electrodes are parallel for a sufficient length for them to be treated (at a distance from the edges) as infinitely long, then a vertical planar section through the electrodes and surrounding area allows a two-dimensional model to be used and applied along the length of the three-dimensional electrode array.

of low field and rapid increases in force very close to the electrode edges. This is not preferable, and in many ways causes the problems associated with coaxial wire electrodes that the isomotive geometry was developed to avoid.

10.3.2   Modeling time-dependent behavior using analytical methods

It was assumed, at the time of the original work on polynomial electrode geometries, that the electric field rotated in the same manner at all points across the interelectrode volume, which was determined not to be the case by numerical modeling, (as described in Section 10.8). However, phase effects in traveling-wave electrodes were known to have an effect on particles suspended above them, and a number of researchers have attempted to describe these effects analytically.

In order to simplify the problem of determining phase effects, approaches to modeling traveling-wave electrodes have simplified the problem to two dimensions; that is, they consider the traveling electric field to be generated by infinitely long electrodes, across which an infinitesimally thin slice is taken. However, this plane is usually in what we might consider the vertical direction of the electrodes we are simulating, such as the vertical plane intersecting the electrode array in Figure 10.4. Furthermore, since the electrodes are deposited by thin-film methods and are rarely more than 200 nm, but the width of the electrodes is typically 10 µm, simulations often consider the electrodes to be infinitely flat.

The first approach to analytical modeling of traveling-wave effects was developed by Masuda et al.[4] in 1987, and it has recently been reexamined by Morgan et al.[5] This approach is developed from the idea of Fourier series — that is, the idea that any repeating wave form can be reconstructed from the superimposition of an infinite number of sinusoids of different frequencies

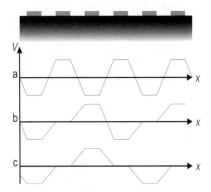

Figure 10.5 A two-dimensional model of interdigitated electrodes, showing the potential along a line through the electrodes. (a) Two phase for dielectrophoresis, (b) three-phase, and (c) four-phase traveling-wave dielectrophoresis. The model can be further simplified by setting the electrode thickness to zero, such that the potential is applied along the interface line between substrate and medium.

and magnitudes. This series of sinusoids is known as a Fourier series. This can be applied to the potential along the plane of the electrodes — which appears as a line in the cross section of our two-dimensional simulation — by considering the potential as being uniform across the electrodes themselves and linearly varying in between them. In their studies, Masuda et al.[4] considered electrodes with three phases applied in sequence (0°, 120°, 240°), whereas Morgan et al.[5] considered a simple dielectrophoretic case (alternating phases of 0° and 180°) and four-phase traveling fields (0°, 90°, 180°, 270°). Applying these different divisions of phase to the electrodes, we expect to see potential profiles along the line through the electrodes similar to those shown in Figure 10.5. These square-topped potentials V along the x-axis can be produced from an infinite Fourier series of the form[5]

$$V(x) = \sum_{i=1}^{\infty} A_i \cos\left(\frac{2i\pi x}{\lambda}\right)$$
(10.7)

where i is the number of the harmonic of the Fourier series, A_i is the magnitude of the ith harmonic, and λ is the wavelength of the potential wave along the electrode array. By analyzing the problem numerically, Morgan and coworkers derived expressions for the values of A_i, and hence the value of the electric field, from which the electric field across the entire two-dimensional plane could be derived.

Another approach was adopted by Wang and colleagues in Houston.[6] This approach employed similar simplifying assumptions to the Fourier model described above, but instead defined the problem according to Green's theorem, which states that within a volume v, two scalar functions ϕ and ψ sharing a boundary surface A can be described by the expression

$$\int_v dv(\phi\nabla^2\psi - \psi\nabla^2\phi) = \oint_A dA\left(\phi\frac{\partial\psi}{\partial n} - \psi\frac{\partial\phi}{\partial n}\right) \tag{10.8}$$

where $\partial/\partial n$ indicates differentiation in a direction normal to the area A. In order to perform this function, the variables ϕ and ψ were chosen to be the potential due to Laplace's equation and the potential due to a point charge at r_0; thus,

$$\nabla^2\phi = 0$$
$$\nabla^2\psi = -4\pi q\delta^{(3)}(r - r_0) \tag{10.9}$$

where $\delta^{(3)}(r - r_0)$ indicates the delta function in three-dimensional space. By solving these equations via Equation 10.8, it is possible to derive expressions for ∇E^2 across the plane intersecting the electrodes. Again, in order to determine the potential due to the electrodes, an assumption must be made about the potential distribution, which is again approximated to a series of sinusoids. However, in this case the potential is expanded into a Taylor series, with approximations made for symmetry and only a few additional terms being used.

10.4   Numerical methods

While analytical methods allow the exact calculation of the electric field gradient at any point in space due to the electrodes used, it is only useful when considering a limited number of simple electrode geometries, generally in two dimensions. However, the majority of electrode configurations do not conform to these simplified models (with the principal exception being the interdigitated arrays described in Section 10.3.2). Even where the electrode structures resemble those derived by analytical means, such as polynomial electrodes, they are usually constructed using thin-film technology and are therefore far from being represented by infinitely long (two-dimensional) models. In fact, the electric field morphology around them is a complex function that varies in magnitude (and often phase) in all three dimensions. These geometries are so complex that it is not feasible (and in some cases, not possible) to derive analytical expressions for the field. In such circumstances, it is necessary for us to employ approximation methods that simplify the problem, allowing us to gain an approximately accurate answer for an arbitrary arrangement of electrodes. We term this *numerical* analysis because we are reducing the problem to a specific set of input numbers at our boundary conditions and calculating the electric field accordingly, whereas with analytical modeling, we use symbolic mathematical representations that hold for any boundary conditions we wish to apply.

The topic of numerical analysis covers a wide variety of mathematical models, some of which are more appropriate to specific problems. In this section we will examine a number of models that have been used for electric field derivation. It is the nature of all numerical models that the solution provided is an approximation to the actual field, rather than an exact solution. By introducing simplifying conditions, such as discretization of otherwise continuous functions, errors are introduced into the solution. The nature of these errors depends on the approximations made; for example, some methods introduce errors near to the electrode surface but are accurate at greater distances; other methods are accurate up to the electrodes but are less accurate farther away. Furthermore, these methods differ widely in the computer resources required to determine them and in the length of time required to process a solution. An understanding is required of the approximations made in a model, the limitations placed on the construction of that model, and the restrictions on determining whether the solution is useful in deciding which method is appropriate for a given problem.

10.4.1 The finite difference method

Originally derived from the work of Gauss,[7] finite difference models are calculated using a regular mesh superimposed across the solution space. At each intersection (*node*) of the mesh, Poisson's equation is approximated to a difference equation relating the potential at the node to the potentials at all the immediately connected nodes, as shown for a two-dimensional example in Figure 10.6. The boundary conditions of the model are the known potentials of electrode surfaces, as indicated in our example for two castellated electrodes. The potential at each node is affected by the estimated potential on those nodes adjacent to it. The boundary conditions (the circumstances imposed on the model that define our particular problem, such as the voltages on the electrodes) are set by those nodes having known values — that is, those nodes of known potential such as those lying on the electrode surface.

Originally, solutions were performed by hand using relaxation methods, where the residual of the sum of potentials acting on a given node is minimized from a series of estimated starting values. However, the advent of high-speed computing has brought about the replacement of this method by an iterative approach, where the solution is reached by repeatedly calculating the unknown potentials as a series of simultaneous equations until the answers converge. Weightings may be introduced into the difference equation to model the effects of permittivity.

The finite difference method was the primary means of numerical field analysis from the 1930s to the 1960s, when it was largely superseded by finite element analysis. However it is still used in contemporary studies. For example, the recent AC resistor-network model of potential calculation postulated by Hölzel[8] shares many of the principles of finite difference analysis; another application was the calculation of potentials in ratchet electrodes by

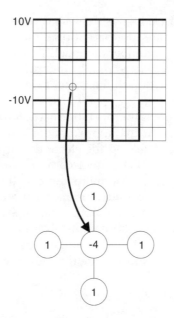

Figure 10.6 In finite-difference simulations, the solution space is divided up into a regular mesh, and the value of the potential at each node is determined by taking the difference between a first-estimate value and its neighbors. The values are determined across the solution space including the boundary conditions (where potentials are imposed, i.e., at electrode edges), and iteratively determined until a satisfactory answer is reached.

Rousselet et al.[9] The advantage of finite difference models is that they are simple to construct using programs or mathematical software packages, and a number of books (e.g., Binns and Laurenson[10]) have been written on the subject.

10.4.2   *The finite element method*

Finite difference models eventually became unfavorable because the models required the application of rigid, regular meshes across the solution space. An alternative method, known as the finite element method, was soon adopted. This method was originally developed for solving mechanical stresses in civil engineering (e.g., Nath[11]) but was later adopted for calculating electric fields (see Zienkiewicz and Taylor[12]). As in the finite difference method, the solution space is divided up into a series of nodes that define the corners of enclosed elements. However, unlike the finite difference method, these elements need only loosely follow a general shape such as a cuboid or tetrahedron. As long as the element has the appropriate number of faces and corners, it does not need to be rigidly shaped. The element is transformed locally onto a rigidly defined master element, and Poisson's equation is approximated linearly across that element. The results are mapped onto a system matrix and solved as a series of simultaneous equations to determine the overall electric field. Boundary conditions are mapped as defined potentials at the appropriate nodes. Since the electric field is approximated to be linear across each element, most finite element models use automated meshing techniques to ensure that the elements are small in regions where the electric field is changing the most, such as at electrode edges. However, this means solutions produced by this method have pseudo-randomly distributed nodes, which can sometimes make direct comparisons of two field patterns for the same electrode array difficult.

This method is now widely used for electric field calculations,[13] mechanical stress,[11] and fluid flow,[14] as well as for electrical, magnetic, and optical problems. Many commercial software suites containing finite element solvers, design packages for defining the system to be modeled, and postprocessing programs for data analysis are available. Examples of the use of finite element software for dielectrophoretic analysis include the determination of ∇E^2 for polynomial and sawtooth geometries,[15,16] interdigitated castellated electrodes,[17] electrodes for field-flow fractionation,[18] and for electrodeless systems for dielectrophoretic attraction by charge deposition[19] as discussed in Chapter 7. It has also been used in the study of electrohydrodynamic fluid flow, as described in Chapter 3.

10.4.3 Boundary element methods

The boundary element method[13] is effectively an extension to the finite element method. The surfaces of the electrodes are discretized into elements, and the potentials at enclosing boundaries surrounding the electrodes are coupled to these elements, where the potential is calculated. This method is accurate but highly computationally intensive and is best applied to functions where far-field considerations are important or where a core element is surrounded by many layers of material with different dielectric properties.[20] This method has also been combined with finite element modeling in some commercial software (e.g., SI Eminence™, Ansoft™ Inc.) where a finite element model is used to determine the solution space and a boundary element model is then used, taking the previous model as its core element, to determine far-field effects. It is also used in determining properties of an enclosed volume from data taken at the outer boundary, such as determining the properties of the human body by examining the electric signals transmitted to electrodes on one part of the skin due to an input potential applied at another part.[21]

10.4.4 The Monte Carlo method

As with the finite difference model, the Monte Carlo method[10,22,23] involves the superimposition of a mesh across the solution space, with a series of difference equations relating to the potential at the nodes. However, unlike the finite difference model, coefficients relating to the difference equations are interpreted as probabilities of a fictitious particle moving from one node to its neighbor. By evaluating the random walk of the particle from a given node to a known boundary, it is possible to determine the most probable value of the potential at the original node. Permittivity, charge density, and other factors may be included in the probability equations.

The method produces a slowly converging result, as the random nature of testing requires extensive recursion in order to "settle" with any degree of confidence. It is common for calculation times to be greater than those for methods such as finite difference by factors of 20,[10] and, as such, it is generally not the method of choice for the majority of field calculations. However, there are applications for which the Monte Carlo method is well suited, those applications being related to the random nature of the analysis. There are many examples in this book that must consider the random motion of particles in electric fields, particularly due to the action of Brownian motion that itself introduces a random walk factor to the collection of particles. For example, the Monte Carlo method has been employed in studies of particle motion in ratchet electrodes,[24] or the diffusion of particles held within dielectrophoretic traps.[25,26] Note however, that the calculation of the electric field may be calculated by some other method and then applied as a boundary condition to the analysis of the motion of the particle.

10.4.5 The method of moments

This method differs from the others presented here in that while it imposes a mesh across the charge-bearing electrodes, it does not impose a grid across the sample space. Referred to as the charge density or moment method, it differs from the other methods represented here in that the electric field is calculated based on Coulomb's law:

$$E = \sum_i \frac{Q_i}{4\pi\varepsilon r_i^2} r_i \qquad (10.10)$$

where E is the electric field at a point due to i charges of magnitude Q_i a distance r_i away along unit vector r_i. This principle was used by Maxwell to calculate the charge across a square area by dividing it into smaller regions across which the charge was approximately uniform.[27] If the surfaces of the electrodes are divided into sufficiently small subareas, the charge across these subareas can be assumed to be uniformly distributed. The charge on each subarea can be calculated by determining the contribution a unit charge on a given subarea makes to the potential at all other subareas. By solving against the known potentials on the electrodes, the charge distribution may be derived. A similar process is subsequently used to calculate the contributions of the charges on the electrodes to the potential at any arbitrary point.

The method of moments is generally applied to electric field problems where charge distribution is required, such as capacitance calculations,[28] since the process is used to calculate the distribution of charge accumulated on an arbitrary shape. The method also has been widely used in dielectrophoresis research, including studies of dielectrophoretic forces on castellated electrodes,[29,30] quadrupole electrodes,[31–34] octopole electrode field cages,[35,36] and traveling wave dielectrophoresis in both interdigitated[6,37–39] and polynomial[40] electrode geometries. It is simpler to implement than most other methods described here but is very computationally intensive.

10.5 Finite element analysis

As described in the previous section, most models introduced a numerical approximation by discretizing the solution space into a series of evaluated nodes. The finite element method is similar, but it divides the solution space between the nodes into elements across which Poisson's equation is approximated. The element approach has many advantages over the imposition of meshes, as described in the previous section. Here, the procedures that underlie construction of a finite element model are described.

10.5.1 Local elements and the shape function

Consider the electrode geometry shown in Figure 10.7. We may partition it into a number of elements, such as the rectangular elements shown in Figure 10.7; the nodes of the elements are the points where we wish to

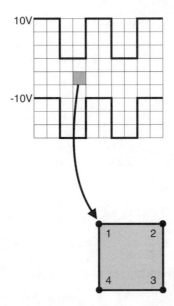

Figure 10.7 In finite-element simulations, a mesh (which may or may not be regular) divides the solution space into elements; the potential across each element is determined, and all the elements are then recombined to determine the potential across the whole array.

evaluate the potential. The elements may follow any mathematically defin-
able shape, and here we will consider both triangular and square elements.
Taking triangular elements first and examining a single element (the local
element, such as the shaded area in Figure 10.7b), it is possible to approximate Poisson's equation (Equation 10.1) linearly across it using an equation
in potential variable u:

$$u \approx U = \alpha_1 + \alpha_2 x + \alpha_3 y \tag{10.11}$$

As we would wish to examine the potential at all three nodes, we may apply
this equation to all three nodes (locally numbered 1 to 3);

$$U_1 = \alpha_1 + \alpha_2 x_1 + \alpha_3 y_1$$
$$U_2 = \alpha_1 + \alpha_2 x_2 + \alpha_3 y_2 \tag{10.12}$$
$$U_3 = \alpha_1 + \alpha_2 x_3 + \alpha_3 y_3$$

This may be written in matrix form, i.e.,

$$\begin{pmatrix} U_1 \\ U_2 \\ U_3 \end{pmatrix} = \begin{pmatrix} 1 & x_1 & y_1 \\ 1 & x_2 & y_2 \\ 1 & x_3 & y_3 \end{pmatrix} \begin{pmatrix} \alpha_1 \\ \alpha_2 \\ \alpha_3 \end{pmatrix} \tag{10.13}$$

Solving Equation 10.13 for α we find

$$\alpha_1 = \frac{1}{2A} \begin{vmatrix} U_1 & x_1 & y_1 \\ U_2 & x_2 & y_2 \\ U_3 & x_3 & y_3 \end{vmatrix}$$

$$\alpha_2 = \frac{1}{2A} \begin{vmatrix} 1 & U_1 & y_1 \\ 1 & U_2 & y_2 \\ 1 & U_3 & y_3 \end{vmatrix} \tag{10.14}$$

$$\alpha_3 = \frac{1}{2A} \begin{vmatrix} 1 & x_1 & U_1 \\ 1 & x_2 & U_2 \\ 1 & x_3 & U_3 \end{vmatrix}$$

where A is the area of the triangle, and

$$2A = \begin{vmatrix} 1 & x_1 & y_1 \\ 1 & x_2 & y_2 \\ 1 & x_3 & y_3 \end{vmatrix} \tag{10.15}$$

If we define variables a, b, c such that for node 1

$$a_1 = x_2 y_3 - y_2 x_3$$
$$b_1 = y_2 - y_3 \qquad (10.16)$$
$$c_1 = x_3 - x_2$$

and similarly for nodes 2 and 3, then Equations 10.13 and 10.15 may be expressed as

$$\alpha_1 = \tfrac{1}{2A}\left(a_1 U_1 + a_2 U_2 + a_3 U_3\right)$$
$$\alpha_2 = \tfrac{1}{2A}\left(b_1 U_1 + b_2 U_2 + b_3 U_3\right)$$
$$\alpha_3 = \tfrac{1}{2A}\left(c_1 U_1 + c_2 U_2 + c_3 U_3\right) \qquad (10.17)$$
$$2A = a_i b_i x_i + c_i y_i$$

This may be generalized for node i using Equation 10.11,

$$UU = \tfrac{1}{2A}\left[(a_1 + b_1 x + c_1 y)U_1 + (a_2 + b_2 x + c_2 y)U_{21} + (a_3 + b_3 x + c_3 y)U_3\right]$$
$$= \sum N_i U_i \qquad (10.18)$$

where U_i is the potential at node i. N_i is the *shape function*, and, for triangular elements, it is written as

$$N_i = \frac{a_i + b_i x_i + c_i}{2A} \qquad (10.19)$$

which is a transformation into a coordinate system known as area coordinates,[13] effectively normalizing the dimensions of the element. This coordinate transformation is an important concept and forms the basis for the examination of square elements, as discussed later.

10.5.2 The Galerkin method

Let us suppose that across all the elements i, the potential u is such that

$$u = \sum_i N_i a_i \qquad (10.20)$$

where a_i is a set of parameters that will need to be determined. It is possible to insert this into Equation 10.1, which may be expressed in the form,

$$\nabla .k\nabla u + Q = 0 \qquad (10.21)$$

Substituting Equation 10.20 into Equation 10.21, we obtain the expression

$$R = \left(\nabla .k\nabla \sum_i N_i a_i + Q \right) = 0 \qquad (10.22)$$

where $R = 0$ for an exact solution of Equation 10.20 and is otherwise an indication of the error introduced due to the linear approximation used in Section 10.5.1. It is possible to force R to equal zero over the solution space Ω by satisfying the equation

$$\int_\Omega w_i \mathbf{R} d\Omega = 0 \qquad (10.23)$$

where w is a set of *weighting functions*. In order to determine Poisson's equation over the solution space, we substitute Equation 10.21 (Poisson's equation across a single element) into Equation 10.23. Considering the effect of the potential function across a single element U on a single node j, this result is given by

$$\int_\Omega w_j [\nabla .k\nabla U + Q] d\Omega = 0 \qquad (10.24)$$

If we expand this to consider the effect for all nodes, we find

$$\int_\Omega w_j [\nabla .k\nabla U + Q] d\Omega = \sum_{e=1}^n \int_{\Omega_e} w_j [\nabla .k\nabla U + Q] d\Omega_e = 0 \qquad (10.25)$$

Effectively, the integral across the solution space Ω has been replaced by the sum of the integrals across the elements Ω_i. If we integrate Equation 10.25 by parts, we obtain the following expression:

$$\int_\Omega w_j [\nabla .k\nabla U + Q] d\Omega = -\int_\Omega \nabla w_j k\nabla U d\Omega + \int_\Omega w_j Q d\Omega + \int_\Gamma w_j k \frac{\partial U}{\partial n} \partial \Gamma = 0 \qquad (10.26)$$

where Γ is the surface surrounding the solution space Ω. The weighting functions w may take a wide range of values. In the Galerkin approach, we define the weighting functions as being the shape functions of the elements over which the integration is taking place; hence

$$\mathbf{R}_i^{(e)} = \left[\int \nabla N_i . k \nabla N_j d\Omega_e \right] U_j + \int_\Omega N_i Q d\Omega_e \qquad (10.27)$$

where $\mathbf{R}_i^{(e)}$ is the residue at node i due to element e. This can be expressed in the form

$$\mathbf{R}_i^{(e)} = k_{ij}^{(e)} u_j + f_i^{(e)} \qquad (10.28)$$

where

$$k_{ij}^{(e)} = \int_{\Omega_e} k \left(\frac{\partial N_i}{\partial x} \frac{\partial N_j}{\partial x} + \frac{\partial N_i}{\partial y} \frac{\partial N_j}{\partial y} \right) d\Omega_e$$

$$f_i^{(e)} = \int_{\Omega_e} N_i Q d\Omega_e$$

$$(10.29)$$

These integrals are easily performed over the triangular shape functions described earlier:

$$\int_{\Omega_e} N_1^a N_2^b N_3^c d\Omega_e = 2A \frac{a!\, b!\, c!}{(a+b+c+2)!} \qquad (10.30)$$

where

$$k_{ij}^{(e)} = \int_{\Omega_e} \frac{k\left(b_{ij}b + c_i c_j \right)}{4A^2} d\Omega_e = \frac{k\left(b_{ij}b + c_i c_j \right)}{4A}$$

$$f_i^{(e)} = \int_{\Omega_e} \frac{\left(a_i + b_i x + c_i y \right)}{2A} Q d\Omega_e = \frac{A}{3} Q$$

$$(10.31)$$

These results may be organized into the matrix form of Equation 10.28; the local system matrices for a single element are

$$\begin{bmatrix} R_1 \\ R_2 \\ R_3 \end{bmatrix} = \frac{k}{4A} \begin{bmatrix} b_1^2 + c_1^2 & b_1 b_2 + c_1 c_2 & b_1 b_3 + c_1 c_3 \\ & b_2^2 + c_2^2 & b_2 b_3 + c_2 c_3 \\ sym. & & b_3^2 + c_3^2 \end{bmatrix} \begin{bmatrix} u_1 \\ u_2 \\ u_3 \end{bmatrix} + \frac{A}{3} Q \begin{bmatrix} 1 \\ 1 \\ 1 \end{bmatrix} \qquad (10.32)$$

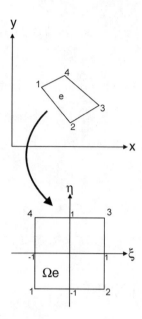

Figure 10.8 A quadrilateral element e is mapped onto a square master element Ωe.

10.5.3 Quadrilateral elements

The principles described above may be applied to elements of other shapes. If we consider quadrilateral elements rather than triangular elements, the linear approximation across the element is now an equation in four, rather than three, unknowns;

$$u \approx U = \alpha_1 + \alpha_2 x + \alpha_3 y + \alpha_4 xy \tag{10.33}$$

This transforms to a shape function N_i where

$$N_i = \tfrac{1}{4}\left(1 + \xi_i \xi_0\right)\left(1 + \eta_i \eta_0\right) \tag{10.34}$$

where ξ_0 and η_0 may take the values ± 1 according to how the shape function has been mapped onto the local element. By considering ξ and η as axes, this mapping process can be seen as a transformation of the element onto a master element, as shown in Figure 10.8. The conformal mapping of one shape to another varies the lengths of the edges by a factor defined by the Jacobian of the transformation;[41]

$$|J| = \begin{vmatrix} \dfrac{\partial x}{\partial \xi} & \dfrac{\partial x}{\partial \eta} \\[2mm] \dfrac{\partial y}{\partial \xi} & \dfrac{\partial y}{\partial \eta} \end{vmatrix} \tag{10.35}$$

from which we obtain the following relationships:

$$\frac{\partial \xi}{\partial x} = \frac{1}{|J|}\frac{\partial y}{\partial \eta} \qquad \frac{\partial \xi}{\partial y} = \frac{-1}{|J|}\frac{\partial x}{\partial \eta}$$

$$\frac{\partial \eta}{\partial x} = \frac{-1}{|J|}\frac{\partial y}{\partial \xi} \qquad \frac{\partial \eta}{\partial y} = \frac{1}{|J|}\frac{\partial x}{\partial \xi} \qquad (10.36)$$

Here we define the master element, to which the element is transformed, as follows:

$$x = \sum x_i U_i(\xi, \eta) \qquad y = \sum y_i U_i(\xi, \eta) \qquad (10.37)$$

We require these expressions to map out the potential U across the master element. This is expressed in terms of the potential function across the master element $U_i^{(e)}$

$$U_i = U_i^{(e)}\big(\xi(x,y), \eta(x,y)\big) \qquad (10.38)$$

from which we find, using the chain rule,

$$\frac{\partial U_i}{\partial x} = \frac{\partial U_i^{(e)}}{\partial \xi}\frac{\partial \xi}{\partial x} + \frac{\partial U_i^{(e)}}{\partial \eta}\frac{\partial \eta}{\partial x} \qquad \frac{\partial U_i}{\partial y} = \frac{\partial U_i^{(e)}}{\partial \xi}\frac{\partial \xi}{\partial y} + \frac{\partial U_i^{(e)}}{\partial \eta}\frac{\partial \eta}{\partial y} \quad (10.39)$$

Substituting Equation 10.37 into Equation 10.39, we obtain the following result:

$$\frac{\partial U_i}{\partial x} = \frac{1}{|J|}\left[\frac{\partial U_i^{(e)}}{\partial \xi}\sum_j y_j \frac{\partial U_j^{(e)}}{\partial \eta} - \frac{\partial U_i^{(e)}}{\partial \eta}\sum_j y_j \frac{\partial U_j^{(e)}}{\partial \xi} \right]$$

$$\frac{\partial U_i}{\partial y} = \frac{1}{|J|}\left[\frac{\partial U_i^{(e)}}{\partial \xi}\sum_j x_j \frac{\partial U_j^{(e)}}{\partial \eta} + \frac{\partial U_i^{(e)}}{\partial \eta}\sum_j x_j \frac{\partial U_j^{(e)}}{\partial \xi} \right]$$

$$(10.40)$$

We now have an expression for the potential U across the original element expressed entirely in terms of the master element dimensions. We wish to determine the potential across the element according to Poisson's equation, so from Equation 10.28 we need to solve for k and f. Since this is a highly complex integral, it is appropriate to approximate a solution using the Gaussian quadrature principle, where the integral of a function g across a square area may

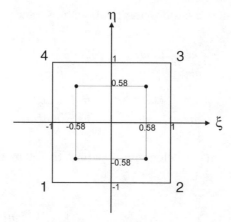

Figure 10.9 The location of the Gaussian quadrature points (at the center of the large dots) within a quadrilateral master element.

be determined by summing the value of the function at four specific points. Across the master element, this is described by Equation 10.41

$$\int g(\varsigma, \eta) d\xi d\eta = \sum_{1}^{n} g(\xi_l \eta_l) w_l + E \tag{10.41}$$

where l is the set of coordinates shown in Figure 10.9, E is the quadrature error (which equals zero if sufficient quadrature points are taken, as is the case here), n is the number of Gaussian quadrature points taken across the element, and w is the set of weighting functions (equal to the shape function of the element, as expressed in Equation 10.34). The values of k_{ij} and f_i are determined using this approximation on the integrals described in Equations 10.29 and are assembled into a local matrix as described in Equation 10.32. This principle may be extended to general element shapes, such as cuboids or other three-dimensional elements, with little difficulty.

10.5.4   Assembling the elements

We have formulated a means of approximating the potential across an element by mapping it to a master element. We now examine the method of assembling the elements in such a way as to determine the overall potential distribution across the whole solution space. This is performed by assembling the local matrices, of the form expressed in Equation 10.32, into system-wide equivalent matrices. For example, consider the two-element solution space (that is, the entire solution space is composed of two triangular elements) shown in Figure 10.10. From Equation 10.32, the local system matrices are defined in terms of the nodes of the element. For element 1, the system is represented as follows:

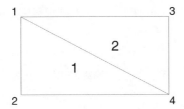

Figure 10.10 A complete solution space (points 1, 2, 3, 4) can be divided into two triangular elements, 1 (corners 1, 2, 4) and 2 (corners 1, 3, 4).

$$\begin{bmatrix} R_1 \\ R_2 \\ R_4 \end{bmatrix} = \begin{bmatrix} k_{11} & k_{12} & k_{14} \\ & k_{22} & k_{24} \\ sym & & k_{44} \end{bmatrix} \begin{bmatrix} u_1 \\ u_2 \\ u_4 \end{bmatrix} + \begin{bmatrix} f_1 \\ f_2 \\ f_4 \end{bmatrix} \tag{10.42}$$

Similarly, we can construct the system matrices for element 2:

$$\begin{bmatrix} R_1 \\ R_4 \\ R_3 \end{bmatrix} = \begin{bmatrix} k_{11} & k_{14} & k_{13} \\ & k_{44} & k_{43} \\ sym & & k_{33} \end{bmatrix} \begin{bmatrix} u_1 \\ u_4 \\ u_3 \end{bmatrix} + \begin{bmatrix} f_1 \\ f_4 \\ f_3 \end{bmatrix} \tag{10.43}$$

The system matrices for the whole solution space are the result of merging both of the above according to nodal positions. The R term may be set to zero, as is the condition defined by the Galerkin method in Equation 10.22. Hence:

$$\begin{bmatrix} k_{11}^{(1)} + k_{11}^{(2)} & k_{12}^{(1)} & k_{13}^{(2)} & k_{41}^{(1)} + k_{41}^{(2)} \\ k_{12}^{(1)} & k_{22}^{(1)} & 0 & k_{24}^{(1)} \\ k_{13}^{(2)} & 0 & k_{33}^{(2)} & k_{43}^{(2)} \\ k_{41}^{(1)} + k_{41}^{(2)} & k_{24}^{(1)} & k_{43}^{(2)} & k_{44}^{(1)} + k_{44}^{(2)} \end{bmatrix} \begin{bmatrix} u_1 \\ u_2 \\ u_3 \\ u_4 \end{bmatrix} = - \begin{bmatrix} f_1^{(1)} + f_1^{(2)} \\ f_2^{(1)} \\ f_3^{(2)} \\ f_4^{(1)} + f_4^{(2)} \end{bmatrix} \tag{10.44}$$

Note that $k_{23} = k_{32} = 0$. This is due to the arrangement of nodes such that no element contains both node 2 and node 3, and hence there is no interaction between these two nodes. Large numbers of nodes thus lead to potentially very sparse matrices, which may be exploited during the solution phase, as is described later.

10.5.5 Applying boundary conditions

Boundary conditions upon the system may take one of two forms: the Dirichlet boundary condition where a potential at a given boundary (such as an electrode surface) and hence the potential at those nodes located on that surface is defined, and the Neumann boundary condition that states that across the surface Γ that encloses solution space Ω, the following condition must apply:

$$\frac{\partial u}{\partial g} = \textbf{constant} \tag{10.45}$$

where g is an axis normal to the surface at the given point. This condition exists across the whole surface and need not be integrated into the procedure. Indeed, consideration of this condition is important, and such an imposition of symmetry of potential at the outer boundary may be exploited in simulating symmetrical systems. Consideration must also be taken when simulating systems that involve changing phase relationships, since this symmetrical boundary will negate any phase relationships at the boundary.

However, it is necessary to define the known system potentials that comprise the Dirichlet boundaries. Having determined the system matrices, we now apply the initial conditions to the system. In the case of the solution of Equation 10.1, these take the form of known node potentials. Consider the case of a three-node system, as shown in Equation 10.46.

$$\begin{bmatrix} k_{11} & k_{12} & k_{13} \\ k_{21} & k_{22} & k_{23} \\ k_{31} & k_{31} & k_{33} \end{bmatrix} \begin{bmatrix} u_1 \\ u_2 \\ u_3 \end{bmatrix} = \begin{bmatrix} f_1 \\ f_2 \\ f_3 \end{bmatrix} \tag{10.46}$$

If it is known that potential $u_3 = U_3$, then Equation 10.46 may be rewritten as

$$\begin{bmatrix} k_{11} & k_{12} & 0 \\ k_{21} & k_{22} & 0 \\ 0 & 0 & 1 \end{bmatrix} \begin{bmatrix} u_1 \\ u_2 \\ u_3 \end{bmatrix} = \begin{bmatrix} f_1 - k_{13}U_3 \\ f_2 - k_{23}U_3 \\ U_3 \end{bmatrix} \tag{10.47}$$

In general, for the boundary condition of potential $u_i = U_i$ where U_i is known, the following procedure is followed:

1. Subtract $k_{ij}U_i$ from $k_{ij}U_i$, where j is the list of all nodes, excluding i.
2. Assign $k_{ii} = 1$.
3. Set all other values in row and column i to 0.
4. Set $f_i = U_i$.

Note that for a large number of boundary conditions, matrix **k** becomes increasingly sparse due to procedure number 3. As mentioned previously, sparsity may be exploited during the solution phase in minimizing the quantity of memory required for storing the matrix during the solving process.

10.5.6 The solution process

The solution of the potentials is thus performed by solving Equation 10.28 across solution space Ω, with the residue set to zero,

$$\mathbf{u} = \mathbf{k}^{-1}\mathbf{f} \tag{10.48}$$

This calculation produces a vector $\mathbf{u}$ containing the values of the potential at all nodes. There is a wide variety of well-documented methods for implementation of this function, such as Gaussian elimination[41] or incomplete Choleski-conjugate gradients.[13] A number of commercial software tools are available for performing this equation, including the NAG™ libraries for use within FORTRAN, or the software suites described in the next chapter. The advantages of these methods are discussed elsewhere.[13] A consideration when performing such an implementation is that matrix $\mathbf{k}$ is largely sparse (i.e., contains a large quantity of zero-value elements) and is also symmetrical. It may be expedient to use a solution method such as ICCG, which takes these features into account if computer storage space is at a premium. Having determined the potentials across Ω, it is a relatively straightforward procedure to determine the electric field by calculating ∇u across the nodes. This may be performed as a postprocessing function, using mathematical tools such as MATLAB™ (The Math Works, Inc.) or Mathematica™ (Wolfram Research), which are commercially available.

At this stage, it is worth noting a possible source of error when designing with finite element models. The Neumann boundary condition, in which the gradient of the field across the boundary of the solution space is zero, effectively creates "ghost" electrodes in mirror image of the electrodes being simulated. These are reflected at the outer boundary of the solution space in all directions. This has the effect of raising the magnitude of the potential at the boundary and setting the electric field across it to zero, both of which create errors in the final solution. Further care must be taken where a phase direction is studied (such as the traveling-wave simulations of Chapter 8), since the reflected electrodes at either end of the electrode array will have ghost traveling waves running counter to those on the actual electrode simulated. To avoid this, the electrodes and region of study should be isolated from the boundary by large, empty elements. These introduce a large distance between the real and ghost electrodes but increase simulation time. Alternatively, it is sometimes possible to make use of symmetry in the electrode geometries by placing the Neumann boundary at the time of symmetry and only simulating half of the problem space.

10.6 The method of moments

In contrast to the method described above, the method of moments (or charge density method) does not discretize a finite solution space to derive

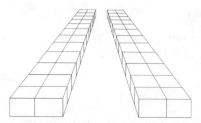

Figure 10.11 In simulations using the method of moments, the electrode surfaces (but not the intervening space) are discretized into subareas, each having an associated potential V and charge q, with the potential taken relative to 0 V at infinity.

the electric field across it. Instead, the surfaces of the electrodes generating the electric fields are divided into a series of charges whose contributions to the electric field according to the principles of Coulomb are described in Equation 10.10 (see also Chapter 2).[27–40,43–46] The model derives its solution by the examination of unrelated points at which the potential is determined, rather than adopting the more abstract concept of a finite solution space, and is consequently simpler to define.

10.6.1 Calculating charge density

Consider the system electrodes, which have been divided into n conducting electrode subareas as shown in Figure 10.11; we can consider each subarea to be an independent electrode to which potentials V_n are applied, relative to 0 V at infinity. Each carries a surface charge q_n. Owing to the superposition principle, the potential on any subarea is related to the charge on itself and all other subareas by a parameter dependent on the distance between the subareas, material properties, and so forth:

$$V_1 = p_{11}q_1 + p_{12}q_2 + \ldots + p_{1n}q_n$$

$$V_2 = p_{21}q_1 + p_{22}q_2 + \ldots + p_{2n}q_n$$

$$V_3 = p_{31}q_1 + p_{32}q_2 + \ldots + p_{3n}q_n \qquad (10.49)$$

$$\cdot$$

$$V_n = p_{n1}q_1 + p_{n2}q_2 + \ldots + p_{nn}q_n$$

where p_{ij} is a parameter that couples the potential on subarea i due to charge j. These expressions may be written in the form,

$$
\begin{pmatrix} V_1 \\ V_2 \\ \\ V_n \end{pmatrix} = +
\begin{pmatrix}
p_{11} & p_{12} & & p_{1n} \\
p_{21} & p_{22} & & p_{2n} \\
\\
p_{n1} & p_{n2} & & p_{nn}
\end{pmatrix}
\begin{pmatrix} q_1 \\ q_2 \\ \\ q_n \end{pmatrix}
\qquad (10.50)
$$

and hence:

$$V = PQ \tag{10.51}$$

It is possible to simulate electrode systems by representing each electrode as a single point charge.[44] However, the more complex and detailed simulations presented here require that the physical dimensions of the electrodes be represented in the model and that the charge density vary across the surface of the electrodes. In order to account for this charge distribution, an approximation is made here. By dividing the surfaces of the electrodes into subelectrode areas, the charge distribution across the subarea is considered uniform. This approximation becomes more accurate as the number of subareas is increased.[44]

Element p_{ij} of matrix P is the potential on subarea i resulting from a unit charge on subarea j in the absence of any other charges. If subarea j is sufficiently small, then the potential at a point (chosen as the midpoint of subarea i) is determined by integrating over subarea dA_j a distance r_{ij} from such a unit charge, thus,

$$p_{ij} = \frac{1}{4\pi\varepsilon_0\varepsilon_m} \int \frac{\sigma_j |dA_j|}{|r_{ij}|} \tag{10.52}$$

where ε_m is the relative permittivity of the medium surrounding the electrodes, which is assumed to be homogeneous. Since subarea j holds unit charge,

$$\sigma_j = \frac{1}{A_j} \tag{10.53}$$

From Equations 10.52 and 10.53 we obtain an integration over subarea j:

$$A_j p_{ij} = \frac{1}{4\pi\varepsilon_0\varepsilon_m} \int \frac{|dA_j|}{|r_{ij}|} \tag{10.54}$$

If we assume that subarea j is rectangular and is in relation to the midpoint of subarea i in the axes imposed as shown in Figure 10.12 then, provided the rectangle j does not cross the x or y axes, the result of the above integration is given by[44]

$$4\pi\varepsilon_0\varepsilon_m A_j p_{ij} = \left| I(x_2,y_2,z_1) - I(x_2,y_1,z_1) - I(x_1,y_2,z_1) + I(x_1,y_1,z_1) \right| \tag{10.55}$$

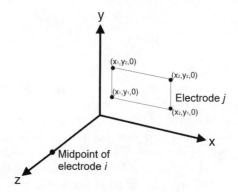

Figure 10.12 A coordinate system is imposed in order to determine the effect on electrode subarea *i* of the charge on electrode subarea *j*.

where

$$I = x.\sinh^{-1}\left\{\frac{y}{\sqrt{(x^2+z^2)}}\right\} + y.\sinh^{-1}\left\{\frac{x}{\sqrt{(y^2+z^2)}}\right\} + z.\tan^{-1}\left\{\frac{z}{xy}\sqrt{(x^2+y^2+z^2)}\right\}$$

(10.56)

This can be coded in this manner, or through the use of logarithmic identities[47,48]

$$I = x\ln\left|\frac{y}{\sqrt{(x^2+z^2)}}\sqrt{1+\frac{y^2}{\sqrt{(x^2+z^2)}}}\right| + y\ln\left|\frac{x}{\sqrt{(y^2+z^2)}}\sqrt{1+\frac{x^2}{\sqrt{(y^2+z^2)}}}\right|$$

$$+z.\tan^{-1}\left\{\frac{z}{xy}\sqrt{(x^2+y^2+z^2)}\right\}$$

(10.57)

If the rectangle *j* lies across either the *x* or *y* axes, it must be divided into two (or four) separate subareas partitioned by the axes and calculated as before.

Having determined the values of matrix **P**, it is possible to calculate the charges on each subarea by reorganizing Equation 10.51 and solving for charge **Q**:

$$\mathbf{Q} = \mathbf{P}^{-1}\mathbf{V}$$

(10.58)

where **V** is the potential applied to the subareas. Alternatively, by considering the relationship described by Equation 10.53, we may express Equation 10.58 in terms of charge density, rather than charge:

$$\sigma = \mathbf{P}'^{-1}\mathbf{V}$$

(10.59)

where σ is the charge density vector and $p'_{ij} = A_j p_{ij}$, as calculated using Equations 10.55–10.57. The solution of Equation 10.59 may be performed using one of the methods described in Section 10.5.5. Having derived the charge-density distribution across the electrode surfaces, it may be saved for future use. Any subsequent calculations involving a previously simulated electrode array may reuse an existing charge matrix for calculating the potential and electric field.

10.6.2 Calculating the potential

If we consider an electrode array that has been subdivided into a system of charges across the electrode surfaces, there are a number of methods that may be employed for determining the potential due to these charges. One possible method is to apply Coulomb's law (Equation 10.10) and derive the electric field directly by calculation of the contributions of each charge to the electric field. However, this method may be inefficient to implement directly due to the requirement to consider vector computations.

An alternative approach is to consider the expression given in Equation 10.59, which states that for a given set of points in space, the potential ψ is dependent on the charge density across the subareas of the electrodes, and on a charge-potential relationship matrix

$$\psi = \sigma \mathbf{P}' \tag{10.60}$$

If this is expressed in the form of Equation 10.52, we can express the relationship of the potential at a point charge n a distance s away at its center in the following form:

$$p'_n = \frac{1}{4\pi\varepsilon} \int \frac{|dA_n|}{|s_n|} \tag{10.61}$$

Substituting Equation 10.61 into Equation 10.60, we can derive an expression for the potential at a general point k:

$$\psi_k = \sum_{n=1}^{N} \sigma_n \int \frac{|dA_n|}{|s_n|} \tag{10.62}$$

where N is the total number of subareas and s_n is the distance from the center of subarea n to the point k. The advantage of this implementation is that the code required to perform Equation 10.59 is the same as that required for Equation 10.54, allowing both to be performed by the same subroutine. This program is implemented in such a way as to provide the results in the form of two-dimensional matrices at locations $\mathbf{K}$, which may then be loaded into a mathematical postprocessing system such as MATLAB. Furthermore, it is a straightforward task to adapt the calculation of the potential so as to calculate potentials at points displaced a small, regular distance from the points of study along the axes centered on the sample point. This allows calculation of the electric field at the series of points, by calculating field gradients based on differences between potentials at these extra points. Such a method increases computation time but gives high levels of accuracy by minimizing the distance from sample point to extra nodes. Having calculated the electric field in vector form in this manner, the data may be saved in the form of matrices Ex_K, Ey_K, and Ez_K. Since the electrode model in the method of moments requires a fixed charge density, it is unable to perform true AC analysis. However, by discretizing the AC cycle into a number of static frames, it is possible to analyze phase effects across a cycle in great depth.

The method of moments does not employ the concept of a defined solution space and thus does not restrict the choice of sample points to be examined, or create ghost electrodes. However, the system of discretization of the electrodes creates approximation errors of a different nature. The approximation of expressing the charge across a subarea as if it were a single point creates errors in close proximity to the electrode surfaces, where the difference between charge focused at set points, rather than distributed equally across the subarea surfaces, becomes more apparent. Within a distance from the electrodes approximately equal to the width of adjacent subareas, the potential will vary according to whether it is facing a point charge or a subarea boundary.

Another limitation to the moments method is that due to the nature of the model considering a direct-line relationship between the sample point and the point charges, the introduction of heterogeneous dielectric media between these points (or into the model generally) is a complex problem that potentially increases computation time to make the method unsuitable for any problems of this nature. Thus this method is most appropriate to the simulation of electrodes suspended in a medium such as the aqueous solution of basic electrokinetic experiments.

10.7   Commercial versus custom software

The steady but rapid increase in computer power has meant that, in the last few years, a number of powerful PC-based electric field solvers have become available. Many of these will be directly applicable to the analysis of electric field for dielectrophoresis. However, some will not, and for more unconventional analyses such as the determination of phase-based effects, it is likely that a few packages will have the features necessary to even allow other packages such as MATLAB (described in the next chapter) to perform the appropriate analyses.

For simple analysis of dielectrophoretic force, it is important that the electric field solver not only be able to determine the electric field, but also the electric field gradient. The majority of commercially available field solvers use finite element analysis (FEA) to determine the electric field across a three-dimensional volume, and so they should be able to determine the gradient in three dimensions. However, the facility to calculate this might not be available. Similarly, there may not be a function to multiply the electric field by itself — a prerequisite for determination of dielectrophoretic force.

If these facilities are not available, a second option is to export the electric field data and perform these calculations in another, mathematics-based software package. In order to do this, the data must be exported in a format that can be easily understood by the receiving package. For example, it is relatively straightforward to analyze data exported in the form of a regularly ordered matrix of two-dimensional or three-dimensional points at which the electric field vectors are measured. However, most finite element solvers use a technique called adaptive meshing to ensure the greatest density of elements is in those areas where the electric field is changing most rapidly. Therefore, the representation of the field in the software is as a series of seemingly random nodes. In order to overcome this, some software packages allow the super-position of a regularly ordered mesh across which the field can be calculated, even if the points on that mesh do not correspond to calculated nodes.

This feature is of particular importance where the solver is required to determine the time-dependent electric field for analysis of electrorotation and traveling-wave dielectrophoresis (as described in the next section). Since these are complex functions that exist outside of the mainstream applications of most FEA electric field solvers, it falls upon the user to perform the analysis in another software package such as MATLAB. In order to do this, the ability to export data in a regular mesh is vital; the analysis described below is based on the assumption that the magnitude of the electric field can be determined at *exactly the same place* in a number of different simulations.

With these constraints in mind, it may be preferable to write one's own code for the performance of electric field modeling; the procedures outlined in the previous sections for construction of moments and finite element solvers is straightforward, both resulting in programs of about 6 kilobytes in length when coded in FORTRAN. These solvers lack the bells and whistles of commercial programs, such as a computer-aided design package for the definition of the problem, but have the advantage of exporting the data in a form tailored exactly to the needs of the user, such as formatting prior to postprocessing (literally after processing, this refers to calculations per-formed on the electric field calculations after the field has been determined).

10.8   Determination of dynamic field effects

10.8.1   The nature of the dynamic field

The above descriptions of processes by which the electric field can be deter-mined provide us with a useful basis for calculating the electric field. However, we may on many occasions need to determine the *time-dependent* nature of the dynamic electric field. If we wish to determine the traveling-wave dielectro-phoretic force or electrorotation torque in a given electrode geometry, then we need to model the manner in which the electric field changes over time.

Since the phase relationships are consistent from one cycle of the electric field to the next (assuming that a single frequency electric field is applied), we principally need to observe the variations in electric field across the solution space for a period of one cycle. We can do this by *discretizing* the electric field into a series of steps and using those steps to calculate the variation in magnitude and phase across the whole cycle. By modeling discrete time intervals, we are effectively sampling the signal; when we have all the samples, we can analyze them to determine the complete effect of the time-variance of the field.

For example, we might wish to consider the electric field varying around a traveling-wave electrode array, such as the one described in Chapter 8 or the nanomotor in Appendix A. In sampling the electric field, we need to ensure the electric field can be modeled with sufficient accuracy to be able to determine the precise form of the electric field. From the sampling theorem,[48] we need to make at least two samples per cycle, although that would give very little of the information of the magnitude and phase that we require. A greater number of samples leads to more accuracy but a greater number of simulations being required. We can take advantage of the fact that the magnitudes of the electric field in the second half of the cycle are equal but opposite to those of the first half of the cycle (assuming all potentials vary sinusoidally) to simulate only the first half of the signal and invert this to generate the data for the second. We can then use postprocessing methods to determine the exact magnitude and phase of the applied field, using methods described in detail in the next chapter.

From our work in Chapter 2, we have established that a time- and phase-variant electric field can be written in the form:

$$\mathbf{E} = E_x(t)\mathbf{i} + E_y(t)\mathbf{j} + E_z(t)\mathbf{k}$$

$$= E_{x0}(x,y,z)\cos(\omega t + \phi_x(x,y,z))\mathbf{i}$$

$$= E_{y0}(x,y,z)\cos(\omega t + \phi_y(x,y,z))\mathbf{j} \tag{10.63}$$

$$= E_{z0}(x,y,z)\cos(\omega t + \phi_z(x,y,z))\mathbf{k}$$

Similarly, the dipole moment takes the form

$$\mathbf{m} = m_x(t)\mathbf{i} + m_y(t)\mathbf{j} + m_z(t)\mathbf{k}$$

$$= 4\pi\varepsilon_0\varepsilon_m r^3$$

$$\begin{pmatrix} \left(E_{x0}(x,y,z)\operatorname{Re}[K(\omega)](\omega t + \phi_x(x,y,z)) - \operatorname{Im}[K(\omega)]\sin(\omega t + \phi_x(x,y,z))\right)\mathbf{i} \\ +\left(E_{y0}(x,y,z)\operatorname{Re}[K(\omega)](\omega t + \phi_y(x,y,z)) - \operatorname{Im}[K(\omega)]\sin(\omega t + \phi_y(x,y,z))\right)\mathbf{j} \\ +\left(E_{z0}(x,y,z)\operatorname{Re}[K(\omega)](\omega t + \phi_z(x,y,z)) - \operatorname{Im}[K(\omega)]\sin(\omega t + \phi_z(x,y,z))\right)\mathbf{k} \end{pmatrix}$$

$$\tag{10.64}$$

and these can be combined to determine the time-dependent force:

$$\mathbf{F}(t) = (\mathbf{m} \cdot \nabla)\mathbf{E}$$

$$= F_x(t)\mathbf{i} + F_y(t)\mathbf{j} + F_z(t)\mathbf{k} \tag{10.65}$$

$$= \sum_{a=x,y,z} \left(m_x(t)\frac{\partial E_x(t)}{\partial a} + m_y(t)\frac{\partial E_x(t)}{\partial a} + m_z(t)\frac{\partial E_x(t)}{\partial a} \right)$$

Averaging across a number of samples, we can determine the time-averaged force value: for example, for samples taken every 10° of phase of the applied signal,

$$F_\alpha = \frac{1}{36}\sum_i \left(m_x(i)\frac{\partial E_\alpha(i)}{\partial x} + m_y(i)\frac{\partial E_\alpha(i)}{\partial y} + m_z(i)\frac{\partial E_\alpha(i)}{\partial z} \right) \tag{10.66}$$

where i corresponds to each of the 36 instants in time.

10.9 Example: simulation of polynomial electrodes

In order to examine both the process, and indeed the requirement, for simulation, we will examine the electric field around a planar electrode array with electrodes shaped according to the principles of the polynomial electrode geometry described earlier in this chapter.

The principal use of electrorotation electrode arrays is to allow the rate of rotation to be measured for individual objects, which allows the determination of the imaginary part of the Clausius–Mossotti factor, and hence the dielectric properties of each particle. However, it was established early in the development of the technique that the rate at which particles rotate within the electrode area is not consistent, preventing the direct comparison of cells spread across the entire electrode chamber. In order to address this, a number of attempts were made at modeling the electric field and its variation,[2,8,31,32,50] including the use of analytical, moments, and finite-difference methods.

In order to examine the method of analysis, we will consider here a time-dependent simulation of the electric fields due to a four-electrode polynomial geometry addressed by four sinusoidal voltages, of equal magnitude, phased 90° apart.

10.9.1 Simulations

The results presented here were obtained using a simulation model based on the method of moments; the electrode structures were each represented by 150 subareas of equal size, making 600 subareas across the whole simulation. The simulation was used to determine the potential (scalar) and

Figure 10.13 The polynomial electrode geometry simulated; the dotted square indicates the area across which the solution was determined.

electric field (vector) in a regular mesh of 40×40 points arranged in a square, defined by the area in the center of the electrode array and with the inner tips of the electrodes touching the midpoints of the square sides, as shown in Figure 10.13.

The geometry of the polynomial electrodes in the x-y plane was defined by a simplified version of Equation 10.5, such that

$$\left|x^2 - y^2\right| = k^2 \qquad (10.67)$$

where $2k$ defines the distance between opposing electrode tips. Since the simulation is of electrodes fabricated using standard photolithography, they were defined as being 0.2 μm thick with the spacing between electrode tips chosen to be 400 μm. The subareas were chosen to be square with sides 20 μm and were uniformly distributed over the electrode surface to a distance of 200 μm back from the electrode tips. Four sinusoidal voltages of peak potential $V_o = 10$ V were assigned to the electrodes, with 90° phase difference between adjacent electrodes. To simulate the sinusoidal voltages, the electrodes were assigned phase advance values ϕ of 0°, 90°, 180°, and 270°, with the potentials described by the equation

$$V = V_o \cos(\omega t + \phi) \qquad (10.68)$$

so that the potentials of the electrodes for the first simulation were 10 V, 0 V, –10 V, and 0 V; at the second simulation (10° advanced from the first) they were 9.8 V, –1.7 V, –9.8 V, and 1.7 V, and so on. Calculations were performed for the first 18 steps and reversed to determine the next 18. When all 36 results had been determined, the three electric field vectors were stored in three-dimensional matrices representing two dimensions of space (the plane in which the results are obtained) and one of time. A MATLAB function was then used to fit a sine wave to the data points, allowing the amplitude to be matched at each point to a sine wave with its own amplitude and phase relationship to ωt. These were then compared to determine the magnitude and phase distribution across the simulation plane. The 18 calculations were performed using FORTRAN 77, and the results were then processed using MATLAB to produce plots of the electrical potential in the plane of the electrodes as well as of the magnitude (E_x, E_y, E_z) and phase (ϕ_x, ϕ_y, ϕ_z) of the rotating field.

10.9.2   Simulation results

We have established in previous chapters that the electrorotational torque acting on a sphere of radius r is given by

$$\Gamma = -4\pi\varepsilon_m r^3 \operatorname{Im}[K(\omega)]E^2 \qquad (10.69)$$

However, we know from Chapter 2 that this is only true for the case where the electric field rotates in a circular pattern — that is, the phase difference between the x and y vectors is 90°. To explain this, consider viewing a rotating disk with a nail inserted at one point on its circumference. If viewed from the side, the nail would appear to move up and down as the disk rotates; if viewed from overhead, it would appear to move side to side. However, when it appears to be at one end of its travel in one view, it appears to be at the *center* of its travel in the other view. Hence, the position of the nail in the two views (and the phase of the motion in the two axes) is 90° out of phase. If the nail appeared to be at the end of its travel in both directions simultaneously, this would indicate that it was not moving in a circle, but back and forth along a diagonal line; the phase difference is not 90°, but is either 0° or 180°. Hence the motion is not circular. Transposing this to our electric field, if the phase difference is not 90°, the electrorotation torque is diminished, and if it equals 0° or 180°, it disappears, leaving only dielectrophoresis.

In practice, the electrorotation torque is given by the general equation

$$\Gamma = -4\pi\varepsilon_m r^3 \operatorname{Im}[K(\omega)]E_x E_y \sin(\phi_x - \phi_y) \qquad (10.70)$$

where E_x and E_y are the magnitudes in the x and y (orthogonal) directions, and ϕ_x and ϕ_y are the phase relationships between those two vectors and some reference phase. However, it is only when $\phi_x - \phi_y = 90°$ and $E_x = E_y$ that Equations 10.69 and 10.70 are synonymous. We can define an effective value of electric field to allow this to be the case, where

$$E_{\text{eff}}^2 = E_x E_y \sin(\phi_x - \phi_y) \qquad (10.71)$$

This indicates the reason why the previous models had not correctly predicted the distribution of torque across the interelectrode plane; it had been

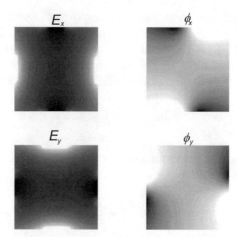

Figure 10.14 The magnitude (left column) and phase (right column) of the electric field in the x and y directions (top and bottom rows, respectively) across the area shown in Figure 10.13. The magnitude is greatest (white regions) at the tips of the electrodes arranged along the axis in question, but the phase relationship changes quite markedly near the other electrodes (with white and black indicating 90° phase advance and retardation, respectively, and medium gray representing 0°).

assumed that the conditions of $\phi_x - \phi_y = 90°$ and $E_x = E_y$ had existed at all points across the plane, but simulation indicated that this was not the case.

In order to correctly determine the torque distribution, it is necessary to calculate the magnitudes and phases of the three electric field vectors independently. These are shown in Figure 10.14 for the electric fields in the x and y directions. Analysis of the figure shows the extent to which the electric field varies from its ideal case. For example, the x component of the field has its largest values near the (x-axis) electrode edges and minima at the y-axis electrode edges. As is expected from the symmetry shown in Figure 10.13, a corresponding behavior exists for the y component of the field. Also, ϕ_x deviates from its ideal value of zero (and correspondingly ϕ_y deviates from 90°) increasingly with distance from the central region between the electrodes. It is clear that the ideal situation of two equal E_x and E_y field components of phase 90° apart holds only in a circular region approximately one quarter of the size of the square defined by the electrode tips. The phase difference between E_x and E_y approaches 0° and 180° alternatively between adjacent electrodes, where the resultant field vectors in these regions have vibrational rather than rotational characteristics as described above.

All other factors in Equation 10.70 being constants, we can determine the spatial dependence of the electrorotational torque by determining the field factor E_{eff}^2. Indeed, we can compare the variation in E_{eff}^2 with the variation in torque predicted by models considering only the magnitude of the electric field, and with the original predictions of the analytical model for polynomial electrodes, as well as for results of the rotational torque obtained by experiment. These are shown in Figures 10.15 to 10.18.

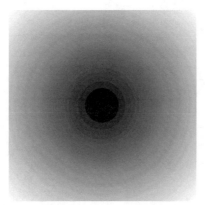

Figure 10.15 The magnitude of the torque predicted across the square in Figure 10.13, determined using the analytical solution in Section 10.3, with lowest torque indicated by the black regions (at the center of the array) and white indicating the highest values (at the interelectrode gaps).

Figure 10.16 The torque in the area at the center of the electrode array in Figure 10.13 determined by finding the square of the electric field, as simulated using the method of moments. The minimum value is still at the center of the array, but the maximum rotation is now predicted to be at the electrode edges.

As can be seen, the increase in the complexity of the model increases the complexity of the predicted output. For example, the simple analytical model (Figure 10.15) indicates that the force increases almost linearly from the center of the array toward the edges and is not influenced by factors such as the position of the electrodes. In Figure 10.16, the result of the numerical model where only the magnitude of the electric field is considered, we see that there is significantly higher electric field strength near the electrode edges. This is because this model considers the electrodes not as two-dimensional structures, but as thin three-dimensional structures. If we then consider the full model including the effects of phase variation, the magnitude of torque appears as shown in Figure 10.17. This shows not only that the magnitude

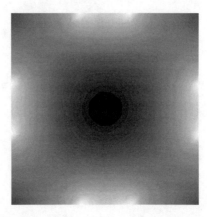

Figure 10.17 The result shown in Figure 10.1 is adapted to compensate for the fact that the torque distribution is not uniform across the interelectrode area, and the torque distribution is again slightly different. The maximum value of torque is still at the electrode edges, but there is a saddle effect where the torque along the line from center to corner rises, reaches a peak, and then diminishes toward the corner. This is due to the phase relationship near the interelectrode gaps causing the electric field to oscillate back and forth rather than rotate.

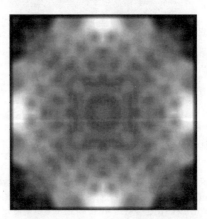

Figure 10.18 A pattern representing the actual rotation rates in the interelectrode gaps for an array of similar geometry, made using over 200 latex beads. As can be seen, the form is similar to that shown in Figure 10.17 with the same saddle effect, peak rate at the electrode tips, and minimum at the interelectrode gaps.

increases at a much higher rate near the electrode tips, but also that it then diminishes toward the corner of the plot, indicating the interelectrode gaps. This is because in these areas, the influence of the other electrode is reduced and the electric field moves monotonically backward and forward between the electrodes.

In order to determine which of these simulations is correct, we can compare the results to a surface generated from the distribution of actual rotation rates in polynomial electrodes, obtained experimentally[51] and shown in Figure 10.18. We see that the pattern closely matches that of the simulation including phase effects, as shown in Figure 10.17; key features to note are the peak rotation at the electrode edges and the drop in rotation toward the corner of the graph. The variation in torque at the center of both graphs is similar, though it appears to be smaller in Figure 10.18 because the magnitude of the torque at the corners is so low as to distort the scale. The reason for the difference in magnitudes between predicted and observed rotation rates in these corners may be due to experimental artifacts. The measurements were taken using ellipsoidal beads, which increased the ease of measurement of the rotation rate but also increased the effects of other factors such as electro-orientation where the magnitudes of E_x and E_y are significantly different (such as in the interelectrode gaps).

As can be seen from this example, the variation in electric field — both magnitude *and* phase — can have significant implications for determining the behavior of particles in the electrode array. Choosing the appropriate model to suit the application is important, as is consideration of the geometry and the task required of it. This is particularly true where numerical values, such as the determination of force or torque, are important. In these cases, electric field simulation is the only method of determining spatial variations in force and torque in all but the simplest electrode geometries.

References

1. Pohl, H.A., *Dielectrophoresis*, Cambridge University Press, Cambridge, 1978.
2. Huang, Y. and Pethig, R., Electrode design for negative dielectrophoresis applications. *J. Meas. Sci. Tech.*, 2, 1142, 1991.
3. Wang, X.-B., Huang, Y., Burt, J.P.H., Markx, G.H., and Pethig, R., Selective dielectrophoretic confinement of bioparticles in potential energy wells, *J. Phys. D: Appl. Phys.*, 26, 1528, 1993.
4. Masuda, S., Washizu, M., and Iwadare, M., Separation of small particles suspended in liquid by nonuniform electric field, *IEEE Trans. Ind. Appl.*, 23, 474, 1987.
5. Morgan, H., Izquierdo, A.G., Bakewell, D., Green, N.G., and Ramos, A., The dielectrophoretic and traveling wave forces generated by interdigitated electrodes: analytical solution using Fourier series, *J. Phys. D: Appl. Phys*, 34, 1553, 2001.
6. Wang, X., Wang, X.B., Becker, F.F., and Gascoyne, P.R.C., A theoretical method of electrical field analysis for dielectrophoretic electrode arrays using Green's theorem, *J. Phys. D: Appl. Phys*, 29, 1649, 1996.
7. Gauss, C.F., Brief an Gerling, *Werke*, 9, 278, 1823.
8. Hölzel, R., Electric-field calculation for electrorotation electrodes, *J. Phys. D: Appl. Phys.*, 26, 2112, 1993.
9. Rousselet, J., Salome, L., Ajdari, A., and Prost, J., Directional motion of Brownian particles induced by a periodic asymmetric potential, *Nature*, 370, 446, 1994.
10. Binns, K.J. and Laurenson, P.J., *Analysis and Computation of Electric and Magnetic Field Problems* Pergamon Press, Oxford, 1963.
11. Nath, B., *Fundamentals of Finite Elements for Engineers*, Athlone Press, London, 1974.

12. Zienkiewicz, O.C. and Taylor, R.L., *The Finite Element Method*, Vol. 1, 5th ed., McGraw-Hill, London, 2000.
13. Binns, K.J., Lawrenson, P.J., and Trowbridge, C.W., *The Analytical and Numerical Solution of Electric and Magnetic Fields*, Wiley, Chichester, 1992.
14. Gallagher, R.H., Zienkiewicz, O.C., Oden, J.T., Cecchi, M.M., and Taylor, C., Eds., *Finite Elements in Fluids*, Wiley, Chichester, 1978.
15. Green, N.G., Morgan, H., and Milner, J.J., Manipulation and trapping of sub-micron bioparticles using dielectrophoresis, *J. Biochem. Biophys. Methods*, 35, 89, 1997.
16. Green, N.G. and Morgan, H., Dielectrophoresis of submicrometre latex spheres. 1. Experimental results. *J. Phys. Chem. B*, 103, 41, 1999.
17. Green, N.G. and Morgan, H., Dielectrophoretic separation of nano-particles, *J. Phys. D: Appl. Phys*, 30, L41, 1997.
18. Rousselet, J., Markx, G.H., and Pethig, R., Separation of erythrocytes and latex beads by dielectrophoretic levitation and hyperlayer field-flow fractionation, *Colloids and Surfaces A*, 140, 209, 1998.
19. Fudouzi, H., Kobayashi, M., and Shinya, N., Arrangement of 100nm scale particles by an electrostatic potential field, *J. Nanoparticle Res.*, 3, 193, 2001.
20. Yu, Q., The new parallel numerical approach for analysis of the larger scale three-dimensional electromagnetic field problems of arbitrary shape, paper presented in the Proceedings of the 2nd International Conference on Computation in Electromagnetics, Nottingham, UK, 315, 1994.
21. Dai, W.-W., Marsili, P.M., Martinez, E., and Morucci, J.P., Using Hilbert uniqueness method in a reconstruction algorithm for electrical impedance tomograpy, *Physiol. Meas.*, 15, 161, 1994.
22. Sobol, I.M., A primer for the Monte Carlo method, CRC Press, Boca Raton, FL, 1994.
23. Madras, M., Lectures on Monte Carlo methods, paper presented to the American Mathematical Society, Providence, RI, 2002.
24. Doering, C.R., Horsthemke, W., and Riordan, J., Nonequilibrium fluctuation-induced transport, *Phys. Rev. Lett.*, 72, 2984, 1994.
25. Sancho, M., Giner, V., and Martinez, G., Monte Carlo simulation of dielectro-phoretic particle chain formation, *Phys. Rev. E.*, 55, 544, 1997.
26. Llamas, M., Giner, V., and Sancho, M., The dynamic evolution of cell chaining in a biological suspension induced by an electric field, *J. Phys. D: Appl. Phys.*, 31, 3160, 1998.
27. Maxwell, J.C., *A Treatise on Electricity and Magnetism*, 3rd ed., Vol. 1, Oxford University Press, Oxford, 1892.
28. Martinez, G. and Sancho, M., Accurate calculation of electrical capacitances of axially symmetric systems including dielectric media, *IEE Proc.* 127, 531, 1980.
29. Pethig, R., Huang, Y., Wang, X.B., and Burt, J.P.H., Positive and negative dielectrophoretic collection of colloidal particles using interdigitated castel-lated electrodes, *J. Phys. D: Appl. Phys.* 25, 881, 1992.
30. Wang, X.-B., Huang, Y., Burt, J.P.H., Markx, G.H., and Pethig, R., Selective dielectrophoretic confinement of bioparticles in potential energy wells, *J. Phys. D: Appl. Phys.*, 26, 1528, 1993.
31. Hughes, M.P., Wang, X.B., Becker, F.F., Gascoyne, P.R.C., and Pethig, R., Computer-aided analyses of electric fields used in electrorotation studies, *J. Phys. D: Appl. Phys.*, 27, 1564, 1994.

32. Hughes, M.P., Computer-aided analyses of conditions for optimizing practical electrorotation, *Phys. Med. Biol.*, 43, 3639, 1998.

33. Hughes, M.P., Morgan, H., Rixon, F.J., Burt, J.P.H., and Pethig, R., Manipulation of herpes simplex virus type 1 by dielectrophoresis, *Biochim. Biophys. Acta*, 1425, 119, 1998.

34. Hughes, M.P. and Morgan, H., Dielectrophoretic trapping of single sub-micrometre scale bioparticles, *J. Phys. D: Appl. Phys.*, 31, 22205, 1998.

35. Schnelle, T., Hagedorn, R., Fuhr, G., Fiedler, S., and Müller, T., Three-dimensional electric field traps for manipulation of cells — calculation and experimental verification, *Biochim. Biophys. Acta,* 1157, 127, 1993.

36. Hagedorn, R., Fuhr, G., Muller, T., Schnelle, T., Schnakenberg, U., and Wagner, B., Design of asynchronous dielectric micromotors, *J. Electrostatics*, 33, 159, 1994.

37. Hughes, M.P., Pethig, R., and Wang, X.-B., Dielectrophoretic forces on particles in traveling electric fields, *J. Phys. D: Appl. Phys.*, 28, 474, 1996.

38. Hagedorn, R., Fuhr, G., Muller, T., and Gimsa, J., Traveling-wave dielectrophoresis of microparticles, *Electrophoresis*, 13, 49, 1992.

39. Fuhr, G., Fiedler, S., Muller, T., Schnelle, T., Glasser, H., Lisec, T., and Wagner, B., Particle micromanipulator consisting of two orthogonal channels with traveling-wave electrode structures, *Sensors and Actuators A*, 41, 230, 1994.

40. Wang, X.-B., Hughes, M.P., Huang, Y., Becker, F.F., and Gascoyne, P.R., Non-uniform spatial distributions of both the magnitude and phase of the AC electric fields determine dielectrophoretic forces, *Biochim. Biophys. Acta*, 1243, 185, 1995.

41. Kreyszig, E., *Advanced Engineering Mathematics*, 5th ed., Wiley, New York, 1983.

42. Prost, J., Chauwin, J.F., Peliti, L., and Ajdari, A., Asymmetric pumping of particles, *Phys. Rev. Lett.*, 72, 2652, 1994.

43. Martinez, G. and Sancho, M., Integral equation methods in electrostatics, *Am. J. Phys.*, 51, 170, 1983.

44. Birtles, A.B., Mayo, B.J., and Bennett, A.W., Computer technique for solving three-dimensional electron-optics and capacitance problems, *Proc. IEEE,* 120, 213, 1973.

45. Reitan, D.K. and Higgins, T.J., Calculation of the electrical capacitance of a cube, *J. Appl. Phys.*, 33, 223, 1951.

46. Reitan, D.K., Accurate determination of the capacitance of rectangular parallel-plate capacitors, *J. Appl. Phys.*, 30, 172, 1959.

47. Harrington, R.F., *Field Computation by Moment Methods*, Macmillan, New York, 1968.

48. Abramowitz, M. and Stegun, I.A., *Handbook of Mathematical Functions*, Dover Publications, New York, 1965.

49. O'Reilly, J.J., *Telecommunications Principles*, Van Nostrand Reinhold, Wokingham, 1984.

50. Gimsa, J., Glaser, R., and Fuhr, G., Remarks on the field distribution in 4 electrode chambers for electrorotational measurements, *Studia Biophysica*, 125, 71, 1988.

51. Hughes, M.P., Archer, S., and Morgan, H., Mapping the electrorotational torque in planar microelectrodes, *J. Phys. D: Appl. Phys.*, 32, 1548, 1999.

chapter eleven

Dielectrophoretic response modeling and MATLAB

11.1 Modeling the dielectrophoretic response

In this book, we have examined the dielectrophoretic responses of colloids, viruses, and proteins. We have seen how we can compare predicted and observed dielectrophoretic responses across the frequency range, and we have used this to determine the dielectric properties of the particles. By modeling the response of more than one type of particle, we have been able to determine the optimum conditions for separating those particles into homogeneous groups in separation devices. In order to achieve these, we have used models of the frequency-dependent Clausius–Mossotti factor to predict the relative polarizability of the particles across the frequency spectrum; we have then expanded that into three dimensions, considering how variations in the conductivity of the suspending medium affect the dielectrophoretic response — and, in particular, the crossover frequency — of the particles. In this chapter, we will examine how this is achieved and examine software-based models for determining these properties.

Since the modeling process requires a not-insignificant amount of calculations to accurately represent the Clausius–Mossotti factor across a range of frequencies and conductivities, it makes sense to automate the process using software. It is possible to write programs directly into code (such as C or FORTRAN) that will perform these calculations, though the user will then still require a method of displaying this information. Since the problem is fairly straightforward, there are a number of computer-based mathematical packages that can be used to both calculate and display the polarizability; these include widely known titles such as Mathematica™, Maple™, and MATLAB™. In order to examine the procedures used in modeling these parameters, this chapter will cite examples using the MATLAB programming language, but the procedures are all fairly commonplace and can easily be adapted to other mathematical languages.

11.2 Programming in MATLAB

MATLAB is a combined programming language and data analysis/presentation computing suite. It is The Mathworks™ Inc.'s software suite for analyzing, manipulating, and displaying matrices (hence its name, derived from Matrix Laboratory). It operates principally from a command line system where commands are typed into a window and executed; it is possible to create programs using lists of these commands, so that they execute as if they had been typed in sequence, by simply typing in the name of the program. An extensive array of operations for data analysis already exists with MATLAB; this includes a large number of functions dedicated to data display.

The MATLAB language is similar to many older computer languages such as BASIC and FORTRAN. It allows the user to write programs for performing a number of operations, including data analysis, which we will be examining in this chapter. In order to distinguish MATLAB programs from conventional text, all material from MATLAB has been written in `courier typeface`. This has also been used within the text when referring to variables used within the programs.

The interface and programming language of MATLAB is fairly intuitive, and it is not difficult to go from mathematical operations to programs easily. One feature of MATLAB worth mentioning before the start of our investigations is that the program uses a colon (`:`) for the *wildcard* operation, which stands for all of the numbers in a given range or all of the numbers within certain limits. For example, "`i = 1:3`" means "`i` equals one to three," while "`i = j(3,:)`" means "`i` equals the numbers in the third row of all columns in matrix `j`."

11.3 Modeling the Clausius–Mossotti factor

As we have learned in previous chapters, the dielectrophoretic force, $\mathbf{F}_{DEP}$, acting on a dielectric sphere of radius r in medium of permittivity ε_m is given by

$$\mathbf{F}_{DEP} = 2\pi r^3 \varepsilon_m \operatorname{Re}[K(\omega)]\nabla E^2 \qquad (11.1)$$

where $\operatorname{Re}[K(\omega)]$ is the real part of the Clausius–Mossotti factor, given by

$$K(\omega) = \frac{\varepsilon_p^* - \varepsilon_m^*}{\varepsilon_p^* + 2\varepsilon_m^*} \qquad (11.2)$$

where p and m are subscripts indicating particle and medium, and complex permittivity is given by

$$\varepsilon^* = \varepsilon - \frac{j\sigma}{\omega} \tag{11.3}$$

Furthermore, we also know that due to viscous drag (Stokes's force), the particle reaches a terminal velocity given by the equation

$$v = \frac{\mathbf{F}_{\mathbf{DEP}}}{6\pi\eta r} \tag{11.4}$$

where η is the viscosity of the medium. Combining these, we obtain

$$
\begin{aligned}
v &= \frac{r^2 \varepsilon_m}{3\eta} \operatorname{Re}\left[\frac{\varepsilon_p^* - \varepsilon_m^*}{\varepsilon_p^* + 2\varepsilon_m^*}\right] \nabla E^2 \\
&= \beta \operatorname{Re}\left[\frac{\varepsilon_p^* - \varepsilon_m^*}{\varepsilon_p^* + 2\varepsilon_m^*}\right] \nabla E^2
\end{aligned}
\tag{11.5}
$$

where β is constant for a given particle and medium. In the previous chapter, we explored methods by which we can determine the local value of ∇E^2. By inserting this in Equation 11.5 we can determine the value of the Clausius–Mossotti factor of a particle merely by determining its velocity at a given point within the electrode array. However, while this approach can be used (as described in Chapter 5) to determine particle properties, only by examining this as a function of frequency can we be more confident in our predictions by enabling us to determine the behavior of the particle as it undergoes dielectric dispersion (as described in Chapters 2 and 4). If we examine the velocity of particles through a particular volume as a function of frequency, then we can treat that local value of ∇E^2 as constant, such that

$$v \propto \operatorname{Re}\left[\frac{\varepsilon_p^* - \varepsilon_m^*}{\varepsilon_p^* + 2\varepsilon_m^*}\right] \tag{11.6}$$

Therefore, the velocity of particles passing through a volume is directly proportional to the value of the real part of the Clausius–Mossotti factor, and by studying the changes in the behavior we can match the model of the dielectric response to fit.

If we are simply trying to determine the properties of homogeneous particles in the absence of surface conduction effects, we need only calculate the Clausius–Mossotti factor for our frequencies of interest. A simple program to achieve this in MATLAB is

```
e0  =  8.85e-12;

sm  =  0.001;

em  =  78*e0;

sp  =  .01;

ep  =  2.55*e0;

f  =  logspace(4,7,1000);

w  =  2*pi*f;

j  =  sqrt(-1);

epc  =  (ep-j*sp./w);

emc  =  (em-j*sm./w);

cm  =  (epc-emc)./(epc+2*emc);

output  =  real(cm);

semilogx(f,output);

xlabel('frequency (Hz)');

ylabel('Re[k(w)]');
```

This 14-line program produces an output such as that shown in Figure 11.1. The function of the program is fairly straightforward. The first five lines set the parameters we need to use; since we cannot use Greek characters, we must use e for epsilon (ε) and s for sigma (σ). Similarly, we cannot use an asterisk to denote a complex number, since the asterisk is used to indicate a multiplication, so we must use another method to indicate it; here I have used epc to represent "ep (complex)." The conductivities are expressed in S m^{-1}.

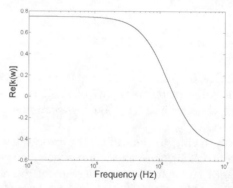

Figure 11.1 The dielectric dispersion of a 100-nm radius colloidal particle, as described in the text.

The sixth and seventh lines generate a range of frequencies (f) and corresponding angular frequencies (w, instead of ω). The `logspace` command simply generates a string of numbers on a logarithmic scale. The three numbers in brackets indicate the order of ten at which the list starts, the order of ten at which it stops, and the number of points in the list — so that here, we are generating a list of 1000 points between 10 kHz and 10 MHz. Line eight defines j as the square root of minus one, using the built-in `sqrt` command. The ninth and tenth lines calculate the complex permittivities of particle and medium by directly expressing Equation 11.3. The presence of a period (.) before the slash (indicating division) shows that the operation is to take place on a list of numbers, or vector — in this case, the list of different frequencies we are examining. Line 11 is the calculation of the Clausius–Mossotti factor cm — compare it with Equation 11.2. Line 12 uses the `real` command to calculate the real part of the Clausius–Mossotti factor, and line 13 produces the graph in Figure 11.1 by using the `semilogx` command; this produces a plot that is logarithmic in only one direction (hence a semilog plot), that direction being the x direction, where the values in the brackets indicate the values on the x – (frequency) and y – (Re[$K(\omega)$]), respectively. Finally, the last lines label the two axes. If this program is saved in a file with a name ending in .m (such as `cmfactor.m`), then it can be run from within MATLAB simply by typing the name of the file, provided the file is in a location of which MATLAB is aware (consult the manual about setting path names for more information about this).

11.4   Determining the crossover spectrum

We now have a program that can plot the Clausius–Mossotti factor as a function of frequency. However, while this measurement can have direct applications to modeling experimental data, such as that collected by collection-rate measurements as described in Chapters 4 and 5, it is not widely used because colloidal particles are also influenced by Brownian motion. A far more common method of particle analysis is to use crossover measurements (as seen in Chapters 4 to 6), where the frequency at which the dielectrophoretic force is zero, is determined for a range of medium conductivities. We can use our existing program for this, but must add to it, and also *nest* it within another program.

In order to operate the program, we need to use two fundamental software operations, the *FOR LOOP* and the *IF* operator. The former executes a piece of code a certain number of times, with a variable being used to count the number of the times the code has been executed. For example, a MATLAB loop such as

```
for i = 1:10

...

end
```

will execute the code between the start and end lines ten times; every time the computer passes around the loop, the value of i is incremented. At the last pass, the value of i is 10; at the end of that time around the loop, the program moved onward from the loop. The second operator is IF, which tests whether a statement is true. If it is, then a piece of code is executed. If not, then another piece of code is carried out. This is expressed in MATLAB thus:

```
if x>1

...

else

...

end
```

which will execute one set of instructions if x is greater than one, and another set of instructions if x is less than one.

We can combine these with our first program in order to calculate the conductivity-dependent crossover spectrum. A third useful programming tool we have is that we can treat programs as commands or instructions, so long as its name is not the same as an existing MATLAB command and it is in a path (part of the computer) that is recognized by MATLAB. We can therefore use this to construct a program structure of the kind shown in Figure 11.2; a first program runs through a list of conductivities and for each,

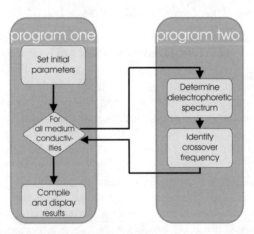

Figure 11.2 A flow diagram indicating the method by which the crossover spectrum can be calculated. The first program established a parameter set, and then uses a FOR loop to examine all parameters within that set, one at a time, via a second program. This enables the determination of the crossover frequencies of all the conductivities examined. The first program compiles each of these results as the second program completes them, and when the whole list is completed, the first program compiles them and displays them.

it calls a second program, not unlike our program described in Section 11.3, which determines the Clausius–Mossotti factor for that particular conductivity and determines the frequency at which the crossover takes place. This is then stored when control returns to the first program, giving a list of crossover frequencies corresponding to the starting list of conductivities; plotting one against the other gives the crossover spectrum.

The first program — we will call this `crosslist.m` — generates a list of 40 medium conductivities (using `logspace`, as before) called `smlist`, and then loops through the list of 40 using a `for j = 1:40` command. This executes a second program — called `fullmodel.m` — which determines the crossover frequency and stores it in a variable called `crossover`. When `fullmodel` has been executed, the value in `crossover` is saved as a number in a list of crossover frequencies called `crossoverlist`. When all 40 conductivities have been determined, `loglog` produces a log–log plot (logarithmic on both axes) of the crossover frequencies versus medium conductivity. Labels are added; note the use of the ^ symbol, which turns the following character into a superscript such that Sm^-^1 appears as Sm^{-1}; however, in the context of a mathematical operation, ^ means to the power of. The final line uses the `axis` command to define the maximum and minimum values shown along the x and y axes of the displayed graph.

```
smlist = logspace(-3,-1,40);

for i = 1:40;

sm = smlist(i);

fullmodel;

crossoverlist(i) = crossover;

end;

loglog(smlist,crossoverlist);

xlabel('Conductivity (Sm^-^1)');

ylabel('Crossover frequency (Hz)');

axis([1e-3 1e-1 1e5 1e7]);
```

The second program, called `fullmodel.m`, is presented below and has been divided into a number of parts for ease of description. Note that the parts are separated by comments, indicated by a percentage mark %; MATLAB ignores anything after these until it sees a semicolon or end of line. The first part of our program looks similar to the program in Section 11.3, since it calculates the frequency-dependent crossover spectrum, as before. Note however that there is no longer a statement of the value of sm (medium conductivity), since that was established within `crosslist.m` just before `fullmodel` was called; this means that the same version of fullmodel can be used for all values of medium conductivity. Note also that the lines from the end of the program, displaying the Clausius–Mossotti factor, have been removed because we do not want the program to display each Clausius–Mossotti factor for all 40 conductivities.

Part two of the program is used to examine the real part of the Clausius–Mossotti factor to detect the crossover frequency. It uses a variable called marker, which is originally set to zero. The program then uses a for loop to examine all 1000 numbers in the Clausius–Mossotti factor. It also uses an if statement to test if the value of $Re[k(\omega)]$ is greater than 0; if it is, then it makes marker equal to the number of the value in the list. When $Re[k(\omega)]$ drops below 0, the condition is not met and marker is no longer updated; therefore, when all values have been tested, it contains the number in the list of the highest sample where $Re[k(\omega)]$ is greater than 0, indicating that the following value is negative and that between them is the crossover frequency.

Part three of the program determines the crossover frequency more accurately within the frequency band between the two values either side of the crossover frequency — those being the values in the marker[th] value in the list (the positive value) and the one after it (the negative value). By approximating the change in $Re[k(\omega)]$ to a straight line between these values, the program calculates the exact value at which $Re[k(\omega)] = 0$ and records it in crossover. If no value of crossover frequency exists (because $Re[k(\omega)]$ is only positive or negative through the entire sweep of frequencies being investigated), the program returns a value 0.01, indicating a very low crossover frequency. Since we are using logarithmic scales, we cannot display zero or negative numbers. Similarly, if the crossover is greater than the maximum frequency examined, the program returns NaN; this stands for "not a number," and will be ignored by subsequent processing.

```
% part one

e0 = 8.85e-12;

em = 78*e0;

sp = .01;

ep = 2.55*e0;

f = logspace(4,9,1000);

w = 2*pi*f;

j = sqrt(-1);

epc = (ep-j*sp./w);

emc = (em-j*sm./w);

cm = (epc-emc)./(epc+2*emc);
```

```
output  =  real(cm);

%  part  two

marker  =  0;

for  k  =  1:1000

if  output(k)>0

marker  =  k;

end

end

%part  three

if  marker<1000

if  marker>0

xlower  =  f(marker);

vlower  =  output(marker);

xupper  =  f(marker+1);

vupper  =  output(marker+1);

ratio  =  vupper/(vlower-vupper);

crossover  =  xlower+ratio*(xupper-xlower);

else  crossover  =  0.01;

end

else  crossover  =  NaN;

end
```

The output of these programs for the set of electrical characteristics from our first example is shown in Figure 11.3. As can be seen, the crossover frequency is mapped as a function of conductivity, exhibiting a relatively stable value at lower conductivities but dropping sharply at higher conductivities. This actually drops to 0.01Hz in our model, due to the restriction on not being able to display negative numbers. Since the lower threshold on our graph is 1 MHz, it appears to go to zero. Using NaN where there is no crossover means that the line effectively disappears in midgraph, which, while perhaps more correct, is less effective in displaying the dielectrophoretic response.

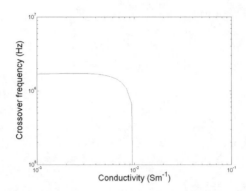

Figure 11.3 A dielectrophoretic crossover spectrum for the programs in the text, using the method shown in Figure 11.2.

11.5 Modeling surface conductance effects

The model has now been developed to determine the dielectric response according to basic Maxwell–Wagner relaxation, but we know from the analysis in Chapter 4 that for particles of the size described here, surface charge effects dominate dielectrophoretic behavior. In order to account for these effects, it is necessary to adapt our `fullmodel.m` program to calculate them.

We learned in Chapter 4 that the effective particle conductivity is given by the expression

$$\sigma_p = \sigma_{pbulk} + \frac{2K_s^i}{r} + \frac{2K_s^d}{r} \tag{11.7}$$

where σ_{pbulk} is the conductivity of the particle interior, r the radius, and K_s^i and K_s^d the conductance through the Stern and diffuse double layers, respectively. We can elaborate further; K_s^d is given by the expression

$$K_s^d = \frac{\left(4F^2cz^2D^d\left(1+3m/z^2\right)\right)}{RT\kappa}\left(\cosh\left[\frac{zq\zeta}{2kT}\right]-1\right) \tag{11.8}$$

where

$$m = \left(\frac{RT}{F}\right)^2 \frac{2\varepsilon_m}{3\eta D^d} \tag{11.9}$$

and where D^d is the ion diffusion coefficient for the ionic species (counterion) in the diffuse layer, z the valence of the counterion, F the Faraday constant, k Boltzmann's constant, R the gas constant, q the charge on the electron, T the temperature, κ the inverse Debye length, c the electrolyte concentration (mol m^{-3}), ζ the ζ potential, and η the viscosity. Similarly, conductance in the Stern layer can be expressed thus:

$$K_s^i = \frac{u\mu_s^i\Lambda}{2z^iF\mu_m^i}$$ (11.10)

where Λ is the molar conductivity of the bulk suspending medium, u is the surface charge density of the particle, μ_s^i and μ_m^i are the mobilities of the ion species in the Stern layer and bulk medium, respectively, and z^i is the valency of that ionic species. However, since μ_s^i is unknown, we need to use the model to estimate a value of K_s^i and determine μ_s^i from that value.

Both of these affect the effective conductivity of the particle and therefore must be inserted into part one of fullmodel.m where the definitions of particle properties are made. For the sake of simplicity, we will call this part 1a, shown below. Notice that in addition to the numerical constants, we now need to introduce the particle radius for the first time (as the variable r). As before, we cannot write Greek letters, so the ζ potential and the viscosity appear as zeta and neta, respectively. In the example, the radius of the particle is 100 nm and the temperature is assumed to be 300°K (about 26°C, or approximately room temperature).

```
% Part 1a
e0 = 8.854e-12;
ep = 2.55*E0;
em = 78*E0;
Ks = 0.8e-9;
zeta = -0.098;
r = 100e-9;
j = sqrt(-1);
k = 1.38e-23;
T = 300;
e = 1.6e-19;
NA = 6e23;
R = 8.3144;
F = 9.648e4;
z = 1;
```

```
spbulk = 0;

T = 300;

D = 1.94e-9;

NA = 6e23;

neta = 8e-4;

q = 1.6e-19;

lambda = 14e-3;

c = sm/lambda;

N = (c*z*z*NA);

m = ((R*T/F)^2)*(2*Er*E0/(3*neeta*D));

Kapa = 1./(sqrt((E0*Er*k*T)./(2*N*q*q*z*z)));

KDiff1 = (c*4*F*F*z*z*D./(R*T*Kapa)).*(1+(3*m/(z*z)));

KDiff2 = cosh((z*q*zeeta)./(2*k*T))-1;

KDiff = KDiff1*KDiff2;

KStern = Ks;

sp = spbulk+(2*KDiff/rad)+(2*KStern/rad);
```

11.6   Multishell objects

In Chapter 5, we examined how objects consisting of a number of concentric shells could be modeled using an extension of the Clausius–Mossotti equation to describe the interfacial polarizations between each layer. We can model this simply by including each iteration of Equation 5.4. Let us, for example, consider a simple model (discounting surface effects) of an object consisting of a single insulating shell 10-nm thick surrounding a conducting core of radius 100 nm. For a single shelled object, the effective complex permittivity is given by the expression

$$\varepsilon_{1eff}^{*} = \varepsilon_{2}^{*} \frac{\left(\dfrac{r_2}{r_1}\right)^3 + 2\dfrac{\varepsilon_1^* - \varepsilon_2^*}{\varepsilon_1^* + 2\varepsilon_2^*}}{\left(\dfrac{r_2}{r_1}\right)^3 - \dfrac{\varepsilon_1^* - \varepsilon_2^*}{\varepsilon_1^* + 2\varepsilon_2^*}} \qquad (11.11)$$

where ε_1^* and ε_2^* refer to the complex permittivities of the particle and medium, respectively, and r_2 and r_1 refer to the radii of the particle to the outer boundaries of the interior and the shell. Coding this in MATLAB, we obtain the following:

```
e0  = 8.85e-12;

s1  = .1;

e1  = 50*e0;

s2  = 10e-6;

e2  = 7*e0;

em  = 78*e0;

r1  = 100e-9;

r2  = 110e-9;

f   = logspace(3,9,1000);

w   = 2*pi*f;

ce1 = e1-j*s1./w;

ce2 = e2-j*s2./w;

cem = e5-j*sm./w;

cm12 = (ce1-ec2)./(ce1+2*ce2);

ef1 = ce2.*((r2/r1)^3+2*cm12)./((r2/r1)^3-cm12);

cm  = (ef2-cem)./(ef1+2*cem);

output = real(cm);
```

Note that there are now complex conductivities corresponding to both interior materials, represented by ce1 and ce2. The program determines the response by determining the Clausius–Mossotti factor of the interior interface, then substitutes this into Equation 11.11 to determine the effective complex permittivity of the particle before using this with the complex permittivity of the medium to determine the net dielectric response.

In order to determine the best fit to dielectric properties of many-shelled objects, care must be taken since there are often many combinations of parameters of different values that might fit a particular set of experimental data. However, as described in Chapter 5, there is an extensive body of literature describing the biophysical parameters of biological materials, allowing the user to make appropriate decisions regarding the appropriate approximate starting values for estimating the properties of the particle. This can be further augmented, again as described in Chapter 5, by altering the particles using some agent that will only significantly affect one known aspect of the particle structure.

This is further complicated when additional factors such as surface conductance effects are added; all the programs presented here can be combined and altered as appropriate, and the reader is encouraged to experiment. For example, where (as with capsids and certain viruses in Chapter 5, or proteins in Chapter 6) the conductivity of the interior of the particle is proportional to the conductivity of the medium, a line such as sp = sm/4 might be introduced.

11.7 Finding the best fit

When using MATLAB to determine electrical parameters, two approaches may be taken; either the user may manually alter the electrical parameters in a set until a satisfactory match is found, or the computer can be used to find a best fit to the data. In many circumstances, particularly where the approximate parameters are known and need to be refined or where there is a great deal of scatter, best-fit methods may be preferred; alternatively, in circumstances where the processes of charge movement may not be fully understood, a manual method allows far more ability to modify the model itself (rather than just the values within the model).

MATLAB contains functions allowing the determination of the minimum set of parameters (up to five) for a known expression. This is based on the Nelder–Mead simplex method, wherein a solution for the expression for an initial set of parameters is determined. The parameters are then varied slightly and the expression is reevaluated; if the expression reduces the value of the expression, then they are adopted. By repeating this procedure, the values of the parameters are gradually moved as the program iteratively searches for the best match, until an answer within the appropriate tolerance is found.

Where the aim is to determine the best fit to a set of experimental data r_{expt}, then the function to be minimized is the error between the experimental and model data r_{model}, such that the best fit in the following expression is met:

$$\sum_i \left(r_{expt}(\omega_i) - r_{model}(\omega_i) \right)^2 = 0 \qquad (11.12)$$

This finds the lowest error value from the initial starting point, and hence the ideal solution. However, care must be taken where values may exist that are *local minima*; parameter sets that produce a value that has low error but not the lowest that can be found. This is shown in Figure 11.4; as the parameters are swept along the x-axis, the error reduces to a local minimum but then rises before descending to the true minimum value. Note also that the errors between experimental and simulation data may be such that the error might be impossible to minimize to the required tolerance. Where this is the case, it is important to set a maximum number of iterations that the minimization routine will execute before returning an answer.

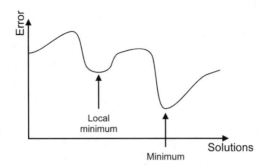

Figure 11.4 When using best-fit algorithms, care must be taken to avoid *local minima*. These are solution sets where any small variation causes an increase in error, and which therefore appear to be the lowest error solution. However, changing the parameter set more extensively may give a more accurate solution (with an even lower error) if the local hump can be overcome.

In most early versions of MATLAB, the `fmins` function was used to determine the set of parameters in a given expression by the Nelder–Mead method; an example of this is shown in the following section. However, more recent versions have also included a newer version, called `fminsearch`, which the readers may wish to investigate if creating their own programs to perform best-fit analysis.

11.8   MATLAB in time-variant field analysis

As seen in the previous chapter, in order to determine the dielectrophoretic force, traveling wave dielectrophoretic force, and electrorotational torque in circumstances where there are more complex phase relationships present, it is necessary to determine the exact magnitude and phase of the electric field across the solution space. The simplest and most accurate method of doing this in circumstances where numerical (rather than analytical) modeling methods are used is to perform simulations at a number of snapshots, each with the phased potentials advanced by a different amount, and then these data are analyzed to determine the exact value at each point.

This can be achieved by using a series of simulations whose phase relationships are equal. When placed into a three-dimensional matrix (with two dimensions representing the plane in which the simulation result is obtained, and the third indicating time), analyzing a single line along the time axis gives an indication of the variation in potential at a single point. The nature of the electric field is such that although the magnitude and phase relationship in the electric field in all three cardinal directions may vary considerably, the form of the electric field in each direction is sinusoidal. Thus, by organizing the data in this way it is possible to use a minimization routine of the type just described to find the magnitude and phase of each of these sinusoids. In order to do this, we need to use a function that determines the error between a sine wave of arbitrary magnitude and phase, and the data generated by the simulation. This can be done using a *function;* similar to the m-files described previously, a function can be treated as if it were a MATLAB command, complete with variable passing (that is, a number within brackets that the function uses). A function to find this phase would look like this:

```
function y = phasefind(ing)

global dummy

mag = abs(ing(1));

phase = ing(2);

errorterm = 0;

for i = 1:36

errorterm = errorterm +
(dummy(i)- mag*cos((((i-1)*10)-phase)/57.3))^2;

end

y = errorterm;
```

This passes a two-value matrix called `ing` to the function, the first being the magnitude of the arbitrary sine wave and the second being the phase advance. For those given values, a sine wave is generated, and the values for the 36 phased simulations are subtracted from that sine wave. The total difference between the arbitrary sine wave and the simulation data at that point is calculated, and that value is returned when the function is used. The command `global` indicates that the value dummy is the one used in the previous part of the program. The function would then be called from within another routine that would assemble the matrices from the data sets. This program would contain parts looking somewhat like this:

```
x = matrixsize(1);

y = matrixsize(2);

for jx = 1:y

for ix = 1:x

a = ix

b = jx

dummy = rex;

m = .01;
```

```
p  =  180;

ing  =  [m;  p];

result  =  fmins('phasefind',ing,options)

magx(ix,jx)  =  abs(result(1));

px(ix,jx)  =  result(2);

dummy  =  rey;

m  =  .01;

p  =  180;

ing  =  [m;  p];

result  =  fmins('phasefind',ing,options)

magy(ix,jx)  =  abs(result(1));

py(ix,jx)  =  result(2);

dummy  =  rez;

m  =  .01;

p  =  180;

ing  =  [m;  p];

result  =  fmins('phasefind',ing,options)

magz(ix,jx)  =  abs(result(1));

pz(ix,jx)  =  result(2);

end

end
```

The first part of this program determines the size of the matrix of data points, such as the 40×40 matrix of data points used to describe the area simulated in the electrorotation example in Chapter 10. The program then loops around two nested loops, with one moving the point of examination from row to row, the other from column to column. For each point in that matrix, starter values of magnitude (m) and phase (p) are combined into a two-element matrix ing. This then uses the fmins command to find the minimum value of error in the routine phasefind described above. This returns another two-element matrix containing the magnitude and phase of the sine wave that best fits the simulation data, which are stored in matrices corresponding to the position in the original simulation array. This returns six matrices, one each for the magnitudes and phases in the x, y, and z directions. Note the use of the term dummy; before the determination of the magnitude and phase for any given electric field data set, the variable dummy is set equal to that data set. By doing this and changing the value contained within the dummy variable, it is possible to use the same function for a number of different data sets. Note also the use of the abs command; this finds the absolute value (i.e., no sign information) of the following variable. The term options contains variables that are important for the minimization routine, specifying the tolerances to which the minimization routine must determine the magnitude and phase before returning the resultant parameters.

11.9   Other MATLAB functions

This chapter has shown how to use some of the functions in MATLAB for data analysis and modeling parameters such as the Clausius–Mossotti factor, as well as some dynamic analysis. However, there is an extensive number of other functions that are of direct benefit to the electromechanic seeking new ways of analyzing data. Of particular use are the functions for two-dimensional data display (e.g., the plots of electric field in Chapter 4, or rotational torque in Chapter 10), three-dimensional data display (see the magnitude and phase plots of the electric field over traveling wave electrodes in Chapter 8, or the protein crossover responses in Chapter 6), and vectors (such as the plot of traveling wave forces in Chapter 8); these were performed by importing the data from experiments or simulation programs and converting them into data matrices for analysis. Similarly, there are a number of tools for data analysis, such as filters and methods for producing planes of data from scattered points. Again, the surface of experimentally acquired protein crossover data in Chapter 6 gives an example of this. Finally, there are functions for animating data output, which can be of tremendous benefit when analyzing dynamic electric fields, though obviously examples of this cannot be included in this book.

Note also that the MATLAB language is not the only tool for performing analyses such as these; there are a number of commercially available alternatives, or the programs can be coded directly into a programming language. This is left to the user's discretion (and wallet) to decide.

appendix A

A dielectrophoretic rotary nanomotor: a proposal

A.1 Electrokinetic nanoelectromechanical systems

As has been stated throughout this book, electrokinetic forces have a wide range of applications on the nanometer scale. However, they may also have potential as a method of actuation in nanometer-scale motor systems. There is a great deal of interest in the production of nanomechanical systems (NEMS), electromechanical equivalents of larger moving devices, but replicated on the nanometer scale. At present, most NEMS research is hypothetical (e.g., Drexler[1]), since electromechanical fabrication technology is only now beginning to be capable of constructing on the micrometer scale. However, we have the theory available to at least model how such devices might behave and how they might be actuated. In this appendix, we will examine the possibilities of developing a dielectric motor[2] using dielectrophoresis as a means of motor actuation and stabilization.

The rotary motor is perhaps one of the simplest NEMS devices. Rotary motor devices were among the first powered machines, and one of the first challenges to microengineers laid down by Richard Feynman in his talk "There's Plenty of Room at the Bottom"[3] was the construction of an electric motor with a maximum length of no more than 400 µm, for which a $1000 prize was offered. There even exists a biological nanomotor in nature; flagellate bacteria move by rotating their corkscrew-like tails (flagella), powered by a protein-based rotary motor.

In order to examine how AC electrokinetics might contribute to this field, we shall examine the construction and performance of a hypothetical AC dielectric nanomotor. The idea of electrokinetically actuated motors is not new; in his follow-up talk, "Infinitesimal Machinery,"[4] Feynman described an idea for an electrostatically actuated, synchronous motor whose operation is similar in principle to the stacked ratchet particle separator described in Chapter 8. The motor described was eventually built, tested, and compared to an electrokinetic version using electrorotation to drive the rotor and dielectrophoresis to hold it in place, by Fuhr et al. in 1992[5] (see also Hagedorn

Figure A.1 A schematic of the nanomotor described here. Four truncated-prism electrodes (the stator) surround a central nanofiber (rotor). The electrodes are energized by potentials in quadrature, such that the phase of the wave on each electrode leads its clockwise neighbor by 90°.

et al.[6] and Müller et al.[7]). The motors demonstrated used a variety of flat rotors trapped within quadrupolar electrode arrays; both rotor disks and stator electrodes were built using conventional planar photolithographic techniques, and the rotors were of the order of 100 μm across.

We can take this concept and shrink it down to the nanometer scale, considering how such a device would perform if the rotor were replaced by a nanofiber of some kind. The proposed nanomotor, in order to keep the dimensions low, should ideally consist of more than just thin disks in an electric field generated by thin electrodes. Since we are speaking of a hypothetical device, we can examine the performance of a three-dimensional structure.

Such a device might be constructed as shown in Figure A.1. The rotor is a nanofiber enclosed within four electrodes that together form a quadrupolar electrode array. Since the rotor is to be held in place by negative dielectrophoresis, it will need to be constructed of an insulating material and surrounded by a more polarizable material (such as water). The shape of the electrodes is, to a degree, arbitrary, but simulation studies of planar electrodes[8] have shown that electrodes of a truncated pyramid design generate a relatively high torque per unit applied voltage, and a large field gradient near the electrode edges (due to the presence of sharp corners between the electrode faces).

A.2 Calculation of motor performance

Unlike the ellipsoids we have examined so far, there is no analytical solution to the induced dipole moment in a rod held perpendicular to an applied electric field. In previous chapters we have seen that, where it is necessary to determine the dielectrophoretic forces on rod-shaped particles such as tobacco mosaic viruses,[9] we can approximate the force by considering the rod as being a prolate ellipsoid of similar dimensions. However, we can

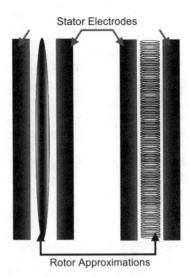

Stator Electrodes

Rotor Approximations

Figure A.2 In order to assess the performance of the motor, we can use two approximations for the cylindrical motor, either a prolate ellipsoid (left) or a stacked series of parallel flat disks (right).

improve our approximation in the case considered here: when the rod is suspended with its axis orthogonal to the plane of the rotating electric field, the dipole moment is not concentrated at a point on the shortest circumference of the ellipsoid, but would in fact consist of charge accumulated along the length of the rod. We can take advantage of this by considering the rod as consisting of a large number of very thin disks held within the field and stacked one upon the other. We can then integrate along the rod to determine the total dielectric response. The flat disk is the limit for zero thickness of an oblate ellipsoid, for which an analytical solution does exist. These possibilities are shown schematically in Figure A.2.

However, the approximation only holds if the axis of rotation of the disk (and hence, the rod) is indeed at 90° to the plane of rotation of the electric field; we can use the prolate ellipsoid model to confirm that the dielectric dispersion along the length of the rod (above which the rod will align perpendicular to the field) is approximately half the frequency of the dispersion along the minor axis if the long axis is double the length of the short one. Since we are principally concerned with the situation where even the short axis has undergone dispersion and is being repelled, the approximation holds and we may proceed. In order to ensure this is the case, the motor would need to be much longer than its width, with axis-to-length ratios of 10:1 or more.

It is known from Chapter 2 that the electric field induced in an ellipsoid exposed to electric field vector E applied along axis x is given by

$$\mathbf{E}_x = \frac{\varepsilon_m^* \mathbf{E}}{\varepsilon_m^* + A_x \left(\varepsilon_p^* \varepsilon_m^* \right)} \tag{A.1}$$

where A_x is the depolarization factor along axis a, and the subscripts m and p refer to the medium and particle, respectively. As described in Chapter 5, there are three depolarization factors A_x, A_y, A_z, one for each axis.

If we consider the case of an electric field applied along the plane of an infinitesimally thin disk (let this be along axis x), then the lengths across axes a and b are equal, and along axis c tends to zero. We can assign these dimensions values a, a, and δc, respectively. It can be shown[10,11] that the depolarization factors A_x, A_y and A_z will have values of 0.5, 0.5, and approximately zero.

The induced polarization P per unit volume is given by the expression:

$$\mathbf{P} = \left(\varepsilon_p^* - \varepsilon_m^*\right)\mathbf{E}_x \tag{A.2}$$

where the induced dipole moment for our example is given by

$$\mu = \tfrac{4}{3}\pi a^2 \delta c \mathbf{P} \tag{A.3}$$

From this, the polarizability can be determined thus,

$$a = \frac{\mu}{\mathbf{E}} \tag{A.4}$$

Combining Equations A.1, A.3, and A.4, we obtain

$$\alpha = \frac{\tfrac{4}{3}\pi a^2 \delta c\left(\varepsilon_p^* - \varepsilon_m^*\right)\varepsilon_m}{\varepsilon_m + \tfrac{1}{2}\left(\varepsilon_p^* - \varepsilon_m^*\right)} \tag{A.5}$$

$$\alpha = \tfrac{8}{3}\pi a^2 \delta c\varepsilon_m \frac{\left(\varepsilon_p^* - \varepsilon_m^*\right)}{\left(\varepsilon_p^* + \varepsilon_m^*\right)} \tag{A.6}$$

We can integrate this along axis c to determine the net polarizability of a rod composed of many such circular elements, giving a net polarizability:

$$\alpha = \int_0^h \tfrac{8}{3}\pi a^2 \varepsilon_m \frac{\left(\varepsilon_p^* - \varepsilon_m^*\right)}{\left(\varepsilon_p^* + \varepsilon_m^*\right)} dc \tag{A.7}$$

$$\alpha = \tfrac{8}{3}\pi a^2 \varepsilon_m h \frac{\left(\varepsilon_p^* - \varepsilon_m^*\right)}{\left(\varepsilon_p^* + \varepsilon_m^*\right)} \tag{A.8}$$

Thus the total induced dipole moment p$_{eff}$ due to applied field E is given by

$$\mathbf{P}_{\mathbf{eff}} = \tfrac{8}{3}\pi a^2 h \varepsilon_m \frac{\left(\varepsilon_p^* - \varepsilon_m^*\right)}{\left(\varepsilon_p^* + \varepsilon_m^*\right)}\mathbf{E} \qquad (A.9)$$

From Chapter 2, we know that electrorotational torque is given by the expression:

$$\Gamma = \tfrac{1}{2}\left[\mathbf{P}_{\mathbf{eff}} \times \mathbf{E}\right] \qquad (A.10)$$

Then the net rotational torque along the rod is given by the expression

$$\Gamma = -\tfrac{8}{3}\pi a^2 h \varepsilon_m \ \text{Im}\left[\frac{\left(\varepsilon_p^* - \varepsilon_m^*\right)}{\left(\varepsilon_p^* + \varepsilon_m^*\right)}\right]E^2 \qquad (A.11)$$

The dielectrophoretic force is given by the expression

$$\mathbf{F} = \tfrac{1}{2}\left[\mathbf{P}_{\mathbf{eff}} \cdot \nabla \mathbf{E}\right] \qquad (A.12)$$

$$F = \tfrac{4}{3}\pi a^2 h \varepsilon_m \ \text{Re}\left[\frac{\left(\varepsilon_p^* - \varepsilon_m^*\right)}{\left(\varepsilon_p^* + \varepsilon_m^*\right)}\right]\nabla E^2 \qquad (A.13)$$

Note that for both dielectrophoresis and electrorotation, the frequency response of the motor is governed by a relationship between the complex permittivities of the particle and medium that is slightly different from the Clausius–Mossotti factor:

$$\left(\frac{\varepsilon_p^* - \varepsilon_m^*}{\varepsilon_p^* + \varepsilon_m^*}\right) \qquad (A.14)$$

In order for the rotor to remain suspended by dielectrophoretic force, it is important that the rotor has a lower complex permittivity than the medium at the frequency of operation. This means that it cannot be used in either a vacuum or in an ordinary gas, as these have near-unity permittivities and very low conductivities. This in turn means that the motor can only be operated in a liquid, which potentially limits the applications to which the motor can be put. Another possibility may be that the rotor could be suspended within a low-pressure plasma (conducting gas), which may enable

it to be repelled due to the conductivity of the gas increasing the polarizability of the plasma by a sufficient amount to cause the rotor to be repelled.
If the motor were to be developed further, this is a possibility that might be
explored, since it would dramatically decrease the viscous drag (and hence
increase the maximum velocity of the rotor), allowing it to be used in nonaqueous applications such as in space science.

A.3 Theoretical limits of motor performance

If we are to suspend our rotor in water, then its ultimate performance will
be limited by the fact that we cannot apply a larger electric field across the
chamber than the dielectric breakdown field of water (at which point current
will conduct across the chamber), of about 20 MV m^{-1} (rms). However, the
electric field is also closely related to the torque we can gain from the device.

We can explore the scaling factors involved in the performance of the
motor from Equation A.11. For a given frequency — that is, a constant value
of the factor expressed in Equation A.14 — the torque is proportional to the
following variables:

$$\Gamma \propto a^2 h E^2 \tag{A.15}$$

If we consider the rotation chamber as enclosing a central region across
which the electric field is approximately uniform and equal to the potential
on opposing electrodes divided by the distance between those electrodes,[8]
then we can replace the E term with:

$$E = \frac{V}{2ka} \tag{A.16}$$

where k is the ratio between the diameter of the rotor and the distance between
opposing electrodes. Substituting Equation A.16 into Equation A.15 produces

$$\Gamma \propto \frac{hV^2}{k^2} \tag{A.17}$$

This indicates that the torque generated by the motor is not related to the
diameter of either the rotor or stator, only the length of the rotor and the
ratio of rotor and stator size. However, the maximum value of E is limited
by the dielectric breakdown condition. If this is applied, then for a fixed,
maximum threshold value of E dictated by Equation A.16 for defining maximum field for any given set of dimensions, we find the limits on motor
performance are

$$\left.\begin{array}{l} \Gamma \propto a^2 h \\ V \propto ka \end{array}\right\} \tag{A.18}$$

indicating that the key factor in torque generation is the volume of the rotor, and the value of the applied potential is proportional to the diameter of the stator chamber.

However, in nanomotor design these are conflicting criteria — it is preferable to have a large torque, small rotor, and low driving voltage. Perhaps one strategy toward design is to define a supply voltage, which in turn defines a value of a, and then use the variation in h to adjust maximum motor torque. The value of k should be as near unity as possible — i.e., the rotor should fill as much of the chamber as possible — but work by Hughes[8] indicates that the maximum ratio is about 1.5:1.

An important consideration here is that we cannot exceed the dielectric breakdown voltage of the medium — this is the electric field that, when applied, causes the material (in this case, the suspending medium) to become conductive. This places a constraint on the electric field that can be applied across a given gap, and hence the maximum torque that can be generated in accordance with Equation A.18. A graph of the maximum torque generated by a motor with 1-μm long rotor shaft is shown in Figure A.3, with the values that break this rule removed from the graph.

In order to examine the frequency-dependent parameters of torque and force generation, we can consider a numerical example. Consider then an electrode chamber (stator) 150 nm in diameter driven by a 1 V_{rms} rotating potential. The rotor at the center of the chamber is an insulating nanofiber 100-nm in diameter 1-μm long and has an internal relative permittivity of 2.55 and net effective conductivity (including both surface and internal

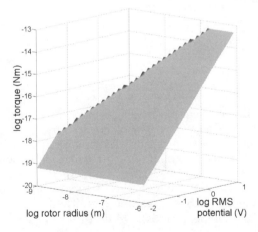

Figure A.3 A graph of the variation in torque developed by a nanomotor 1 μm long, with rotor radius a and applied root mean square (rms) potential V, in a stator chamber 1.5 times wider than the rotor. As can be seen, the torque generated increases with applied potential and is independent of rotor radius; however, the limitation that the electric field in the chamber cannot exceed the electrical breakdown conditions of water restricts the maximum torque that can be developed by small rotors. Combinations of values that break this condition are not illustrated.

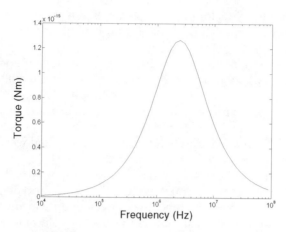

Figure A.4 The frequency-dependent electrorotational torque induced in a 100-nm diameter, 1-μm long rotor under the conditions described in the text.

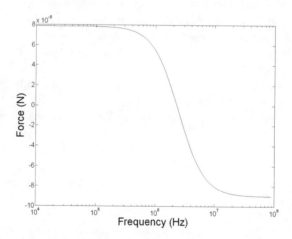

Figure A.5 The dielectrophoretic forces acting between a 100-nm diameter, 1-μm long rotor and the stator electrodes under the conditions described in the text. At frequencies above 6 MHz, the rotor experiences negative dielectrophoretic force and will be electrostatically suspended in the center of the chamber, removing the need for bearings.

components, as discussed in Chapter 4) of 10 mS m^{-1}. The rotor is suspended in water (conductivity 1 mS m^{-1}, relative permittivity 78). By analyzing this motor using a finite element model, we find that the electric field strength across the volume occupied by the rotor is approximately 1.5×10^7 V m^{-1} and that the electric field gradient between the rotor and the electrodes is 1.33×10^{21} V^2 m^{-3}.

If we substitute these figures into Equations A.11 and A.13, we can determine the frequency spectrum for both the torque and stabilizing force exerted on the motor. These are shown in Figures A.4 and A.5, respectively. As can be seen, the peak torque generated by the motor is 1.3×10^{-15} Nm in a frequency window between 2 and 3 MHz. The peak force propelling the rotor to the center of the motor is approximately 9 nN at frequencies above approximately 6 MHz. Combining these, we can see that the optimum condition for the motor is to generate torque between 0.01 and 1×10^{-15} Nm by varying the applied frequency between 5 Mhz and 50 MHz. Control over torque below 0.1×10^{-15} Nm can of course be achieved by the application of higher frequencies. Under these operating conditions, the stabilizing dielectrophoretic force does not drop below 5 nN.

For any moving object in a liquid, there will be a maximum velocity at which the particle can move (its terminal velocity). Unlike ellipsoids, it is not possible to derive an analytical solution for the viscous drag on a rotating cylinder, and we cannot use our flat-plane approximation since the effect of drag occurs across the whole surface of the shape. However, we can still use the approximation of a prolate ellipse to obtain an order-of-magnitude assessment. The steady state rotation rate of the ellipse can be found by calculating the viscous frictional torque of the particle. It depends on the rotation rate of the ellipsoid, and steady state rotation rate is given by

$$R(\omega) = \frac{\Gamma_e}{2\pi D_f} \tag{A.19}$$

where D_f is the hydrodynamic resistance to rotation of the particle around the z-axis, given by[12]

$$D_f = 2v_c\eta\left\{\frac{\left(a_0^2 + b_0^2\right)}{a_0^2 A_x + b_0^2 A_y}\right\} \tag{A.20}$$

were η is the viscosity of the medium, v_c is the volume of the object, a_0 and b_0 are the x and y dimensions of the particle, respectively, and A_{0x} and A_{0y} are the depolarization factors of the particle along the x- and y-axes. However, this equation is dependent on the rotor being able to move freely without viscous drag effects from close proximity to other objects, whereas the rotor here is nanometers away from the electrodes that form the stator. Second, the equations do not take into account any heating effects due to very high rotation rates. If we use Equations A.19 and A.20 to determine the maximum rotation rate of the rotor by approximating it to a 50-nm minor radius and 500-nm major radius, we arrive at a value of maximum rotation rate of approximately 500 million revolutions per second, assuming that the flow around the surface is laminar (i.e., not turbulent). This is a very high value and completely unrealistic — rotation at this rate would almost certainly cause the suspending medium to evaporate. However, this *does* indicate that terminal velocity due to friction would not impair the use of the motor. This would not be a major problem for fluid applications, which would comprise the majority of applications to which such a motor might be put. In such cases, the ability to generate more torque (and hence to move larger masses at slower speeds) will be more important than motor rotation rate.

As can be seen from Figure A.4, at low frequencies the rotor experiences dielectrophoretic attraction to the rotor. Although this does not play a role in torque generation, it may be useful in parking the rotor so that, when the motor is not is use, neither the rotor, or the device to which it is attached is displaced by, for example, the action of Brownian motion. In this manner it could be seen as some form of brake, though conventional braking is not necessary, since the action of viscous drag will mean that following the removal of the applied field, the rotor will come to a halt in fractions of a second.

A second application of positive dielectrophoresis in the motor is in its construction. As we have seen in Chapter 4, the application of first positive dielectrophoresis then negative dielectrophoresis can be used to trap single 93 nm particles in field nulls, and a similar regime could be used here in order to attract a rotor, constructed separately to the stator, into the center of the stator. This is important since the rotor cannot be constructed from the same material as the stator; the former needs to be insulating (in order to ensure that conditions exist to allow negative dielectrophoresis), whereas the stator electrode must be conducting. Therefore, both must be constructed of different materials, separately, and brought together during assembly.

A.4   Digital electronic control of torque generation

In conventional electrorotation applications, electrodes are usually arranged such that three or more electrodes are used to generate a rotating electric field by phasing a 360° sinusoid spatially around the electrode array; for example, a chamber with three electrodes would require a generator that provided 0°, 120°, and 240° phase shifts to the electrodes. A four-electrode array would require 0°, 90°, 180°, and 270° phase shifts to successive electrodes. This sequence of phase shifts creates the effect of a continuously rotating sinusoidal electric field. The greater the number of electrodes, the more complicated the generation system required, but the less distortion observed in the rotating sinusoid. In the literature, the most common number of electrodes used is four, due to the relative ease of phase shifting 90° using devices such as all-pass filters.[13]

It has been demonstrated by Gimsa et al.[14] that square-wave, rather than sinusoidal, signals can be used effectively in the activation of electrodes for electrorotation; the effect is equivalent to the summation of torques induced by the fundamental and higher frequency partials. This scheme is not widely used since the majority of electrorotation work is concerned with the determination of the frequency-dependent properties of complex particles such as cells. However, where the purpose is the manipulation of particles in a controlled way, square wave excitation has been shown to be highly effective. In the case of the motor described here, the higher frequency

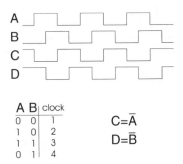

Figure A.6 The proposed motor could be driven using four square wave signals, 90° phase shifted. These are equivalent to the output from two binary counters sequenced as shown in the table, with C and D being the inverse of A and B (or in logic terms, $\bar{A}$ and $\bar{B}$).

components assist by slightly increasing the net dielectrophoretic force holding the rotor in place. The torque is largely unaffected since only the lowest components above the fundamental are responsible for generating torque (as the torque generated diminishes rapidly at applied field frequencies a decade above the resonant peak), and any variation from the torque-frequency characteristic shown in Figure A.4 can be accounted for via a frequency-torque lookup table.

The principal advantage of the application of square wave signals over sinusoids is that it allows direct control of the generation of electric signals by a computer. The motor presented here further aids this integration, since the voltages required for motor activation are sufficiently small to be generated directly by computer devices integrated around the motor itself. In a complete electronic system including both the motor and control circuitry, a controlled sequence of pulses, with duration controlled by a system clock, could be used to achieve effective control over the motor. Variation in frequency can be achieved by altering the number of system clock pulses in the electrode on and electrode off states, that is, by varying the length of a series of 1s followed by an equal number of 0s. As shown in Figure A.6, where there are four phase-shifted square wave signals, A and B are the reverse of C and D, with one having a high signal when the other is low. Therefore, we can produce this circuit using only the A and B signals that are cycled through in the order shown in the table, with C and D constructed by inverting A and B.

We can construct such a device using the circuit shown in Figure A.7. Opposing phases (0°/180° and 90°/270°) can be achieved with NOT gates. If a further NOT gate is applied to the signals powering opposing electrodes, then the direction of rotation is reversed; a switch to connect the additional inverter into the circuit (the circle containing an arrow) allows for changing direction. The torque developed by the motor is well defined by Equation A.11 and shown in Figure A.4, from near zero to near peak, allowing straightforward control of frequency via a lookup table. A further advantage is that since there

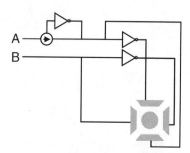

Figure A.7 The circuit by which the motor control could be implemented. The outputs of A and B are inverted to produce C and D, and an additional inverter can be switched on one line to reverse the direction of the motor.

is no requirement for the generation of multiple phase-locked analogue sinusoidal signals, any number of field-generating electrodes may be used. Four have been considered here, but this is arbitrary; the larger the number of electrodes used, the less distortion occurs in the rotation sinusoid and the lower the potential between adjacent electrodes. The example of torque generation in Section A.3 uses voltages equivalent to a controlling signal with $V_{OFF} = 0$ V and $V_{ON} = 2.8$ V, similar to the core voltages of modern microprocessors.

A.5 Nanomotor applications

As stated previously, one of the original feats of microengineering set by Feynman was the construction of a micrometer-scale electric motor, work that we can trace down a hypothetical lineage to the motor presented here. But we could ask — what use is such a device? When seeking applications, we are immediately restricted by the requirement that the rotor be submerged in water in order for dielectrophoresis to overcome the need for bearings. While this appears limiting, consider that the majority of cases where a motor may be needed on this scale may well be in aqueous solutions. In gas or vacuum environments, conventional electrostatic interactions (such as Coulombic attraction/repulsion) can be used to move objects, so that motors of this kind are not required.

In seeking ways of using our device, we can look for parallels in nature. Although most locomotion in nature is performed by the contraction of muscles and such, there does actually exist a form of naturally occurring rotary motor of the kind described here. These motors — consisting of nanoscale rotors and stators — are the method by which bacteria such as *Escherichia coli* move around. The bacteria have long, corkscrew tails extending from their bodies, and these are rotated by means of a bacterial rotary motor in a manner similar to a boat's propeller.

The *E. coli* motor is a complex arrangement of proteins surrounding a protein rotor.[15] The whole arrangement (excluding the propeller) is about 100 nm long and 50 nm in diameter. The motor is driven by the movement of protons from the outside of the cell to the inside, attracted by a 100 mV

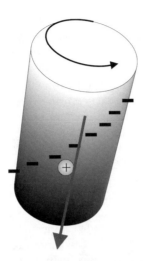

Figure A.8 A schematic of one suggestion of the function of a molecular motor. A proton is propelled along the stator, moving along a diagonal stripe of negative charge. As the charge moves forward, Coulombic interactions between the proton and negative charge induces a torque in the rotor, causing it to rotate. It is known that there are approximately 2000 torque-generating units, corresponding to approximately 2000 charge stripes and corresponding proton pathways.

potential difference. However, the process by which this works is not understood. Theories exist regarding the interaction between the protons and charges on the rotor; one theory suggests that the movement of protons across the membrane and through the protein stator attracts stripes of charge on the rotor causing a torque to be induced by Coulombic attraction; this is shown schematically in Figure A.8. Other theories exist where the torque-generating proteins change shape in the presence of the proton, with these conformational changes in stator proteins causing the movement of the rotor from one unit to the next. It has even been suggested that the motor may operate using the principles of thermal ratchets, which are described in more detail in Chapter 8. However, it is known that, whatever mechanism exists for imparting torque, there are approximately 2000 of them on a single motor. This means that the motor works like an electric stepper motor; rather than imparting continuous rotational motion, each torque generator is only responsible for turning the rotor through a fraction of a degree per step.

Although we do not know how the bacterial motor operates, we are able to measure its mechanical performance. There are a number of ways in which this can be done, including the tethering of the rotor to a surface and measuring the rotation rate of the cell body[16] and by using dielectrophoresis to form a balance force to measure the output of the motor.[17] The total torque produced by a bacterial motor is approximately 1.5×10^{-15} Nm, comparable with our motor described above.

It has been speculated that bacterial motors could be harnessed in order to meet nanotechnological needs, with researchers attaching microengineered rotors to swimming bacteria.[18] As such, it is worthwhile to compare the applicability of the dielectric motor to the same purpose. The biological nanomotor is somewhat (though not vastly) smaller than the dielectrophoretic motor, and is very efficient; the torque generated per unit charge (in this case, the protons) is very high. However, the dielectrophoretic motor also has advantages; as we have seen, the device is easy to integrate into computer control, whereas the bacterial motor is far more complex and offers only a small number of constant rotation rates. Similarly, the bacterial motor is designed to fit in the bacterial membrane and is therefore difficult to position and function within a machine or other device, whereas the dielectrophoretic motor can be fabricated by conventional semiconductor methods.

Since such motors exist in nature, perhaps one application might be the propulsion of independent devices within, for example, the blood stream or a water supply. Perhaps the most commonly described ideal of nanotechnology is that of ultrasmall robots traveling the blood stream and performing clinical functions from within the body. The locomotion of micrometer-scale devices around the body for clinical applications was described by Feynman.[3,4] Although the idea has formed the basis of many flights of fancy — particularly in the description of such robots performing delicate surgery on DNA within cells, or eliminating cancer cells one-by-one, there may be more realistic applications to which swimming robots could be put. For example, localized drug delivery or arterial plaque removal could be achieved by relatively simple swimming machines, without a requirement for onboard computing. The machines would simply move when exposed to an external radio field used to power the devices, which could be modulated to switch simple tasks (such as releasing the drug or activating the plaque removal device, probably also actuated using the same kind of motor). Positional and directional control could be applied merely by moving and changing the orientation of the control coil, in the way that a magnet can be made to move across a sheet of paper by moving another magnet on the other side of the sheet.

A.6 *The way forward?*

In this book we have journeyed from the plains of theories a hundred years old, climbed the mountain of nanomechanics, traveled through the analysis of simple particles and through the manipulation of viruses and molecules and construction on the nanoscale and particle separation, and we have reached the pinnacle of the mountain of what can be achieved today. In this Appendix, we have followed an ancient proverb: "when you reach the top of the mountain, keep climbing." We do not know where we go to from here, but the climb goes on.

In this book, we have examined how the manipulation of particles might revolutionize everything from electronics to medicine — and demonstrated that such advances are perhaps not as disparate as they appear. This field is expanding so rapidly that the direction of research cannot be predicted. Some applications, such as those in medicine and biotechnology, are taking place within a well-established field; others, such as nanotechnology and nanoengineering, are, at best, in their infancy. As such, when some future author comes to write on the impact of electrostatics in nanotechnology, we can barely imagine what might be written, whereas the study of colloids, viruses, and proteins builds on decades of electric methods of biological and chemical analysis of cells and other biological materials. However, we can be sure that, whatever the applications are, the future looks exciting.

References

1. Drexler, K.E., Nanosystems: *Molecular Machinery, Manufacturing and Computation*, Wiley, New York, 1992.
2. Simpson, P. and Taylor, R.J., Characteristic rotor speed variations of a dielectric motor with a low-conductivity liquid, *J. Phys. D: Appl. Phys.*, 4, 1893, 1971.
3. Feynman, R., There's plenty of room at the bottom, 1960; reprinted in *J. Microelectromech. Syst.*, 1, 60, 1992.
4. Feynman, R., Infinitesimal machinery, 1983; reprinted in *J. Micromech. Syst.*, 2, 4, 1993.
5. Fuhr, G.R., Hagedorn, R., and Gimsa, J., Analysis of the torque frequency-characteristics of dielectric induction-motors, *Sensor. Actuat. A-Phys.*, 33, 237, 1992.
6. Hagedorn, R., Fuhr, G., Müller, T., Schnelle, T., Schnakenberg, U., and Wagner, B., Design of asynchronous dielectric micromotors, *J. Electrostatics*, 33, 159, 1994.
7. Müller, T., Gimsa, J., Wagner, B., and Fuhr, G., A resonant, dielectric micromotor driven by low AC-voltages (<6V), *Microsyst. Technol.*, 3, 168, 1997.
8. Hughes, M.P., Computer-aided analyses of conditions for optimizing practical electrorotation, *Phys. Med. Biol.*, 43, 3639, 1998.
9. Morgan, H. and Green, N.G., Dielectrophoretic manipulation of rod-shaped viral particles, *J. Electrostatics*, 42, 279, 1997.
10. Hasted, J.B., *Aqueous Dielectrics*, Chapman & Hall, London, 1973.
11. Jones, T.B., *Electromechanics of Particles*, Cambridge University Press, Cambridge, 1995.
12. Kakutani, T., Shibatani, S., and Sugai, M., Electrorotation of non-spherical cells: theory for ellipsoidal cells with an arbitrary number of shells, *Bioelectrochem. Bioenerg.*, 31, 131, 1993.
13. Hölzel, R., A simple wide-band sine-wave quadrature oscillator, *IEEE Trans. Inst. Meas.*, 42, 758, 1994.
14. Gimsa, J., Pritzen, C., and Donath, E., Characterisation of virus-red-cell interaction by electrorotation, *Studia Biophys.*, 130, 123, 1989.
15. DeRosier, D.J., The turn of the screw: the bacterial flagellar motor, *Cell*, 93, 17, 1998.
16. Berry, R.M. and Berg, H.C., Torque generated by the flagellar motor of *Escherichia coli* while driven backward, *Biophys. J.*, 76, 580, 1999.
17. Hughes, M.P. and Morgan, H., Determination of bacterial motor force by dielectrophoresis, *Biotechnol. Prog.*, 15, 245, 1999.
18. Soong, R.K., Bachand, G.D., Neves, H.P., Olkhovets, A.G., Craighead, H.G., and Montemagno, C.D., Powering an inorganic nanodevice with a biomolecular motor, *Science*, 290, 1555, 2000.

Index

A

AC, *see* Alternating current (AC)
Alternating current (AC), *see also*
        Electrokinetics
   dewatering, 174
   electrokinetics, 10–11, 107, 297
   electrophoretic interaction suppression,
      171
Analytical modeling, 240–246
AND gate, 165–166
Ansoft software, 249
Antifield rotation, 45
Avidin, 140, 145

B

Backbones, 147
BASIC computer language, 280
Best fit data, 85, 292–293
Binary separation, 177, *see also*
        Dielectrophoretic separation;
        Fractionation
Bioparticles
   anatomy of viruses, 108–109
   charge effects, unexpected, 134–136
   collection rate measurements, 115–117
   crossover measurements, 114, 123–130
   dielectrophoresis, 121–123
   levitation height, 118–120
   manipulating viruses, 10, 107–108
   measuring response, 112–121
   multishell model, 109–112
   nonspherical particles, 130–133
   particle velocity measurements, 120–121
   phase analysis light scattering techniques,
      117–118
   separating viruses, 133–134

   structure examination, 121–123
Blood cell separation, 90
Boltzmann constant
   Brownian motion, 71
   charge movement, 83
   modeling surface conductance effects, 288
   particle trapping size, 99
   thermal ratchets, 187
Boundary conditions, modeling, 259–260
Boundary element method, modeling, 249
Boundary layer, 59
Brownian motion
   Boltzmann constant, 71
   colloids, 10, 51, 72
   dielectrophoretic force, 9
   diffusion, 70–72
   Gaussian properties and principles, 71
   Gouy-Chapman model, 53
   levitation height measurement, 118
   Monte Carlo simulation, 250
   negative dielectrophoresis, 97
   particle size, 77, 79, 80, 98–99, 102
   particle velocity measurement, 120
   protein manipulation, 140
   stacked ratcheting mechanisms, 191
   thermal ratchets, 184, 185
   virus manipulation, 108
   viscous drag, 70
Buckminsterfullerene, 154
Buckyballs, 154
Bulk medium, 52, 84–86, 178
Buoyancy, 70
Buoyancy force, 70

C

Capacitance, 23–25
Capsid particles, 109, 123, 134–136

313